Lecture Notes in Electrical Engineering

Volume 817

The book series *Lecture Notes in Electrical Engineering* (LNEE) publishes the latest developments in Electrical Engineering - quickly, informally and in high quality. While original research reported in proceedings and monographs has traditionally formed the core of LNEE, we also encourage authors to submit books devoted to supporting student education and professional training in the various fields and applications areas of electrical engineering. The series cover classical and emerging topics concerning:

- Communication Engineering, Information Theory and Networks
- Electronics Engineering and Microelectronics
- Signal, Image and Speech Processing
- Wireless and Mobile Communication
- Circuits and Systems
- Energy Systems, Power Electronics and Electrical Machines
- Electro-optical Engineering
- Instrumentation Engineering
- Avionics Engineering
- Control Systems
- Internet-of-Things and Cybersecurity
- Biomedical Devices, MEMS and NEMS

For general information about this book series, comments or suggestions, please contact leontina.dicecco@springer.com.

To submit a proposal or request further information, please contact the Publishing Editor in your country:

China

Jasmine Dou, Editor (jasmine.dou@springer.com)

India, Japan, Rest of Asia

Swati Meherishi, Editorial Director (Swati.Meherishi@springer.com)

Southeast Asia, Australia, New Zealand

Ramesh Nath Premnath, Editor (ramesh.premnath@springernature.com)

USA, Canada:

Michael Luby, Senior Editor (michael.luby@springer.com)

All other Countries:

Leontina Di Cecco, Senior Editor (leontina.dicecco@springer.com)

**** This series is indexed by EI Compendex and Scopus databases. ****

More information about this series at https://link.springer.com/bookseries/7818

Atma Ram Gupta · Nirmal Kumar Roy ·
Sanjoy Kumar Parida
Editors

Power Electronics and High Voltage in Smart Grid

Select Proceedings of SGESC 2021

Springer

Editors
Atma Ram Gupta
Department of Electrical Engineering
National Institute of Technology Kurukshetra
Kurukshetra, India

Nirmal Kumar Roy
Department of Electrical Engineering
National Institute of Technology Durgapur
Durgapur, India

Sanjoy Kumar Parida
Department of Electrical Engineering
Indian Institute of Technology Patna
Patna, India

ISSN 1876-1100 ISSN 1876-1119 (electronic)
Lecture Notes in Electrical Engineering
ISBN 978-981-16-7392-4 ISBN 978-981-16-7393-1 (eBook)
https://doi.org/10.1007/978-981-16-7393-1

This Springer imprint is published by the registered company Springer Nature Singapore Pte Ltd.
The registered company address is: 152 Beach Road, #21-01/04 Gateway East, Singapore 189721, Singapore

To Lord Krishna

Committees

Chief Patron: Dr. Satish Kumar, Director, NIT Kurukshetra
Patron: Dr. Akhilesh Swarup, Dean (P&D) and Professor, EED, NIT Kurukshetra
Co-Patron: Dr. L. M. Saini, Professor and Head, EED, NIT Kurukshetra

General Chairs

Dr. Ashwani Kumar, Professor, EED, NIT Kurukshetra
Dr. Sathans, Professor, EED, NIT Kurukshetra

Organising Secretaries

Dr. Atma Ram Gupta, Assistant Professor, EED, NIT Kurukshetra
Dr. Rahul Sharma, Assistant Professor, EED, NIT Kurukshetra
Dr. Amit Kumar, Assistant Professor, EED, NIT Kurukshetra

Publicity, Technical and IT Management Chair

Dr. Jagan Nath, Officer In-Charge (CCN), NIT Kurukshetra

Technical Program Committee

Dr. Pradeep Kumar, Assistant Professor, EED, NIT Kurukshetra
Dr. Bhanu Pratap, Assistant Professor, EED, NIT Kurukshetra

Dr. Arup Goswami, Associate Professor, NIT Silcher
Dr. Y. P. Verma, Associate Professor, UIET Chandigarh

Finance Management Committee

Dr. Shelly Vadhera, Associate Professor, EED, NIT Kurukshetra
Dr. Sandeep Kakran, Assistant Professor, EED, NIT Kurukshetra

Preface

The Department of Electrical Engineering, National Institute of Technology, Kurukshetra, (Institution of National Importance), Haryana, India, organized an international Online Conference on Smart Grid Energy Systems and Control (SGESC-2021) during March 19–21, 2021. The goal of the conference was to bring together leading academic researchers, scientists, and students worldwide to share their insights and research results, including the role of power electronics, high voltage engineering, renewable energy sources, distributed generation, intelligent control and measurement technologies to help in moving toward the smart grid from the conventional grid. The whole idea of the conference was to share the thoughts and opinions in the relevant fields and transform them into real-time solutions. The conference offered a unique opportunity for all persons of electrical background, as well as related engineering, to learn about the most up-to-date technologies and strategies.

Researchers were invited to submit research papers to the conference based on their original work in the area of power electronics, high-voltage engineering, power systems, renewable energy sources, wind energy, solar energy, smart grids, electric vehicles, energy conservation, and intelligent control and measurement. All of these submissions were confined to a thorough peer-review process appropriate to their tracks. As a result, the articles published here are a collection of high-quality submissions to the literature on evolutionary science and experience, covering a wide variety of trending affairs. Finally, 3 members of the Program Committee reviewed all the 31 approved research papers for this volume entitled *Power Electronics and High Voltage in Smart Grid* and discussed in front of committee members in their respective tracks. There are two more volumes entitled Renewable Energy towards Smart Grid and Control and Measurement Applications for Smart Grid from this conference. In addition, several high-profile keynote speakers addressed the SGESC-2021 community.

It involved an enormous time and energy for all of these and put them together into a cohesive curriculum. Our sincere gratitude goes to all of the authors, session presenters, session coordinators, program committee members, volunteers, and keynote speakers, who contributed to this International Conference. Our special thanks go to Springer as our publication partner to publish all the three volumes of conference proceedings in their prestigious Lecture Notes in Electrical Engineering.

The event would not have been as effective without their support.

Kurukshetra, Haryana, India — Atma Ram Gupta
Durgapur, West Bengal, India — Nirmal Kumar Roy
Patna, Bihar, India — Sanjoy Kumar Parida
March 2021

Contents

About the Editors

Dr. Atma Ram Gupta received his B. Tech. in Electrical Engineering from C.V. Raman College of Engineering, Bhubaneshwar, M. Tech. in Electrical Engineering from National Institute of Technology (NIT), Durgapur, and Ph.D. in Power Systems from NIT Kurukshetra. He is a senior member of IEEE and IEEE Power & Energy Society. He is an Assistant Professor of Electrical Engineering at NIT Kurukshetra. Dr. Gupta's research interests are renewable energy-based D.G. and D-FACTS allocation in the distribution system and high voltage engineering. He has 50 research publications in various reputed international journals, book chapters, and conferences.

Dr. Nirmal Kumar Roy is a Professor in the Department of Electrical Engineering, National Institute of Technology, Durgapur, India. He obtained his M.E. (Elect.) from the Indian Institute of Science, Bangalore, and a Ph.D. (Elect.) from the University of South Australia, Australia. He has 80 publications in journals and conferences of international repute. His research interests include high voltage engineering, magnetic fields in power lines, ICT & IoT enabled high voltage laboratories and lab facilities, rural e-Governance, power quality & energy audit, digital e-Learning, and renewable energy.

Dr. Sanjoy Kumar Parida works as an Associate Professor in the Department of Electrical Engineering at the Indian Institute of Technology (IIT) Patna, India. He did his Ph.D. in electrical engineering from IIT Kanpur, India. His research areas are optimal operation and control of power systems, power system dynamics, wide-area monitoring, control and protection, and micro-grid operation. He is a member of the IEEE, IEEE Power & Energy Society, IEEE Control System Society, and IEEE Power Electronics Society. He has got several publications in various journals and conferences.

Moisture Estimation of Power Transformer Using Transfer Function of Conventional Debye Model

Chandra Madhab Banerjee, Arijit Baral, and Sivaji Chakravorti

Abstract The moisture present in the paper insulation of a power transformer is considered to be an important parameter in the field of insulation diagnosis. Over time, deterioration in insulation conditions increases the moisture content. In the present analysis, the moisture is estimated using a noninvasive analysis based on the insulation model. Available similar techniques suffer from limitations. Existing techniques are formulated using data collected from a laboratory sample for a specific measurement temperature. This makes them unreliable for real-life power transformers. The discussed method addresses these issues and, in the process, offers a better, reliable insulation diagnosis methodology.

Keywords Dielectric Response · Polarization Current · Conventional Debye Model

1 Introduction

In the present work, a noninvasive way to determine *paper moisture* present in power transformer insulation is investigated. In order to achieve this, the polarization current recorded from a number of real-life in-service power transformers having different power specifications are measured and analyzed. It is reported in [1] that any oil-paper insulation can be represented by the Conventional Debye Model (CDM) and Modified Debye Model (MDM). It is also reported in [2] that the coefficient of the transfer function of CDM/MDM is useful for estimating *moisture* present in solid insulation. The methodology reported in [2] is formulated using the data obtained from laboratory samples whose insulation condition was known beforehand. It is understood that the complexity of insulation present in real-life equipment cannot

C. M. Banerjee (✉) · A. Baral
IIT(ISM) Dhanbad, Jharkhand, India

S. Chakravorti
Jadavpur University, Kolkata, India

A. R. Gupta et al. (eds.), *Power Electronics and High Voltage in Smart Grid*, Lecture Notes in Electrical Engineering 817,
https://doi.org/10.1007/978-981-16-7393-1_1

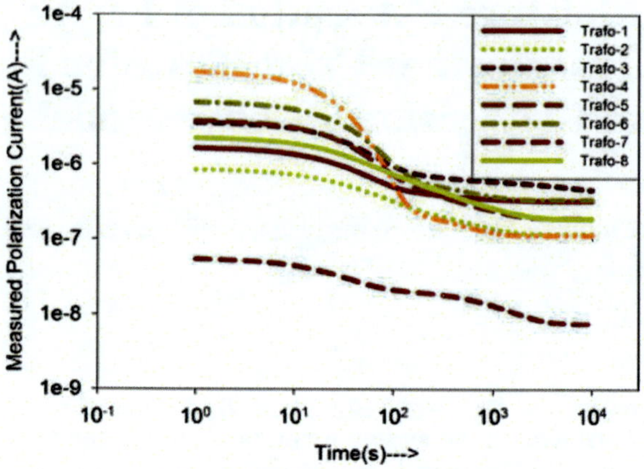

Fig. 1 Measured polarization current profile

be replicated in a laboratory. Further, the power rating and scale of oil-paper insulation of laboratory samples differ significantly from that used in actual equipment. Hence, the characteristics of oil-paper insulation manufactured in the laboratory will differ significantly from that which is present in complex real-life in-service systems. Available literature [3, 4] show that the characteristic of such laboratory sample and real-life unit indeed differs due to the above-mentioned reasons. The above discussion suggests that methodologies developed using data obtained from laboratory sample may not always yield accurate when applied to real-life equipment. This is also illustrated later in the paper.

In the present analysis, the transfer function's coefficient (obtained from CDM) is used in *moisture* estimation. The profile of polarization current and hence its analysis is affected by temperature and aging [5]. Therefore, it is imperative that the effect of temperature and aging on any condition monitoring methodology is also investigated. Hence, the influence of these two key influential factors on the proposed methodology is also discussed in the paper. The profiles of measured polarization current corresponding to the eight power transformers that are used in the present work is shown in Fig. 1. It should be mentioned here that the PDC measurement temperature of the tested units was equal to 35 °C.

The power rating of the tested real-life power transformers is shown in Table 1.

2 Description of Modified Debye Model and Its Transfer Function

Conventional Debye Model, CDM (shown by Fig. 2), can be used to represent any oil-paper insulation [1]. Each branch of CDM contains a single Resistance (R), Capacitance (C) element. In MDM (a structure derived from CDM), a single branch parameter is further segmented into multiple numbers of R–C elements to emulate the

Table 1 Rating of power transformer

Transformer name	Power rating
Trafo-1	>100 MVA/220 kV
Trafo-2	>250 MVA/15.75 kV
Trafo-3	>200 MVA/21 kV
Trafo-4	>250 MVA/15.7 kV
Trafo-5	>150 MVA/21 kV
Trafo-6	>200 MVA/15.7 kV
Trafo-7	>200 MVA/420 kV
Trafo-8	>250 MVA/230 kV

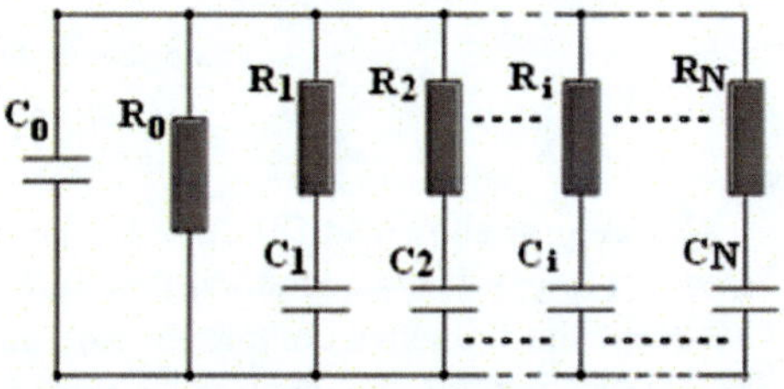

Fig. 2 Conventional debye model

axial temperature gradient present in oil-paper insulation [6]. In Fig. 2, the impedance $Z_i(1,s)$ offered by the CDM's *i*th branch is equal to $R_i(1) + 1/sC_i(1)$. It is generally considered that each series element of CDM characterize the specific dipole group of oil-paper insulation.

The schematic diagram of MDM is displayed in Fig. 3. As illustrated in Fig. 3, the series impedance $Z_i(1,s)$ of CDM is segmented into a number of R–C parameters of MDM. The characteristic response function *Y(s)* of MDM is given by Eq. (1).

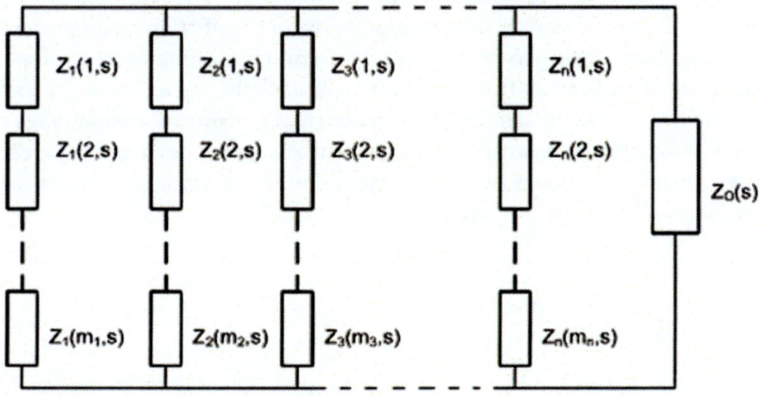

Fig. 3 Structure of modified debye model (MDM)

$$Y(s) = \frac{1}{Z(s)} = \frac{1}{Z_o(s)} + \sum_{j=1}^{N} \frac{1}{Z_j(s)};\ Z_j(s) = \sum_{k=1}^{m_j} Z_j(k, s) \quad (1)$$

In (1), N indicates the number of branches and m_j represent the number of series R–C element of the jth branch.

The CDM/MDM branches with a time constant of less than 100 are reported to be influenced by the characteristics of dipole present in oil [1, 7]. The characteristic of dipoles present in interfacial and solid insulation primarily influences the condition of the solid insulation [1, 7]. Hence, the branches having a time constant greater than 100 characterize dipole groups present in interface and solid insulation [1, 7]. The Transfer Function of CDM is given in (2).

$$TF_D(s) = \frac{(sR_0C_0 + 1)}{R_0} + \sum_{i=1}^{N} \frac{(s \times 1/R_i)}{(s + 1/R_iC_i)} \quad (2)$$

The $TF_D(s)$ shown in (2) can also be expressed as a ratio of two polynomials as shown in (3). It is understood that each branch of the CDM having of series R–C (except the branches containing R_0 and C_0) contributes a real negative pole in $TF_D(s)$ thus making the insulation model an asymptotically stable system.

$$TF_D(s) = \frac{N(s)}{D(s)} = \frac{a_{N+1}s^{N+1} + a_N s^N + \ldots + a}{b_N s^N + b_{N-1}s^{N-1} + \ldots + b_0} \quad (3)$$

As per literature [2], the pole of such transfer function of CDM is useful for estimating *moisture* content present in insulation. However, such relation has been obtained using data recorded from laboratory samples. Due to the complexity and scale of the power transformer, the relation between the transfer function coefficient of CDM and *moisture* may not always give a proper result that is reliable for in situ power equipment. Available works of literature [3, 4] reinforce the fact that relation obtained using sample data cannot be readily applied to data obtained from power transformers owing to its complexity. Furthermore, literature [2] reported a relation between the transfer function coefficient of CDM and *moisture* for a specific temperature for laboratory samples. It is known that performance parameters do not remain constant with the change in measurement temperature and aging. Later in this paper, the change in performance parameters due to the variation of measurement temperature and aging are discussed.

3 Performance Parameter Calculation for Real-Life Power Transformer

In the current investigation, the values of *moisture* of the tested real-life power transformers were measured with the help of IDAX 300. The measured value of *moisture* of the power transformers and their corresponding operational age are shown in Table 2.

It is well known that the numbers of branches of CDM as well as parameters present therein are not unique. In fact, the numbers of branches for a given polarization current are dependent on the curve fitting parameters. Among N number of CDM branches, some number of branches (say b_o) will have a time constant less than 100. These b_0 numbers of branches characterize the dipoles present in bulk oil. Hence, the number of branches ($nb = N{-}b_o$) having a time constant greater than 100 can be assumed to contain information regarding characteristics of dipoles present at the interface and paper. In the present analysis, these nb numbers of branches are used to calculate the transfer function of paper (TF_{PA} (s)) in the form of (4).

$$TF_{PA}(s) = \frac{N(s)}{D(s)} = \frac{a_{nb+1}s^{nb+1} + a_{nb}s^{nb} + \ldots + a}{b_{nb}s^{nb} + b_{nb-1}s^{nb-1} + \ldots + b_0} \quad (4)$$

The coefficient b_{nb} of the transfer function (shown in (4)) is chosen for further analysis. The coefficient value b_{nb} is normalized by a factor 10^{2*nb}. Table 3 shows the value of b_{nb} corresponding to all the tested transformers.

The scatter plot between b_{nb} and *moisture* for ($nb = 3$) is shown in Fig. 4. It can be observed that the b_{nb} of the transfer function of paper (TF_{PA} (s)) maintains a good relationship with *moisture*.

The values of b_{nb} (shown in Table 3) maintain a good relationship with *%pm* in the form of (5).

$$\%pm = Y0 + A \times (1 - \exp(-b_{nb} \times D)) \quad (5)$$

Table 2 Measured *moisture* and operational age of the tested power transformers

Name of transformer	*Moisture*	Operational age *(years)*
Trafo-1	0.8	26.0
Trafo-2	2.3	11.0
Trafo-3	1.8	26.0
Trafo-4	2.3	28.0
Trafo-5	2.3	26.0
Trafo-6	1.8	25.0
Trafo-7	2.4	12.0
Trafo-8	1.2	10.0

Table 3 Calculated value of b_{nb} of power transformer

Transformer name	For $nb = 3$	For $nb = 4$
Trafo-1	76.00	362.00
Trafo-2	139.00	699.00
Trafo-3	106.00	411.00
Trafo-4	190.00	1246.00
Trafo-5	149.00	672.00
Trafo-6	101.00	485.00
Trafo-7	200.00	1600.00
Trafo-8	84.3000	380.00

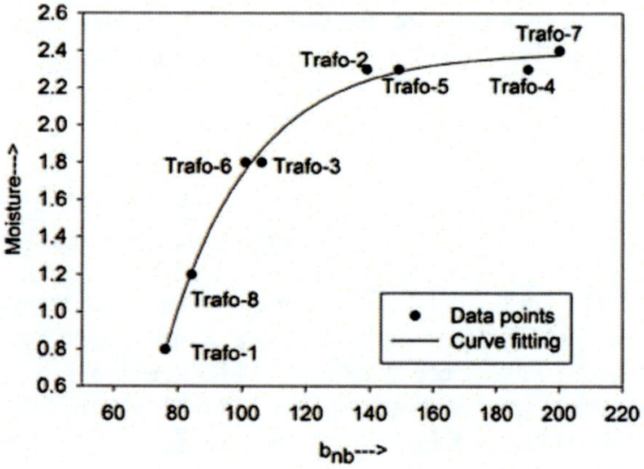

Fig. 4 Curve fitted profile between b_{nb} and *moisture*

The coefficients of (5) are shown in Table 4 for different CDM having the value of $nb = 3$ and 4. It is known that the CDM/MDM branches with a time constant greater than 100 and less than 1000 represent the dipole groups that are located at the oil-paper interface [1, 7]. In literature [2], the value of b_{nb} (for a laboratory sample) is reported to vary between 36 and 226 for a CDM having $nb = 5$. Observing such a low value of b_{nb} for sample (with $nb = 5$), it can be opined that CDM of these samples did not contain any time constant branches having time constant greater than 1000.

Any performance parameter (which is used for condition monitoring of paper insulation) should have the influence of paper insulation. Due to the simple construction of the sample, the value of b_{nb} was minimally affected by the characteristics of solid/paper insulation. Hence, the relation between the transfer function coefficient of CDM and *moisture* of oil paper sample may not always give a proper result that

Table 4 Coefficient of (5) for different values of *nb*

nb of CDM	Y0	A	D
3	−24.7748	27.1665	0.0372
4	−222.0696	224.3785	0.0140

is reliable for field equipment. The above discussion suggests that the relation (5) is much more readily suitable for field equipment. It is also worth mentioning here that (5) is valid only for polarization current measured at 35 °C.

4 Effect of Temperature on Coefficient b_{nb} for Real-Life Transformer

It is understood that it is not always possible to measure polarization current from a power transformer at $T_{re} = 35$ °C. It is reported in literature [8] that the polarization current corresponding to a specific measurement temperature can be used to estimate the polarization current profile at any other temperature with the help of activation energy. It is reported in literature [9] that the value of activation energy (E_a) for in-service unit is near about 111 kJ/mol.

It is also reported in literature [8] that the value of insulation resistance (R_0) and branch resistances (R_i) of CDM are dependent on measurement temperature. However, the geometric capacitance (C_0) and branch capacitances (C_i) do not get significantly affected by measurement temperature [8]. The variation of branch resistance can be estimated based on the value of E_a of the unit. Equation (6) shows the relation that exists between branch elements of CDM and measurement temperature.

$$\left.\begin{aligned} R_i^{T2} &= R_i^{T1} \times \exp\left(\frac{E_a}{K_T} \times \left(\frac{1}{T_2} - \frac{1}{T_1}\right)\right) \quad ; \\ R_0^{T2} &= R_0^{T1} \times \exp\left(\frac{E_a}{K_T} \times \left(\frac{1}{T_2} - \frac{1}{T_1}\right)\right) \\ C_0^{T2} &= C_0^{T1}; \; C_i^{T2} = C_i^{T1} \end{aligned}\right\} \tag{6}$$

In (6), K_T (=8.6173303E-05 eV/K) is the Boltzmann constant. On the other hand, R_i^{T2} and R_i^{T1} exhibit the resistances of the ith branch at T_2 and T_1 K temperature, respectively. In (6), C_0^{T2} and C_0^{T1} symbolize the geometric capacitances at T_2 and T_1 K temperature, respectively. Similarly, C_i^{T2} and C_i^{T1} represent the capacitive property of the ith branch at T_2 and T_1 K, respectively.

Branch elements of CDM for any temperature can be formulated from the branch elements of CDM for any specific temperature. Hence, the polarization current profile corresponding to T_2 K can be calculated using the updated branch elements using (7).

$$\left.\begin{aligned} i_{cal}(T2) &= \sum_{i=1}^{N} M_i \times \exp(R_i^{T2} \times C_i^{T2} \times t) + \frac{V_{in}}{R_0^{T2}} \\ M_i &= \frac{V_{in}}{R_i^{T2}} \end{aligned}\right\} \tag{7}$$

The above methodology is utilized to obtain the polarization current at different temperatures. The calculated polarization current is thereafter utilized to calculate b_{nb} for different temperatures. Table 5 shows the calculated value of b_{nb} at various temperatures ($T \neq T_{re}$) for the tested in-service power transformers. Where T_{re} represents the reference temperature, 35 °C.

Table 5 shows that the value of coefficient b_{nb} reduces as the temperature difference increases. As the measurement temperature decreases, the frequency domain spectroscopy profile gets shifted vertically downwards. As per the equation reported by Zaengl et al. [10], this leads to a decrease in the value of *%pm*. Table 5 shows that temperature has a similar effect on the coefficient of b_{nb}. It is observed by the authors that the variation of b_{nb} due to temperature is exponential in nature (shown in (8)).

$$b_{nb}^{T} = b_{nb}^{t_{re}} \times \exp(-rate \times \Delta temp) \tag{8}$$

Here, b_{nb}^{T} *and* $b_{nb}^{t_{re}}$ are the values of b_{nb} at temperature T and T_{re}, respectively.

Table 5 Value of b_{nb} for different temperatures

Transformer Name	b_{nb} (nb = 4)	b_{nb} (nb = 3)	Δtemp (=T_{re}−T)	Transformer Name	b_{nb} (nb = 4)	b_{nb} (nb = 3)	Δtemp (=T_{re}−T)
Trafo-1	362	76	0	Trafo-5	672	149	0
	307	62	2		562	131	2
	241	48	5		462	95	5
	206	45	7		391	79	7
Trafo-2	699	139	0	Trafo-6	485	101	0
	499	110	2		420	90	2
	251	62	5		299	68	5
	218	55	7		272	54	7
Trafo-3	411	106	0	Trafo-7	1600	200	0
	355	85	2		1178	156	2
	275	73	5		790	101	5
	236	57	7		560	66	7
Trafo-4	1246	190	0	Trafo-8	380	84.3	0
	1075	162	2		273	69.2	2
	883	139	5		174	40.1	5
	774	110	7		150	26.9	7

5 Effect of Aging on Coefficient b_{nb} for Real-Life Transformer

Previously, it was reported that coefficient b_{nb} is useful to determine the health of a power transformer [2]. So, it will be equally interesting to study, how the value of b_{nb} is affected by the operational age of the power transformer. For this analysis, two different operational age groups are defined within the tested transformers. Trafo-2, 7, 8 are assumed to form Group-1 (10–12 years) while Trafo-1,3,5,6 are considered to belong in Group-2 (25–26 years).

It is observed that the rate of decay (*rate*) of b_{nb} with temperature is nearly about the same for a particular age group irrespective of the value of *nb*. The value of *rate* in (8) is shown in Table 6. Figure 5 shows the cluster of *rate* formed by the members of the two age groups. Unfortunately, complete data corresponding to units having ages between 13 and 24 years were not available with the utilities at the time of writing the paper. Hence, the *rate* corresponding to these units could not be determined. The authors are interacting with the utility to access such data. Once available, the analysis of such data will be communicated.

Table 6 Rate of decay (*rate*) of b_{nb} with temperature with aging

Transformer name	Rate	Operational age
Trafo-1	0.080	26
Trafo-2	0.15	11
Trafo-3	0.078	26
Trafo-5	0.080	26
Trafo-6	0.086	25
Trafo-7	0.146	12
Trafo-8	0.143	10

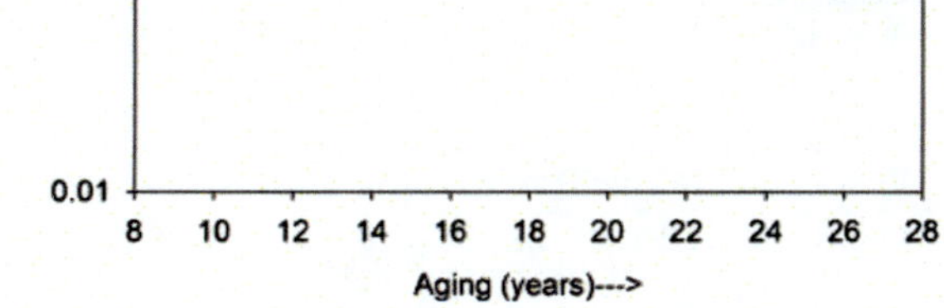

Fig. 5 Cluster of '*rate*' for different aging groups of power transformer

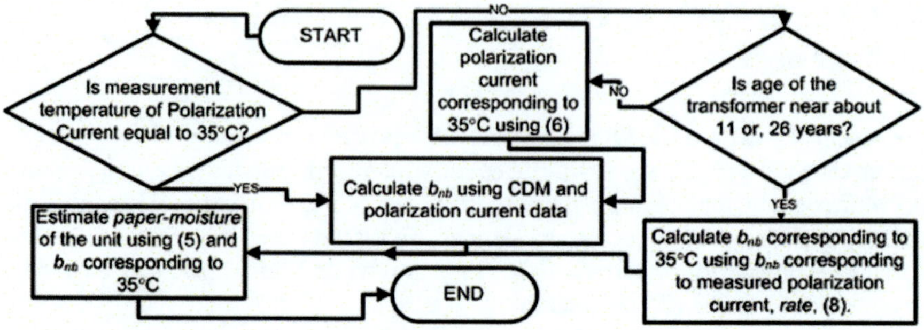

Fig. 6 Flowchart of the proposed methodology

6 Application on Real-Life Unit

To apply the proposed methodology, the polarization current of a unit needs to be measured at any temperature (T). From the polarization current, the value of b_{nb} corresponding to temperature T needs to be evaluated. If the operational age of the unit is near about the operational age of the tested transformers, then b_{nb} corresponding to T_{re}(=35 °C) can be calculated from the *rate* given in Table 6. Thereafter, the *moisture* of any given unit can be predicted using (5). The flowchart of the proposed methodology is shown in Fig. 6.

7 Validation of the Proposed Methodology

In order to validate the proposed methodology, the polarization current of a real-life power transformer at two different temperatures was measured. It is worth mentioning here that moisture was measured only at 35 °C. In Table 7, the polarization current data measured at 30 °C was used to obtain the profile of polarization current at 35 °C and thereafter (5) was used to calculate the paper's moisture. The value of b_{nb} obtained for this transformer is shown in Table 7. The measured and calculated value of *moisture* is also shown in Table 7.

Table 7 Measured and calculated *moisture* and corresponding b_{nb} with temperature having operational age of 26 years

Temperature	*Moisture* measured	b_{nb}	*Moisture* Calculated using (5)	*Moisture* Calculated using (10) [2]	Operational age
30 °C	Not available	58	1.28		26
35 °C	1.2	86		2.31	

Table 7 demonstrates that the present work is better suited in predicting *moisture* by using b_{nb}. It can be further opined that Eqs. (8) along with (5) are capable of identifying the health of oil-paper insulation of real-life units by taking measurement temperature into account.

8 Conclusions

The following conclusion can be made from the above discussion:

1. A well-defined relationship exists between b_{nb} and *paper moisture*. The present paper suggests that this relation can be satisfactorily used for monitoring the health of real-life power transformers.
2. The illustrated methodology is formulated using measured data of real-life in-service power transformers. Hence, the proposed method is applicable to other real-life equipment.
3. The effect of temperature on the reported performance parameter, b_{nb} is also discussed in the present paper for two different operational age groups of transformers.

References

1. Saha TK, Purkait P, Müller F (2005) Deriving an equivalent circuit of transformers insulation for understanding the dielectric response measurements. IEEE Trans Power Deliv 20(1):149–157
2. Baral A, Chakraborty S (2014) Prediction of moisture present in cellulosic part of power transformer insulation using transfer function of modified debye model. IEEE Trans Dielectr Electr Insul 21(3):1368–1375
3. Baral A, Chakraborty S (2014) Condition assessment of cellulosic part in power transformer insulation using transfer function zero of modified debye model. IEEE Trans Dielectr Electr Insul 21(5):2028–2036
4. Goel R, Verma HC, Baral A, Chakraborty S (2017) A novel method to predict moisture in cellulosic insulation of power transformer with improved accuracy using time domain spectroscopy data. In: 6th International conference on computer applications in electrical engineering-recent advances
5. Gubanski SM, et al (2002) Dielectric response methods for diagnostics of power transformers. CIGRE Report of the TF D1.01.09
6. Baral A, Chakraborty S (2013) A modified Maxwell model for characterization of relaxation processes within insulation system having non-uniform aging due to temperature gradient. IEEE Trans Dielectr Electr Insul 20(2):524–534
7. Banerjee CM, Dutta S, Baral A, Chakraborty S (2019) Influence of charging voltage magnitude on time domain dielectric response of oil-paper insulation. IET Sci Meas Technol 13(6):874–882
8. Fofana I, Hemmatjou H, Meghnefi F, Farzaneh M, Setayeshmehr A, Borsi H, Gockenbach E (2010) On the frequency domain dielectric response of oil-paper insulation at low temperatures. IEEE Trans Dielectr Electr Insul 17(3):799–807

9. Lundgaard LE (2007) Ageing of cellulose in mineral-oil insulated transformers. Cigre Task Force D1.01.10 Report
10. Zaengl WS (2003) Dielectric spectroscopy in time and frequency domain for HV power equipment. I. Theoretical considerations. IEEE Electr Insul Magaz 19(5):5–19

Enhanced Operation of Grid-Connected Inverter to Improve Power Quality Issues, Harmonic Compensation with the Proposed Novel Approach

Sudhir Kr Singh, Atul Kumar Kushwaha, Ankit Singhal, and Sourav Diwania

Abstract Increase in power electronic devices/ equipment causes serious power quality concern. Furthermore, the wide use of unbalanced and non-linear loads is responsible for line harmonics and significantly deteriorates power supply quality. Increase penetration of renewable sources injects harmonics into the utility grid. Thus, power quality is a significant issue that requires focus while integrating renewable sources into the grid. The main objective of this paper is to study power management between grid, renewable sources and linear/non-linear loads. Additionally, a simple current controller is proposed which enables the voltage source converter to realize as a shunt active filter. This proposed controller is fixed in voltage source inverter with predefined active and reactive power reference so as to improve grid power quality with reduced THD percentage under unbalanced/non-linear connected at the point of common coupling. This controller is utilized in the outer loop of VSC, where reference currents are generated based on the grid voltage and current with prespecified active and reactive power. Reference signals are then compared with tracked grid side Id and Iq and obtained error is transferred to PI controller for zero steady-state error regulation. Enhanced SRF-PLL is used to generate grid synchronization angle, which is utilized into natural a-b-c to synchronous d-q transformation. The MATLAB/Simulink verifies the capability of the proposed controller.

Keywords Current control · Point of common coupling (PCC) · Proportional integral (PI) controller · Renewable energy sources (RES) · Power quality (PQ) · Voltage source converter (VSC)

S. K. Singh · A. Singhal
KIET Group of Institutions, Delhi-NCR, Ghaziabad, India
e-mail: sudhir.singh@kiet.edu

A. K. Kushwaha
IMS Engineering College, Ghaziabad, India
e-mail: atul.kushwaha@imsec.ac.in

S. Diwania (✉)
Department of Electrical Engineering, Jamia Millia Islamia, New Delhi, India

A. R. Gupta et al. (eds.), *Power Electronics and High Voltage in Smart Grid*, Lecture Notes in Electrical Engineering 817,
https://doi.org/10.1007/978-981-16-7393-1_2

1 Introduction

Small and large-scale RES integrated on transmission as well as distribution network has been significantly enhanced in the past few years [1]. However, due to its intermittent nature and smaller capacity, it is generally connected to a low voltage (LV) distribution network [2]. RES is different from centralized conventional generation due to its uncertain output and mainly depends upon local weather conditions [3, 4]. At LV networks, uses of low power non-linear/unbalanced loads to a large extent cause serious line grid-side harmonics [5]. In the past few years, a decrease in fossil fuel capacity along with the rise in global warming, harnessing, and utilization of natural green sources has been increasing.

The existence of harmonics deteriorates other sensitive linear or non-linear loads leading to higher ohmic and switching losses [6]. Filters are required to minimize harmonic current into the grid. Active filters are preferred over passive filters owing to the higher size of passive filters [7]. Series active filters are used to minimize voltage harmonics whereas shunt active filters minimize current harmonics. Voltage source inverter (VSI) is generally utilized for converting DC into AC using pulse with modulation (PWM). VSI generally acts as a shunt active filter at the distribution network. These filters work in several ways like balancing of power or power management between utility grid and RES, load current harmonics compensation along with compensation of reactive power n [8]. Various control strategies have been proposed to deal with PQ issues in distribution network [9]. Id and Iq Control methods with PI controller has been proposed in [10] in which current harmonics are compensated at PCC. Chilipi et al. [11] has proposed a hysteresis controller which is generally robust and easily implementable. The main problem of hysteresis controller is that it suffers from high switching losses and is difficult to design its output filter.

In [12, 13], the authors suggested utilization of PI controller embedded in VSC which work on synchronous reference frame (d-q frame) which work efficiently for both linear and non-linear loads. In this paper, simple, robust synchronous reference frame SRF PI controller is proposed in which both active and reactive power is specified to generate Id and Iq reference which is compared with tracked grid side Id and Iq and error is transferred to PI controller for zero steady-state error regulation. Further, PI output signal is again transformed back to a-b-c coordinate natural reference frame which is called reference signal U_{ref} and it is compared with triangular pulses at 3 kHz switching frequency and thus six pulses generated and feedback to VSC. A synchronous frame phase-locked loop (SF-PLL) is utilized to synchronize the grid. Generally, a filter is required to connect VSC with the grid in order to compensate the line harmonics produced by inverter PWM. Third-order LCL filter is superior for high power industrial applications. However, it is suffering from resonance phenomena and initiate instability furthermore application of LCL filter in PI-based controllers would not solve the resonance problem effectively. Thus, an LC filter is used here which is more appropriate with the PI controller. PI controller embedded in VSC has been modeled in MATLAB to examine the real, reactive power

Table 1 System parameters in PI-based conventional controller

Proposed system parameters	Numerical values
Inverter Line voltage	520 V
DC Link nominal	700 V
Utility frequency	50 Hz
DC link capacitor	1000
Switching frequency	5 kHz
Inductor (Inverter side)	3 mH
Inductor (Grid side)	1.5 mH
Shunt capacitor	

of utility grid, inverter and loads integrated at PCC. The various system parameters for inverter interfaced DG is mentioned in Table 1.

2 Proposed Controller Description

The proposed system contains RES (PV) which is connected to a grid interfacing voltage source inverter as shown in Fig.1. Here RES is taken as a DC source with VSC coupled to DC link. Generally, PV generation is variable, intermittent and at low DC voltage output, thus power generated from PV needs power conditioning DC-DC converter before DC link. The DC-link capacitor decouples PV and grid and in addition, it independently controls both sides of the DC link. System parameters are specified in Table 1. The system consists of a PV system, non-linear/unbalanced loads, 3-ϕ VSC, and LC filter. The main purpose of grid-connected PV system is to effectively control power between unbalanced loads and to hand over power to the utility grid in case of excessive available power.

3 Proposed Controller Design

The proposed control strategy is mainly based on the current controller embedded in the outer loop of VSC. Enhanced SRF-PLL is used to generate grid synchronization angle, which is utilized into natural a-b-c to synchronous d-q transformation. Grid voltage, grid current are sensed and feedback to reference generation in which two more additional inputs are active and reactive power, thus Idref and Iqref are produced. Now, the obtained reference currents and grid currents Id and Iq are compared which is further passed to the PI controller and output error is utilized to generate 6 control pulses to control VSC output. The proposed method is very easy to utilize, implement, and robust as compared to many proposed controllers discussed in literature, viz., [8] (Fig. 2).

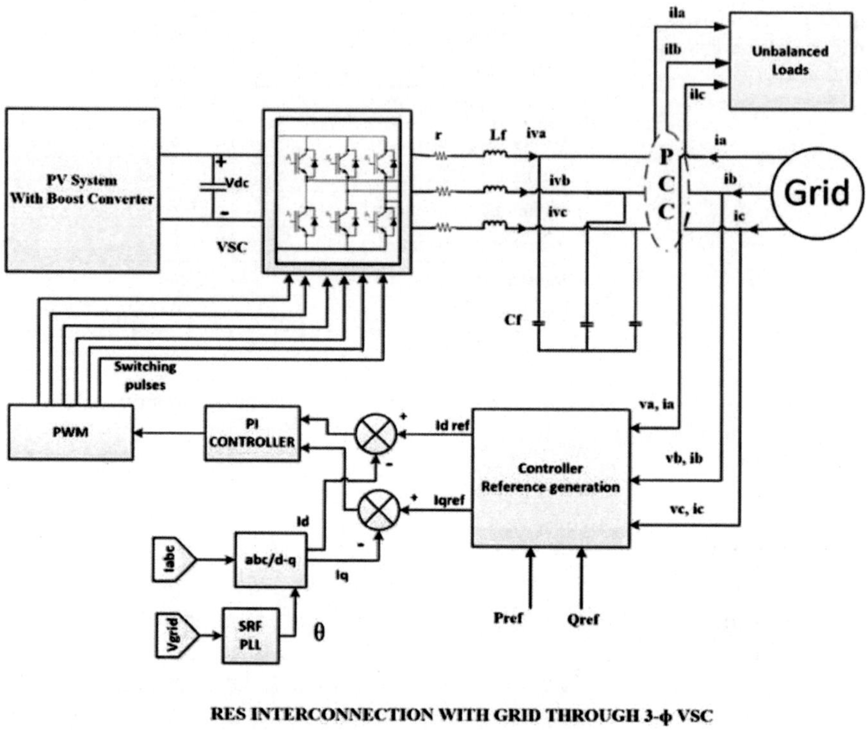

Fig. 1 Schematic of proposed controller embedded in VSC interconnected to the grid

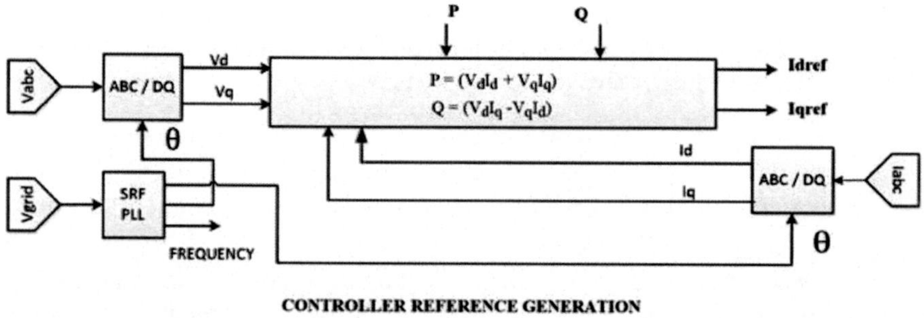

Fig. 2 Controller reference generation with specified active and reactive power

4 Performance Evaluation Through Simulated Results

In order to examine the suggested control strategy, Fig. 1 is simulated using MATLAB. VSC is effectively controlled to accomplish the balanced sinusoidal grid currents at PCC to maintain power factor unity under unbalanced load connected.

Grid voltage and currents (Va, Vb, Vc) and (Ia, Ib, Ic) are illustrated in Fig. 3. Initially grid alone feeds the unbalanced load up to 0.5 s. Thus, the grid current profile seems similar to the load current. At 0.5 s, VSC is connected to grid with the

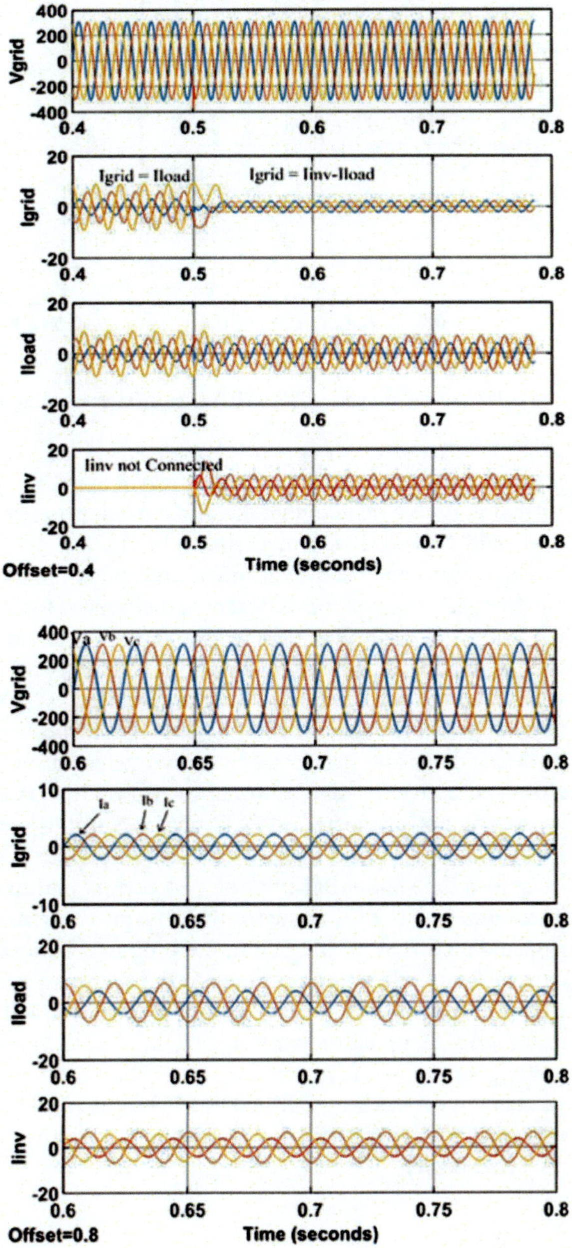

Fig. 3 Simulation results
(a) Utility Grid voltage (V)
(b) Utility Grid current (A)
(c) Load current (A)
(d) Inverter current (A)

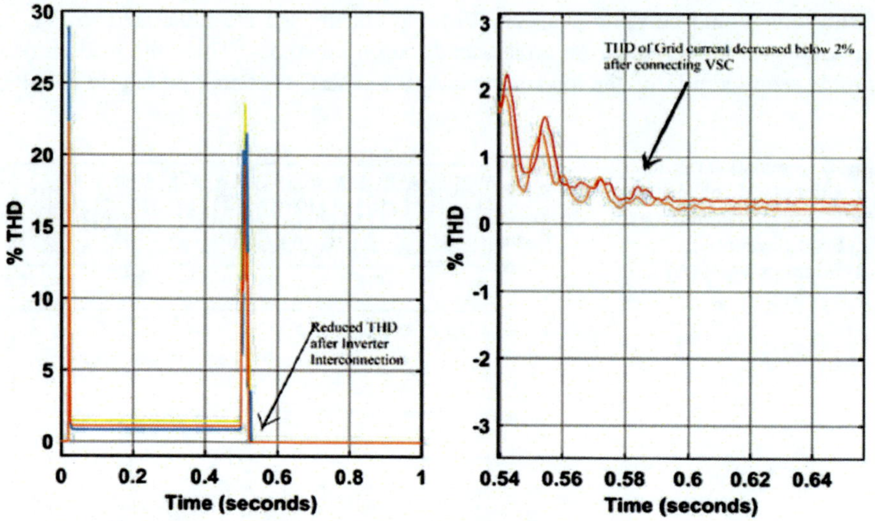

Fig. 4 Reduced Grid current (% THD) after inverter interconnection

proposed controller. Therefore, as represented in Fig. 3, the inverter attempts to inject current in such a manner that the grid current profile transforms from unbalanced to sinusoidal balanced current as shown in Fig. 3.

This discussion follows a similar case study mentioned in literature [8] however, the proposed controller is different and observed better performance as compared to the hysteresis controller. Percentage THD of grid current is shown in Fig. 4, which indicates that after VSC interconnection THD would obtain below 2%.

Grid active and reactive power are specified as Pg and Qg and similarly load and inverter power are specified as Pl, Ql, Pi, Qi, respectively, as portrayed in Fig. 5. The inverter power and grid power positive values are assumed to flow towards PCC. However, active/reactive power absorbed by load is assumed as negative. Initially, inverter is not connected to PCC until 0.5 sec as earlier discussed. Therefore, all load power is supplied by the grid itself before 0.5 s. This is shown in Fig. 5, which illustrates the similar behavior of grid current and load current. At 0.5 s inverter is interconnected to PCC in such a way that grid current will be forced to balanced and sinusoidal current profile due to the proposed controller action. Here, load neutral current is not taken into account. At 0.5 s, the inverter starts injecting power from RES (PV) to feed local load demand and whatever will be the extra has been supplied to the grid. Here, as shown in Fig. 5 only small power remains extra which is fed to the grid after 0.5 s. However, grid reactive power becomes zero and load reactive power demand has been fulfilled by PV. Negative Pg suggests the grid starts receiving power from the inverter. Moreover, the inverter enables the grid to supply/consume only real power.

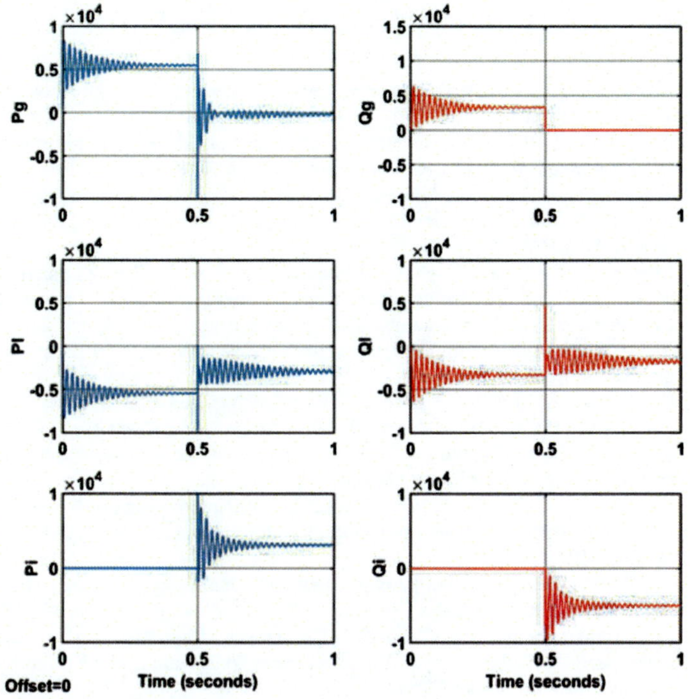

Fig.5 Simulation results: Grid PQ, Load PQ, Inverter PQ(W)

5 Conclusion

This paper illustrates a simple, robust controller, which can be easily installed into an inverter's outer loop and efficiently utilized for power balance and power conditioning. The proposed controller with inverter efficiently operates as an active power filter, inject real power and reactive power to PCC such that grid reactive power will become zero and grid starts working on unity power factor. The proposed controller eliminates any additional power-conditioning device to install which is required to improve PQ. Thus, the VSC effectively works as an Active Shunt Power Filter. The proposed methodology used in this work has been validated using the simulation performed using MATLAB software. Therefore, the proposed study proved the multi-function ability of grid-interfacing inverters. Additionally, when grid power generated from RES is sufficiently high as compared to load demand, the proposed inverter controller fulfills total active/ reactive power load demand along with delivering extra active power in such a way that grid reactive power will become zero.

References

1. Mosobi RW, Gao S (2019) Performance analysis of distributed energy resource-based LV distribution grid. The Institution of Engineers (India) springer
2. Adefarati T, Bansal RC (2016) Integration of renewable distributed generators into the distribution system: a review. IET Renew Power Gener 10(7):873–884
3. Bhandari B, Poudel SR, Lee K-L, Ahn S-H (2014) Mathematical modeling of hybrid renewable energy system: a review on small hydro-solar–wind power generation. Int J Precis Eng Manuf Green Technol 1(2):157–173
4. Nehrir MH, Wang C, Strunz K, Aki H, Ramakumar R, Bing J, Miao Z, Salameh Z (2011) A review of hybrid renewable/alternative energy systems for electric power generation: configurations, control, and applications. IEEE Trans Sustain Energy 2(4):392–403
5. IEEE Std 519–1992 IEEE recommended practices and requirements for harmonic control in electrical power systems. IEEE, New York, NY
6. Enslin JHR, Heskes PJM (2004) Harmonic interaction between a large number of distributer power inverters and the distribution network. IEEE Trans Power Electr 19(6):1586–1592
7. Singh B, Al-Haddad K, Chandra A (1999) A review of active filters for power quality improvement. IEEE Trans Ind Electr 46(5):960–971
8. Singh M, Khadkikar V, Chandra A, Varma RK (2011) Grid interconnection of renewable energy sources at the distribution level with power-quality improvement features. IEEE Trans Power Deliv 26(1):307–315
9. Afonso J, Couto C, Martins J (2000) Active filters with control based on the p-q theory. IEEE Ind Electr Soc Newslett 47(3):5–10
10. Nagaraj C, Manjunatha Sharma K (2016) Improvement of harmonic current compensation for grid integrated PV and wind hybrid renewable energy system. In: IEEE 6th International Conference on Power Systems (ICPS)
11. Chilipi R, Al Sayari N, Al Hosani K, Beig AR (2016) Control scheme for grid-tied distributed generation inverter under unbalanced and distorted utility conditions with power quality ancillary services. IET Renew Power Gener 10(2):140–149
12. Yao Z, Xiao L, Guerrero JM (2015) Improved control strategy for the three-phase grid-connected inverter. IET Renew Power Gener 9(6):587–592
13. Guo XQ, Wu WY (2010) Improved current regulation of three-phase grid-connected voltage-source inverters for distributed generation systems. IET Renew Power Gener 4(2):101–115

Design of a Power Converter for Domestic Induction Heating Driven from PV Module

S. A. Mizbah Farheen, Omsekhar Indela, and S. Sridhar

Abstract The electrical energy extracted from the solar PV unit for domestic induction heating applications is extremely advantageous in many off-grid areas. A cost-effective and highly reliable DC-DC converter to make best utilization of solar power along with a series resonant inverter to provide high-frequency supply with an implicit control strategy is essential to drive the induction heater for domestic applications. This paper proposes a DC-DC converter and a series resonance inverter, which is appropriate for supplying the induction heating load under diverse conditions. The various states of operation have been examined depending on the availability of solar irradiation, status of charge of the battery and the load requirements. Also, a suitable control strategy with the integration of charge controller and MPPT is executed for the proposed converter and asymmetrical control with phase lock loop is implemented for a series resonant inverter. The simulation of this topology under four different operating states is carried out in the MATLAB/Simulink platform.

Keywords Induction Heating (IH) · Series Resonant Inverter (SRI) · Power Converter · Control Strategy · Maximum Power Point Tracking (MPPT)

1 Introduction

There is a rapid rise in energy requirement and heavy intake from renewable resources, which have resulted in solar energy becoming one of the most popular renewable energy. Amidst various applications where the solar panel is being used, a stand-alone PV to battery-powered induction heater is employed in domestic applications. These days induction heating has become an alternative among various

S. A. Mizbah Farheen (✉) · O. Indela · S. Sridhar
Ramaiah Institute of Technology, Bangalore, India

O. Indela
e-mail: omsekhar@msrit.edu

S. Sridhar
e-mail: sridhars@msrit.edu

A. R. Gupta et al. (eds.), *Power Electronics and High Voltage in Smart Grid*,
Lecture Notes in Electrical Engineering 817,
https://doi.org/10.1007/978-981-16-7393-1_3

domestic, medical, and industrial applications because of the merits related to it, like accurate control, cleanliness, safety, efficiency and fast heating. The working principle of IH is a combination of electromagnetic induction and Joule heating. IH load can be modelled by utilizing the equivalent resistance R_{eq} and L_{eq} in series combination. Depending on several variables like shape of the heating coil, distance between the workpiece and IH coil, temperature, frequency, electrical conductivity and magnetic permeability the load parameters are designed.

There exist many converters to drive the Induction Heater from an AC supply [1–4]. Such power drivers are best suited for both industrial and domestic applications. As depicted in Fig. 1. to obtain power from a PV module and a battery in case PV power fails to operate, a DC-DC power converter will have two inputs, one from the PV panel and the other from the bidirectional input port is used [5, 6]. A bidirectional input port is usually preferred for the battery so that it can store the extra energy when the available solar PV power is more and this stored energy can later be used in conditions where the sunlight is not available or when there is partial shading. Therefore, to fulfil the load requirements, to track the maximum power from solar panel depending on the state of charge of the battery, a DC-DC converter is designed [7–9]. Based on the load requirements, Perturb and Observe scheme of MPPT control is utilized for the PV panel [10–13]. Thus, the output obtained from the DC-DC converter is fed to the series resonance inverter, which will provide a high-frequency supply required for the induction heating load. The proposed DC-DC converter can be used to run the motor load [15] or to deliver the grid in the regions, where the availability of solar power is huge [16, 17]. The principle of operation and circuit configuration of a full-bridge SRI are explained in detail in Sect. 2. An asymmetrical control strategy is proposed in Sect. 3. The circuit description along with numerous states of operation for various modes is discussed in segment 4. In Sect. 5, to switch DC-DC converter among various states of operation under different states a suitable control strategy is discussed. The simulation results of DC-DC converter to drive the induction heater from the battery and the solar PV panel are carried out in MATLAB/Simulink environment. Further, results of different states of operation are illustrated in Sect. 6.

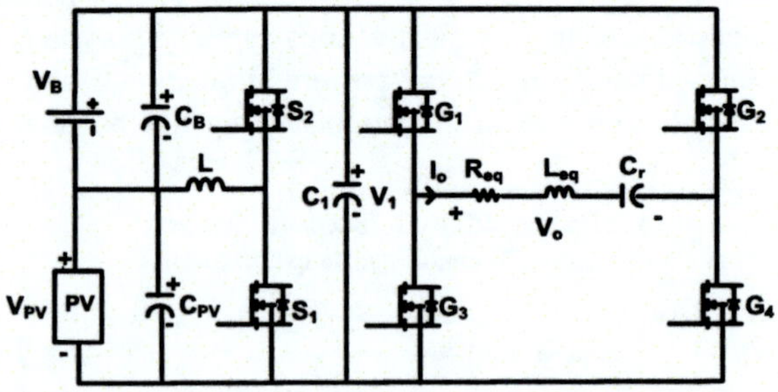

Fig. 1 Proposed power converter

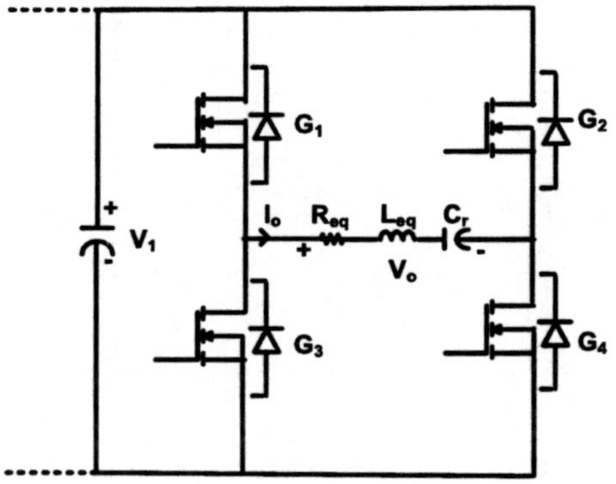

Fig. 2 Series resonant full-bridge inverter

2 Series Resonant Full-Bridge Inverter

A single-phase full-bridge voltage source inverter is utilized for the generation of a high-frequency magnetic induction between the cooking vessel and coil. Figure 2 shows the FBI for low-power induction heating applications. The single-phase voltage source Full-Bridge Inverters (FBI) load circuit consists of an equivalent inductor, equivalent resistor and a resonant capacitor. The DC input voltage V_1 is supplied to the inverter. In order to provide high-frequency current alternatively to the induction coil, switches (G_1, G_4) and (G_3, G_2) are used. When FBI is operated in inductive mode, zero voltage switching can be realized which makes sure that the switches are being operated with a frequency slightly more than the resonant frequency. There are four operating modes under each circuit operation.

Mode 1: As shown in Fig. 3a switches G_1 and G_4 are administered with high-frequency square pulses. The direction of the flow of current is V_1–G_1–V_0–G_4–V_1. During this operation, the output across the load is V_1. Both G_1 and G4 are switched off at the end of this mode.

Mode 2: As shown in Fig. 3b, the diode of G_1 and the diode of G_4 will conduct. Thus, the direction of current is from V_0–G_1–V_1–G_4–V_0. The voltage across the load will be V_1 during this mode.

Mode 3: Fig. 3c shows that this mode begins when G_2 and G_3 will be triggered. The voltage across the load will be—V_1. The direction of load current is V_1–G_2–V_0–G_3–V_1. This mode ends when G_2 and G_3 will be turned off.

Mode 4: As shown in Fig. 3d, the diode of G_2 and the diode of G_3 will conduct. Thus, the direction of the current is from V_0–G_2–V_1–G_3–V_0. The voltage across the load will be–V_1 during this mode.

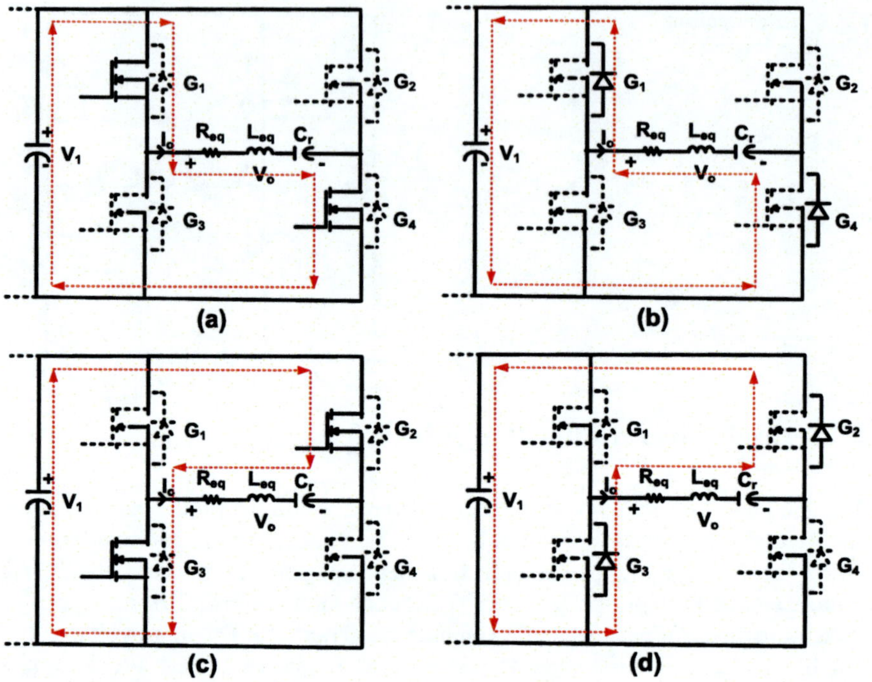

Fig. 3 Operation modes of series resonance full-bridge inverter

2.1 Characteristics of Resonant Tank

For simulation study, an IH load in series with resistance R_{eq} and inductance L_{eq} is considered. The resonant circuit is formed when R_{eq} and L_{eq} and C_r are connected in series. Design parameters of load are explained with the help of the following equations:

The series resonant circuits angular resonant frequency (ω_r) is given by

$$\omega_r = \frac{1}{\sqrt{L_{eq} C_r}} \tag{1}$$

where C_r denotes the resonant capacitor capacitance.

ω_n: normalized angular switching frequency is given as

$$\omega_n = \frac{\omega_s}{\omega_r} \quad (2)$$

The IH load characteristic impedance (Z_0) is

$$Z_0 = \sqrt{\frac{L_{eq}}{C_r}} \quad (3)$$

The IH coil quality factor (Q_L) is given by

$$Q_L = \frac{\omega_r L_{eq}}{R_{eq}} = \frac{1}{\omega_r R_{eq} C_r} = \frac{Z_0}{R_{eq}} \quad (4)$$

The resonant tank circuit equivalent impedance (Z_{eq}) can be expressed as

$$Z_{eq} = R_{eq} + j\left(\omega_s L_{eq} - \frac{1}{\omega_s C_r}\right) \quad (5)$$

The output voltage of the fundamental component can be obtained using the Fourier analysis:

$$V_0 = V_m sin\omega_s t \quad (6)$$

where V_m is the output voltage peak value.

The resonant tank circuit load current (I_0) is described with the following equation:

$$I_0 = I_m sin(\omega_s - \phi) \quad (7)$$

where I_m is the peak value of the output current, Ø is the phase angle.

Therefore, the output power can be expressed as

$$P_{out} = \frac{V_o I_o}{2} cos\phi \quad (8)$$

where $\phi = \frac{\alpha}{2}$ and α is the control angle.

3 Proposed Control Strategy for Series Resonance Inverter

The control strategy that is proposed mainly focuses on two things, the first one is to control the power of full-bridge SRI using the angle alpha (α) of switch G_4 and the second one is to control the frequency of the full-bridge SRI for ZVS operation.

The controller comprises a low-pass filter, phase detector, and voltage-controlled oscillator (VCO), as depicted in Fig. 4. The phase-locked loop (PLL) is used to control the frequency a little above the resonant frequency. In a conventional voltage source inverter, a gate drive signal is in phase with the asymmetrical output voltage signal V_0. Therefore, for phase detection, a gate drive signal is used rather than the output voltage signal. To determine the phase difference, current signal I_0 and voltage signal are compared. Further, using a low-pass filter the output signal obtained from the phase detector is filtered to obtain an average value proportional to the phase difference across the load. Then signal α, and the ramp signal from the phase-locked loop are compared which generates the gate signal for G_4. If the signal α is higher than the ramp signal, the gate pulse of switch G_4 is set to high. The phase detector generates the ramp signal. Thus, the ZVS operation is achieved when the frequency gets automatically adjusted to follow the resonance frequency.

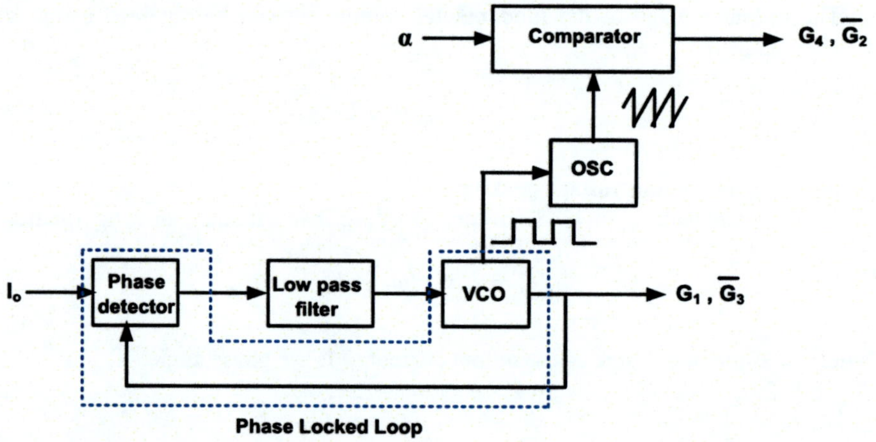

Fig. 4 Control strategy for a series resonant inverter

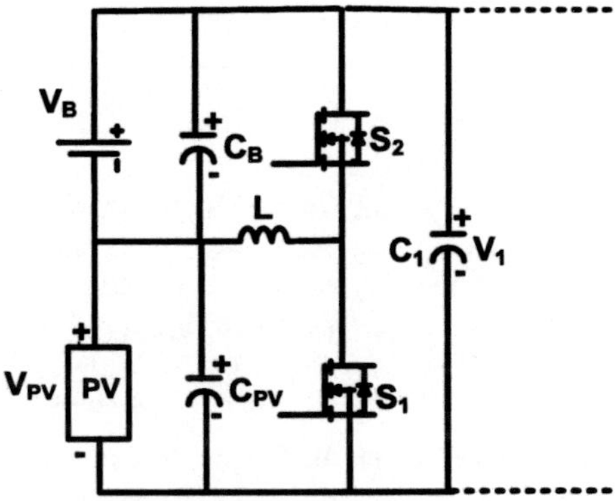

Fig. 5 The proposed power converter

4 Circuit Configuration and Operation States of DC-DC Converter

4.1 Circuit Configuration

In Fig. 5 the circuit configuration of a DC-DC converter is shown. The DC-DC converter uses PV module as a primary source of supply and battery as an auxiliary source of supply. Based on load requirements, climatic conditions and availability of the battery charge level, the converter is able to work under various states of operation. Power MOSFET switches S_1 and S_2 working at a high switching frequency (10 kHz), which are complementary to each other are given to the gate signals. The common inductor L is used to charge the battery from PV power and also supply the load through the switching action of the switches S_1 and S_2 present alongside the output capacitor C_1. The photovoltaic module and the battery system are connected across the capacitors C_{pv} and $C_{B,}$ respectively. When not enough PV power is present to supply the load requirements or during non-availability of sunlight, the battery discharges the current supplying the load when required. A constant current charging method is adopted for charging and discharging of the battery.

4.2 Operating States

Depending on the requirements of the load and the availability of solar power, the analysis of the circuit can be done under four different operating states as discussed below. Further, there are two modes in each switching condition.

Case 1: PV to load operating state $\left(P_{pv} = P_{load}\right)$

In this case, the battery is non-operational or not connected, which results in the PV module supplying power to the load. There are two switching modes in this operating state as depicted in Fig. 6.

Mode I: In this mode, S_1 is switched on while S_2 is off. Inductor L will start charging using the PV power. The capacitor C_1 supplies the load.

Mode II: In mode II, S_2 is switched on while S_1 will remain off. Inductor L will start discharging into the PV source and also charges the capacitor C_1.

Case 2: Battery to load operating state $(P_{batt} = P_{load})$

In this case, when the PV module is unable to supply the power during the night or when it is shaded, the battery will supply the power to load. There are two switching modes in this operating state shown in Fig. 6.

Mode III: In this mode, S_2 is switched on while S_1 is off. The inductor L gets charged by the battery and the output capacitor C_1 supplies the load.

Mode IV: In this mode, S_1 is switched on while S_2 will remain off. Inductor L will start discharging into the battery and also charges the capacitor $C_{1.}$

Case 3: PV and battery to load operating state $\left(P_{pv} + P_{batt} = P_{load}\right)$.

Due to various factors like partial shading or low intensity of solar radiation during the daytime, the load will require power from an additional source. Thus, the load will receive power from the PV panel and battery in this case for the circuit to operate. There are two switching modes in this operating state as illustrated in Fig. 6.

Mode V: In mode V, switch S_1 will be turned on, the inductor L will start charging from the PV source and the battery will supply the load.

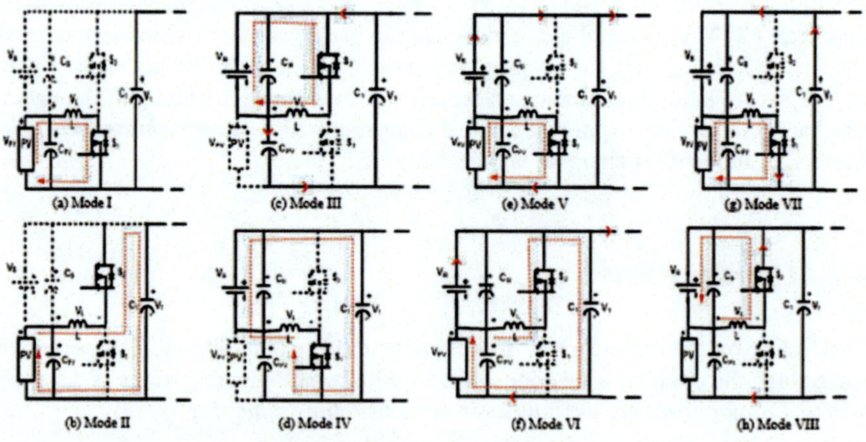

Fig. 6 Operation modes of DC-DC converter

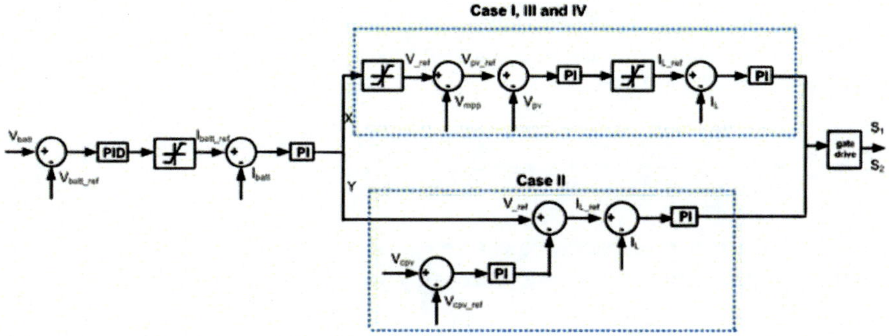

Fig. 7 Closed-loop control for the proposed DC-DC converter

Mode VI: During this mode, S_2 is turned on and switch S_1 turns off. The inductor L will start to discharge into the PV source and also charge the capacitor C_1. The battery will supply the power to the load.

Case 4: PV to load and battery operating state $\left(P_{pv} + P_{batt} = P_{load}\right)$.

During the availability of sunlight that is when the intensity of solar radiation is high and when there is enough power available from the PV panel to satisfy the load requirements, the PV panel will not only supply the power to load to satisfy the load requirements but it will also charge the battery. There are two switching modes in this operating state as illustrated in Fig. 6. Switches S_1 and S_2 operate in complementary mode.

Mode VII: During this mode, S_1 switched on inductor L will start charging from the PV power source V_{pv}. Capacitor C_1 will discharge into the battery and also charge the inductor L. This mode ends when S_1 turns off and S_2 is turned on.

Mode VIII: During this mode, S_1 is turned off and S_2 is turned on. Inductor L, will start discharging the current to the battery, it will also give its energy to the PV source and charge capacitor C_1.

4.3 *Circuit Analysis*

The expressions are derived considering the case of IV operating state. Using the information of the limitation in the ripple contents of current in the inductor for continuous conduction mode and of inductor current ($\triangle i_L$), the expression for inductor can be calculated using Eq. (9),

$$L = \frac{V_{pv}}{\Delta i_L}(DT_s) \tag{9}$$

where D = Duty ratio, T_s = switching time period. Likewise, for the PV capacitor with the ripple content in the battery current ($\triangle i_B$) and PV voltage ($\triangle V_{pv}$) can be calculated using Eq. (10)

$$C_{pv} = \frac{(i_{pv} - i_L)}{2\Delta V_{pv}}(DT_s) \tag{10}$$

Further, the battery capacitance can be calculated with the ripple content in the battery charging current (Δi_B) and the battery voltage (ΔV_B) is given by

$$C_B = \frac{(i'_B - i_{CB})(1 - D)TS}{2\Delta V_{CB}} \tag{11}$$

Lastly, the output voltage is obtained by taking the limit in the ripple content of the output voltage (ΔV_1) as

$$C_1 = \frac{\mathrm{V_1(DTs)}}{\mathrm{2Z\Delta V_1}} \tag{12}$$

where Z is the load impedance under maximum power consumption.

5 Control Strategy

To maintain a balanced power between the output of the DC-DC converter, battery and PV panel, a suitable control strategy is proposed.

Depending upon different climatic factors, load must be able to maintain the power balance within the system. A control strategy as depicted in Fig. 7 under different cases of operation of the DC-DC converter is discussed.

For Cases–1,3,4: The voltage reference $V_{\_ref}$ obtained from the battery current controller will become positive and also restricted to zero. The signal on comparing with the volage signal V_{mpp} obtained from the MPPT shows that V_{pv_ref} signal is acquired from the MPPT controller. The comparison is done between V_{pv} voltage signal and the reference signal V_{pv_ref}. The resultant is given to the voltage controller. Depending on the state of charge of the battery, the controller switches into any of the cases-1,3 or 4. Whenever the available solar energy is not sufficient to fulfill load requirements, the controller operates in case-3 condition and whenever the solar radiation intensity is high and also if available solar power is enough to fulfill the load requirements, then the controller operates in case 1 condition. The difference of the inductor current i_L and inductor reference current i_{L_ref} is provided to the current controller, resultant signals are given to a gate driver circuit to generate the required gating signal. When the PV cannot fulfill load requirements alone, the battery starts to discharge and the controller operates in case 4.

For Case–2:The capacitor voltage V_{cpv} and the reference capacitor voltage V_{cpv_ref} are compared and given to a PI controller. This resultant signal is compared with $V_{\_ref}$ signal obtained by the battery current controller, then output i_{L_ref} is compared with i_L and given to PI controller. The resultant is given to the gate driver circuit which generates gating pulses for the switches.

6 Simulation Results

Detailed simulations for the closed-loop control of the DC-DC converter under different operating cases are carried out based on the specifications mentioned in Table 1.

An MSX60 Solarex PV panel of rating 60 W with specifications as shown in Table 2 is implemented in MATLAB/Simulink environment. Using Perturb and Observe method, MPPT is attained. At 1000 W/m^2 a maximum of 60 W is delivered from the PV panel. A 13 V, 10 Ah, lead-acid battery is employed as a secondary source of power supply. Angle α is varied from 0° to 180° for controlling the induction heater output power. The simulation results of the DC-DC converter under dissimilar states of operation are obtained for the closed-loop control.

Table 1 Circuit parameters

Parameters	Ratings
Battery	13 V, 30 Ah
PV capacitor, C_{pv}	6 µF
Battey capacitor , C_B	6 µF
Inductor, L	10 mH
Output capacitor , C_1	1 mF
Switching frequency of the DC-DC converter, f_s	10 kHz
Switching frequency of SRI, f_{sw}	40 kHz
Resonance frequency, f_r	38 kHz
Load resistance, R_{eq}	7.29 Ω
Battery 100	13 V, 30 Ah

Table 2 Specification of Solarex MSX60 PV panel

Parameters	Ratings
Maximum Power, (P_{max})	60 W
Voltage at P_{max}, (V_{mpp})	17.1 V
Current at P_{max}, (I_{mpp})	3.5 A
Short-circuit current (I_{sc})	3.8 A
Open-circuit voltage (V_{oc})	17.5 V
Maximum Power, (P_{max})	60 W

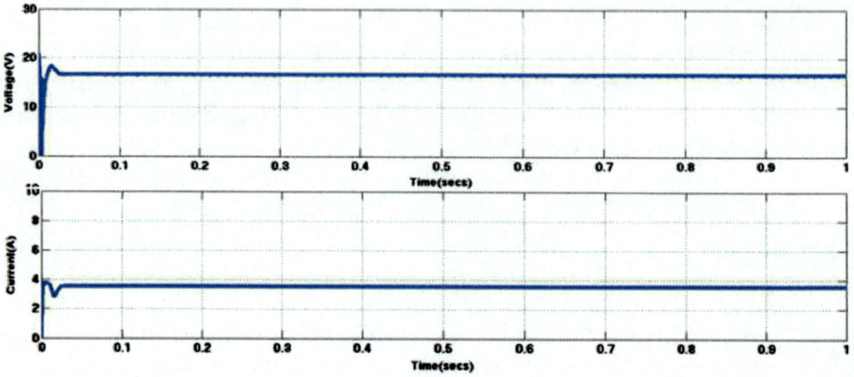

(a)

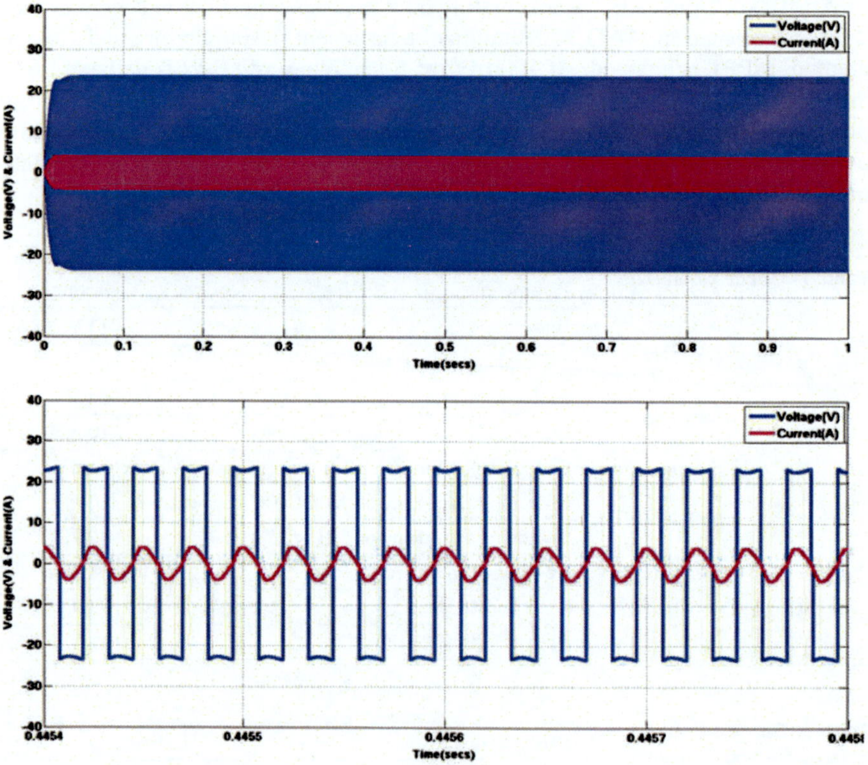

(b)

Fig. 8 Simulation results for case 1 **a** PV V and I **b** Load V and I

Table 3 Simulation results for case 1

Parameters	PV module	Load
Voltage	16 V	24 V
Current	3.2 A	4 A
Power	15.12 W	48 W

Table 4 Simulation results for case 2

Parameters	Battery	Load
Voltage	13 V	13 V
Current	1A	2 A
Power	13 W	12.87 W

Table 5 Simulation results for case 3

Parameters	PV module	Battery	Load
Voltage	10 V	13 V	23 V
Current	2 A	1.8 A	3.5 A
Power	20 W	23.4 W	39.84 W

Case 1: In case 1 condition, power is directly supplied from the PV panel to the load without any involvement of the battery as the available solar power is sufficient for meeting the load requirements. The simulation results are represented in Table 3 and Fig. 8, respectively.

Case 2: In this case, that is, during night time when the sunlight is not available, the load will receive power from the battery. Table 4 and Fig. 9 illustrate the simulation results for this case. In this case, the battery will discharge the current to the load which can be indicated by the positive value of the battery current.

Case 3: In case 3 operation, when the available sunlight is not enough to fulfil the load demands, the battery will start to discharge which can be indicated by the battery's current positive value to satisfy the load demands. Thus, the load, in this case, will receive power from the PV panel and battery, respectively. Simulation results are shown in Table 5 and Fig. 10.

Case 4: In this case, the available sunlight will meet the load requirements as well as charge the battery. The results of the simulation carried out are shown in Table 6 and Fig. 11.

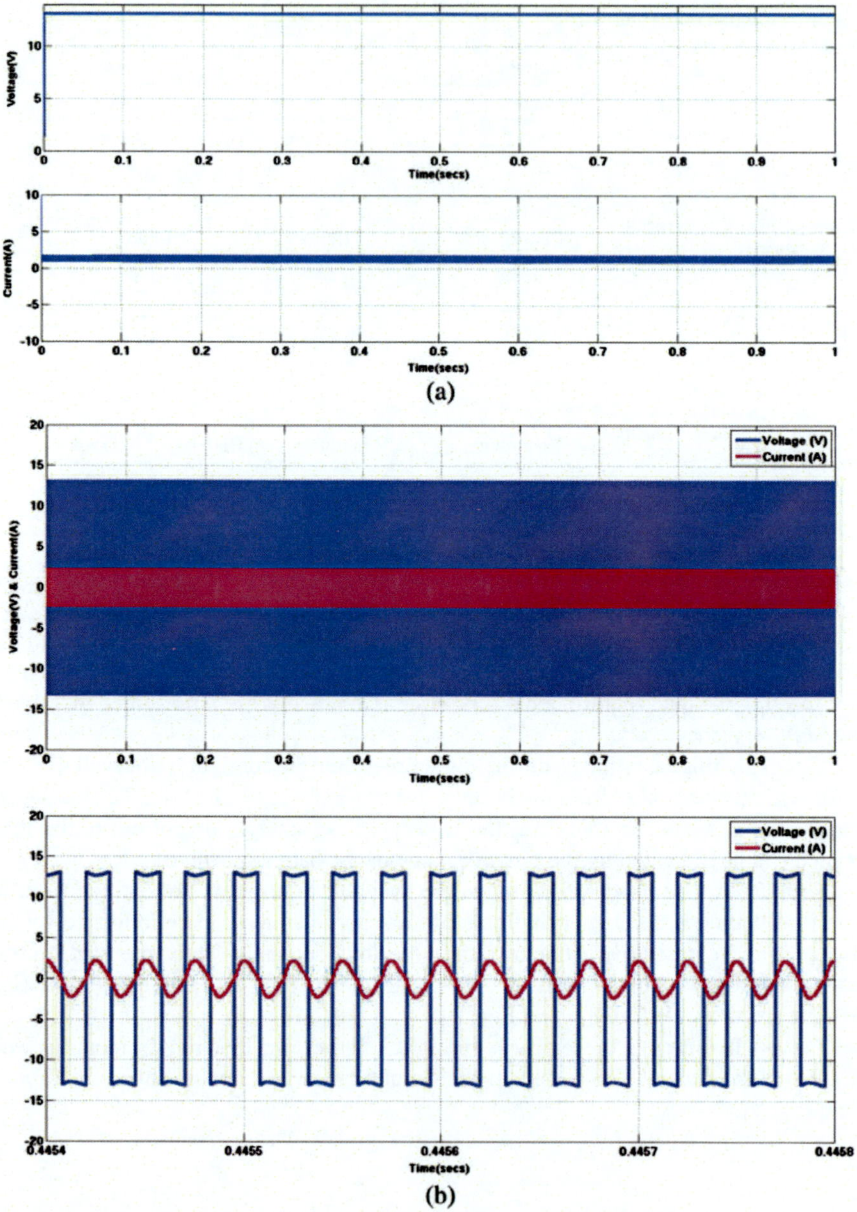

Fig. 9 Simulation results for case 2 **a** Discharging V and I of battery **b** Load V and I

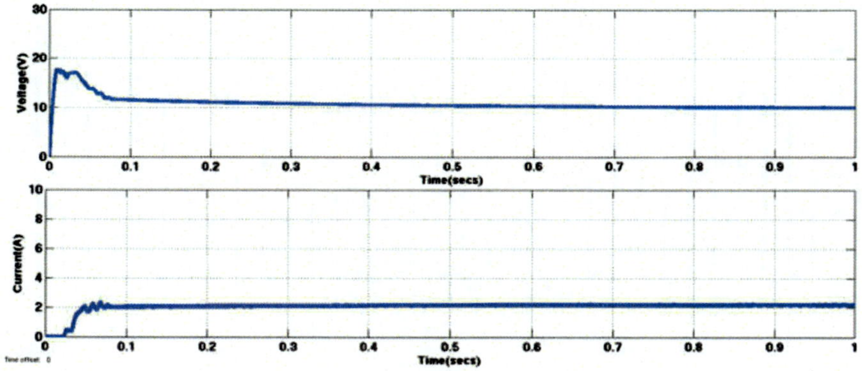

(a)

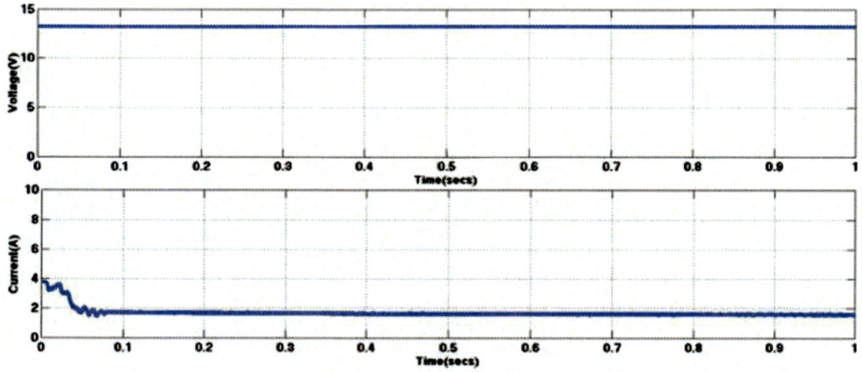

(b)

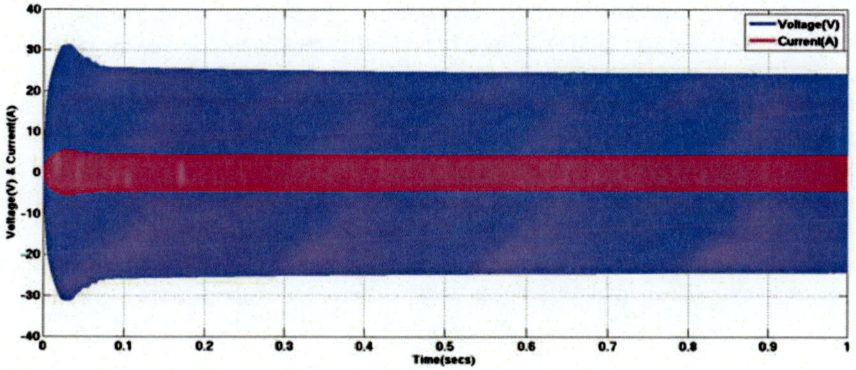

Fig. 10 Simulation results for case 3 **a** PV V and I **b** Battery charging V and I **c** V and I

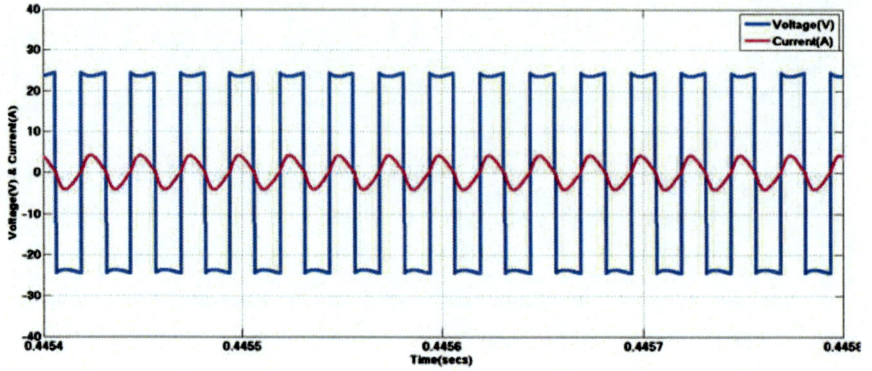

(c)

Fig. 10 (continued)

Table 6 Simulation results for case 4

Parameters	PV module	Battery	Load
Voltage	10 V	13.14 V	25 V
Current	4 A	1 A	4 A
Power	40 W	13 W	25 W

7 Conclusion

To obtain a cost-effective and highly reliable system along with solar PV panel and the battery for domestic induction heating applications, in this paper a DC-DC power converter with a suitable integrated control strategy is implemented. The asymmetrical control strategy method is utilized to control output power (P_o) of the induction heater under four different operating cases depending on the availability, partial availability, and absence of sunlight. The working modes of the DC-DC converter under different operating states are explained in detail. The output voltage obtained from the power converter is then given to the series resonant inverter to provide high-frequency supply to the load. The overall power converter is best suited for domestic induction heating applications which will make the best use of solar power available. A suitable integrated control strategy is used to govern the overall system in order to make the converter operate under different operating states as explained.

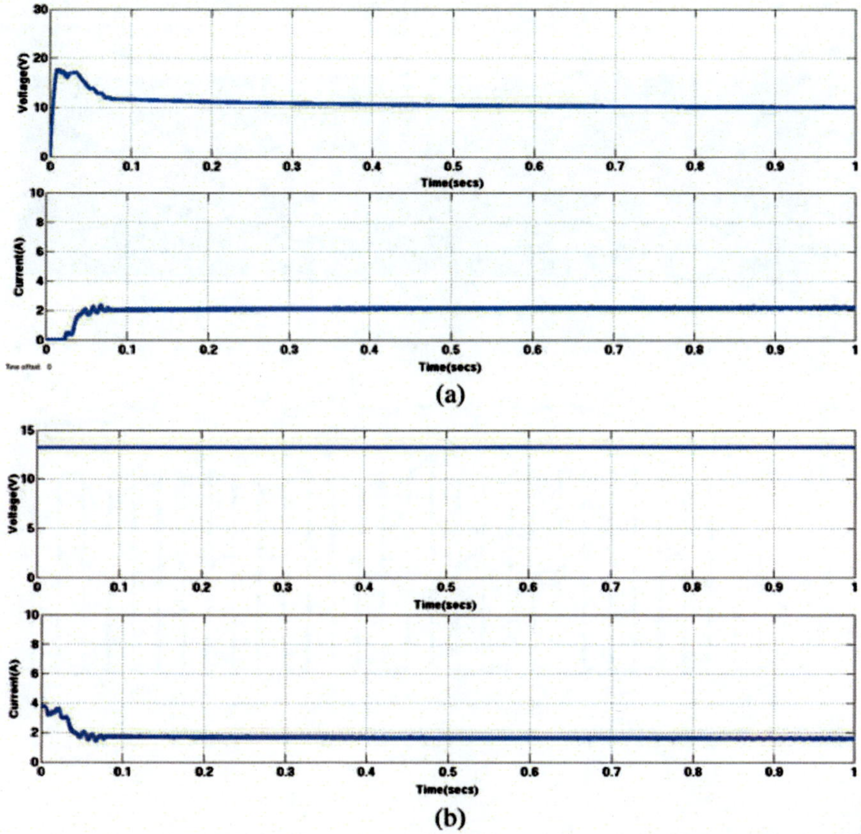

Fig. 11 Simulation results for case 4 **a** PV V and I, **b** Battery charging V and I, **c** Load V and I

The simulation of the overall system is carried out using a MATLAB environment. The output obtained for different cases under different operating states is illustrated. Furthermore, various transitions between various states of operation can be carried out such that the converter is capable of switching into any state of operation. Also, the analysis can be used to supply the grid where there the availability of sunlight is huge and the sudden load changes on the inverter can be studied.

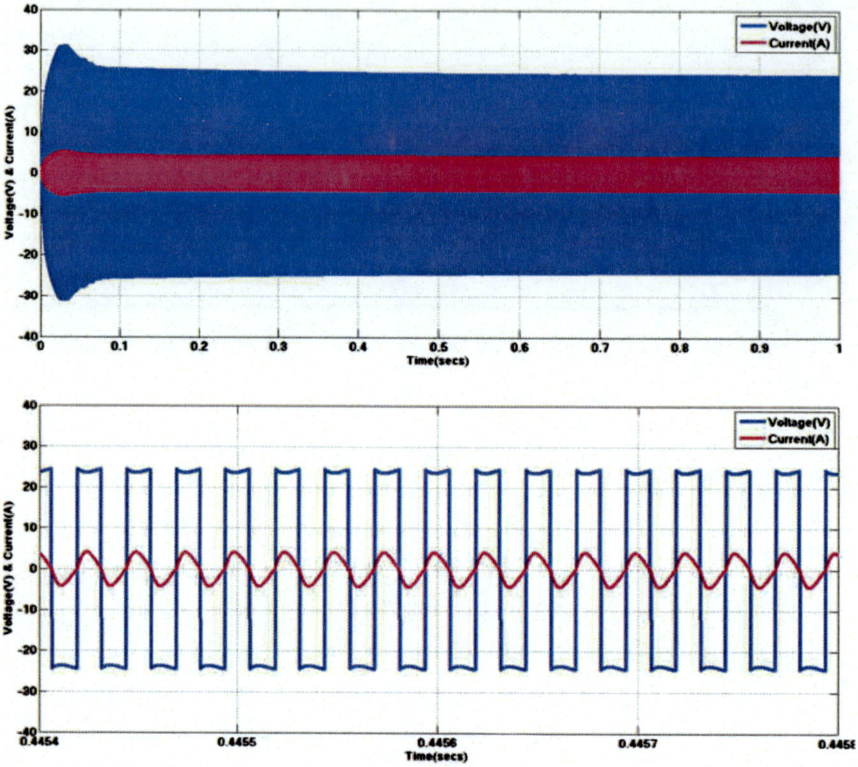

(c)

Fig. 11 (continued)

References

1. Komeda S, Fujita H (2018) A phase-shift-controlled direct AC-to-AC converter for induction heaters. IEEE Trans Power Electron 33(5):4115–4124. https://doi.org/10.1109/TPEL.2017.2712281
2. Meng L, Cheng KWE, Chan KW (2011) Systematic approach to high-power and energy-efficient industrial induction cooker system: circuit design, control strategy, and prototype evaluation. IEEE Trans Power Electron 26(12):3754–3765. https://doi.org/10.1109/TPEL.2011.2165082
3. Sarnago H, Mediano A, Lucia Ó (2012) High efficiency AC–AC power electronic converter applied to domestic induction heating. IEEE Trans Power Electron 27(8):3676–3684. https://doi.org/10.1109/TPEL.2012.2185067
4. Llorente S, Monterde F, Burdio JM, Acero J (2002) A comparative study of resonant inverter topologies used in induction cookers In: APEC. Seventeenth Annual ieee applied power electronics conference and exposition (Cat. No.02CH37335), vol. 2. Dallas, TX, USA, p. 1168–1174. https://doi.org/10.1109/APEC.2002.989392
5. Tayebi SM, Hu H, Abdel-Rahman O, Batarseh I (2018) Design and analysis of a dual-input single-resonant tank LLC converter for PV applications. In: 2018 IEEE applied power electronics conference and exposition (APEC). San Antonio, TX, 476–483

6. Chen G, Jin Z, Deng Y, He X, Qing X (2018) Principle and topology synthesis of integrated single-input dual-output and dual-input single-output DC–DC converters. IEEE Trans Ind Electron 65(5):3815–3825. https://doi.org/10.1109/TIE.2017.2760856
7. Young M (1989) The technical writer's handbook. University Science, Mill Valley, CA
8. Ranjini BR, Omsekhar I, Nayak G (2017) Design and control of power converter for LED lighting driven from the PV module. In: 2017 international conference on intelligent computing, instrumentation and control technologies (ICICICT). Kannur, pp. 61–67. doi: https://doi.org/10.1109/ICICICT1.2017.83425
9. Ranjini BR, Omsekhar I, Nayak G (2017) Integrated control strategy for a power converter used for photovoltaic lighting applications and its transition analysis. In: 2017 7th international conference on power systems (ICPS). Pune, pp. 269–276. https://doi.org/10.1109/ICPES.2017.8387305.
10. Al-Soeidat M, Aljarajreh H, Khawaldeh H, Lu DD, Zhu JG A reconfigurable three-port DC-DC converter for integrated PV-battery system. IEEE J Emerg Select Top Power Electr https://doi.org/10.1109/JESTPE.2019.2941595
11. Al-Soeidat M, Khawaldeh H, Aljarajreh H, Lu D (2018) A compact three-port DC-DC converter for integrated PV-battery system. In: 2018 IEEE international power electronics and application conference and exposition (PEAC). Shenzhen, pp. 1–6 https://doi.org/10.1109/PEAC.2018.8590624
12. Walker G (2001) Evaluating MPPT converter topologies using a Matlab PV Model. J Elect Electr Eng Australia 21
13. Patel H, Agarwal V (2008) MATLAB-based modeling to study the effects of partial shading on PV array characteristics. IEEE Trans Energy Convers 23(1):302–310
14. Hayder W, Abid A, Ben Hamed M (2017) Steps of duty cycle effects in P&O MPPT algorithm for PV system. In: 2017 international conference on green energy conversion systems (GECS). Hammamet, pp. 1–4. https://doi.org/10.1109/GECS.2017.8066203
15. Ferdous SM, Mohammad MA, Nasrullah F, Saleque AM, Muttalib AZMS (2012) Design and simulation of an open voltage algorithm based maximum power point tracker for battery charging PV system. In: 2012 7th international conference on electrical and computer engineering. Dhaka, pp. 908–911
16. Ferreira A, Brito A, Galhardo M, Ferreira L, Macêdo W (2020) Modeling, control and simulation of a small photovoltaic-wind water pumping system without battery bank. Comp Electr Eng 84:106619. https://doi.org/10.1016/j.compeleceng.2020.106619
17. Tahir A, Mohamed EM, Abdulhafid EF, Mohamed F (2020) Grid connected wind energy system through a back-to-back converter. Comp Electr Eng 85:106660, ISSN 0045–7906
18. Li W, Zhang X (2014) Simulation of the smart grid communications: challenges, techniques, and future trends. Comp Electr Eng 40(1):270–288, ISSN 0045–7906
19. Mankour M, Sami BS (2020) Mitigation of commutation failure method in LCC converter based on HVDC systems by mean of modeling and simulation. J Ambient Intell Human Comput. https://doi.org/10.1007/s12652-020-01924-0

Comparative Power Quality Analysis of SRF and UVT Control Based DSTATCOM

Pradeep Kumar

Abstract This paper depicts comparative power quality based analysis of Distribution Static Compensator (DSTATCOM) under SRF and UVT control schemes. The SRF and UVT control schemes supported by DSTATCOM has been implemented under the MATLAB/Simulink platform of power sim categories. DSTATCOM is capable of the current based upgrading of power quality defects in the power network. The SRF and UVT controlled by DSTATCOM are validated through dynamic simulation in a MATLAB/Simulink platform under nonlinear loads. The UVT control based DSTATCOM represents superior performance as compared with SRF control based DSTATCOM by reducing the source current THD.

Keywords Digital Signal Processing (DSP) · Direct Current (DC) · Low Pass Filter (LPF) · Pulse Width Modulation (PWM)

1 Introduction

In the present scenario, inductive loads for instance induction motors, ceiling fan and farming pumps are available in the industry, residence and cultivation, respectively, which draw lagging currents about the voltage. Hence, an adverse effect of burden in reactive power, which results in line losses on the power system and capability of real power transfer in the power system will be decreased. Moreover, nonlinear loads are increasing rapidly due to the enlargement of power electronics. Hence, they create harmonics in the power system and the quality of power gets influenced [1–8]. During abnormal or fault conditions of the distribution system, there will be more prominent power quality difficulties, for instance, voltage dip and interruption. This is a big confront to keep the superiority of power under satisfactory limits [1, 9]. Here, DSTATCOM is utilized for the power quality improvement task of source current THD mitigation. The DSTATCOM has two components, i.e., power circuit and control algorithm. There are several control designs that are

P. Kumar (✉)
NIT Sikkim, Ravangla, Sikkim, India
e-mail: pradeep@nitsikkim.ac.in

A. R. Gupta et al. (eds.), *Power Electronics and High Voltage in Smart Grid*,
Lecture Notes in Electrical Engineering 817,
https://doi.org/10.1007/978-981-16-7393-1_4

accounted in the literature for DSTATCOM control, i.e., IRP theory, PB theory, SRF theory, Lyapunov control and ANN training [10–13]. In this paper, voltage source inverter based DSTATCOM is modelled in MATLAB/Simulink platform with SRF and UVT control schemes for power quality studies. This paper is partitioned into five segments. The paper is started with the introduction as Segment 1. The construction and operation of DSTATCOM have been discussed in Segment 2. Segment 3 describes SRF and UVT control schemes. Discussion about the outcomes of SRF and UVT control schemes based on DSTATCOM have been elaborated in Segment 4. Last Segment 5 summarizes the work in the form of conclusion.

2 DSTATCOM

The STATCOM (Static synchronous compensator) is used in the distribution system that is known as DSTATCOM, which can be adopted for power quality enhancement tasks, for instance, compensation of reactive power, balancing the load and current harmonic compensation. It is one of the best tools to alleviate the entire current-related power quality difficulties.

Various topologies of DSTATCOM available are, i.e., two-wire, three-wire and four-wire configurations. Three-wire configurations are preferable due to the absence of neutral wire and hence saving in conductor materials.

DSTATCOM consists of a VSI attached with a DC bus and its AC parts are fixed in parallel to the end-user loads at PCC. PWM control is used in the VSI. Hence, to alleviate switching ripples, small ripple filters are required. For providing feedback signals, voltage and current sensors are necessary. The control structure is implemented through a DSP kit to provide gate pulse for the switches of the VSI. The system configuration of DSTATCOM is depicted in Fig. 1.

The source resistance and inductance (Rs and Ls) are connected with a nonlinear load of the diode rectifier. A total number of six IGBT switches with anti-parallel diodes are within a VSI. The interfacing inductor Lf (on the AC side of the VSI) compensates harmonic of the compensating currents [14]. UVT/SRF control scheme is fed to PWM generator, which produces pulses for the IGBT switches of VSI [15]. Off–On function of breaker indicates DSTATCOM performance before switching and after switching.

3 Control Methodologies

Control structures are designed for controlling the inverter switches of the DSTATCOM. Two diverse control structures have been adopted in the DSTATCOM which have been discussed below.

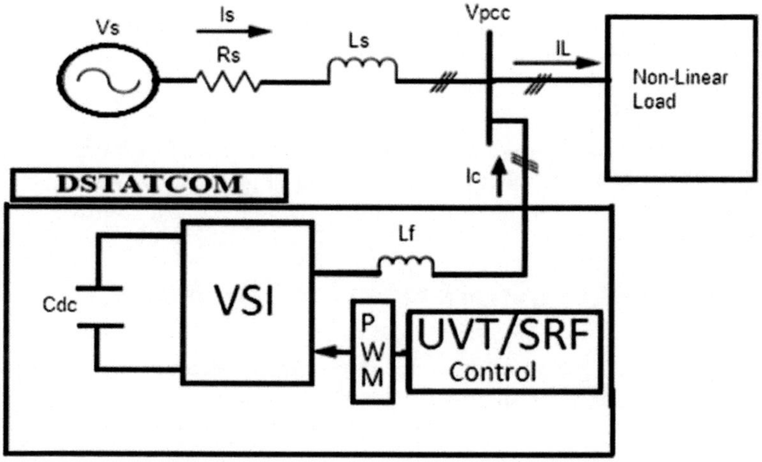

Fig.1 Configuration of DSTATCOM

3.1 Unit Vector Template Method

Phase-locked loop (PLL) is a device that produces unit vectors component of sinθ, cosθ synchronized with three-phase PCC voltages. This unit vectors component from the PLL is employed to calculate 120° shifted unit vectors (u_a, u_b, u_c) as depicted in Fig. 2.

The voltage error signal, which is the difference between desired DC voltage (V^*_{DC}) and filtered DC voltage, has been processed to the PI controller in order to achieve the crest value of desired source currents (I^*_{sm}). The three-phase reference currents on the source side are obtained by multiplying crest value with the unit vectors as mentioned below:

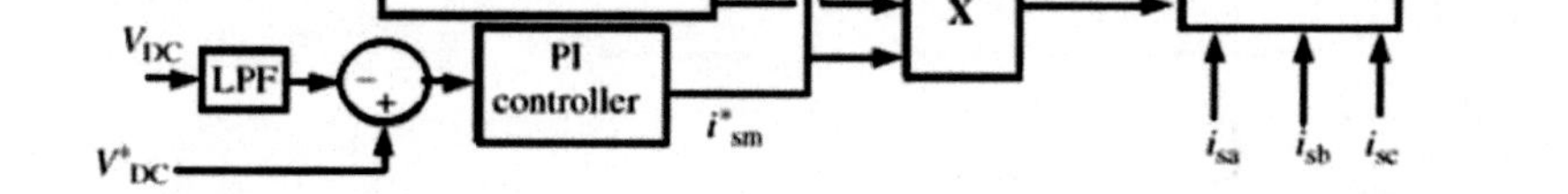

Fig.2 Configuration of UVT

$$\begin{pmatrix} i_{Sa}^* \\ i_{Sb}^* \\ i_{Sc}^* \end{pmatrix} = I_{sm}^* \begin{pmatrix} u_a \\ u_b \\ u_c \end{pmatrix} \tag{1}$$

These reference currents along with the actual currents are processed to the PWM controller. This PWM controller creates switching signals for the VSI.

3.2 Synchronous Reference Frame (SRF) Technique

The SRF (In Fig. 3) generates reference three-phase compensating current by the following mathematical formulation:

$$\begin{bmatrix} i_{Ld} \\ i_{Lq} \end{bmatrix} = {}^{2}\!/_{3} \begin{Bmatrix} \cos\omega t & \cos\left(\omega t - {}^{2\pi}\!/_{3}\right) & \cos\left(\omega t + {}^{2\pi}\!/_{3}\right) \\ \sin\omega t & \sin\left(\omega t - {}^{2\pi}\!/_{3}\right) & \sin\left(\omega t + {}^{2\pi}\!/_{3}\right) \end{Bmatrix} \begin{bmatrix} i_{La} \\ i_{Lb} \\ i_{Lc} \end{bmatrix} \tag{2}$$

$$\begin{bmatrix} \overline{i_{Ld}} \\ \overline{i_{Lq}} \end{bmatrix} = G(s) \begin{bmatrix} i_{Ld} \\ i_{Lq} \end{bmatrix} \tag{3}$$

$\overline{i_{Ld} i_{Lq}}$ = Average d-q axis load current
G(s) = Transfer function of LPF.
For the power factor regulation, K_q is evaluated as

$$K_q = {}^{Q_s^*}\!/_{\overline{Q_L}} \tag{4}$$

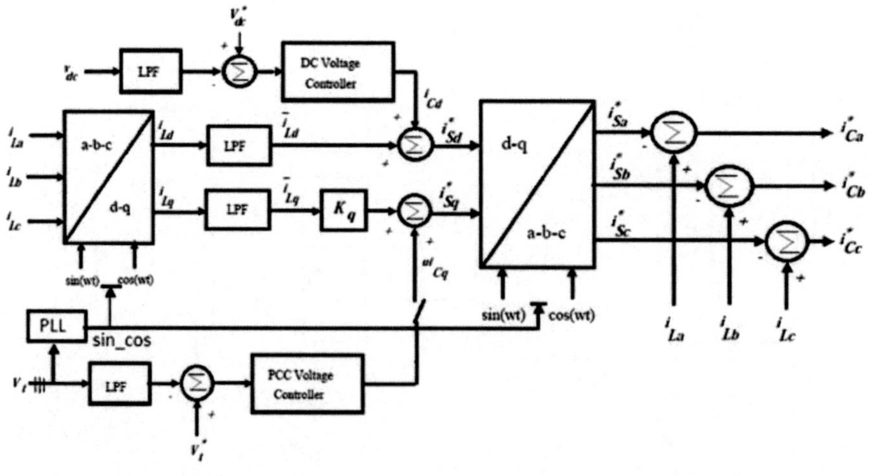

Fig. 3 Configuration of SRF

For power factor correction, $\cos\varphi = 1$, so $\varphi = 0$

$$Q_S^* = VI\sin\varphi = 0$$

The reference supply currents are

$$i_{Sd}^* = \overline{i_{Ld}} + i_{Cd},\ i_{Sq}^* = K_q\overline{i_{Lq}} + ui_{Cq} \tag{5}$$

$$\begin{bmatrix} i_{sa}^* \\ i_{sb}^* \\ i_{sc}^* \end{bmatrix} = {}^2/_3 \begin{Bmatrix} \cos\omega t & \sin\omega t \\ \cos\left(\omega t - {}^{2\pi}/_3\right) & \sin\left(\omega t - {}^{2\pi}/_3\right) \\ \cos\left(\omega t + {}^{2\pi}/_3\right) & \sin\left(\omega t + {}^{2\pi}/_3\right) \end{Bmatrix} \begin{bmatrix} i_{sd}^* \\ i_{sq}^* \end{bmatrix} \tag{6}$$

The reference injecting currents are

$$i_{Ca}^* = i_{La} - i_{Sa}^*,\ i_{Cb}^* = i_{Lb} - i_{Sb}^*,\ i_{Cc}^* = i_{Lc} - i_{Sc}^*$$

4 Outcomes Analysis

Outcomes of two promising control methodologies have been illustrated in three-phase DSTATCOM in MATLAB/Simulink Block set. The power system consists of 415 V line-to-line voltage supply and three-phase diode rectifier as a nonlinear load. DSTATCOM is fixed in shunt with the load at PCC.

4.1 Performances of DSTATCOM Under SRF Technique

Figure 4 shows source currents waveform at before compensation and after compensation in the SRF technique. DSTATCOM is connected at 0.2 s. From the above figure, it is examined that due to nonlinear load, supply current's waveform is distorted before 0.2 s. After compensation, i.e., after 0.2 s, the supply current is approximately sine wave.

Figure 5 depicts the waveform of DC voltage regulation during the period of 0–1 s. Before 0.2 s, DSTATCOM is in off condition and DC voltage is a constant value of 700 V. At 0.2 s, when DSTATCOM is operated, a transient period occurs that causes a sudden increase in the capacitor DC voltage. From 0.3 to 1 s, there is steady state and DC voltage stabilizes to 700 V. Hence, DC voltage regulation is achieved.

Figure 6 shows the THD spectrum of source current before switching off SRF-based DSTATCOM. The THD value is 13.27% which is excess. Figure 7 shows THD

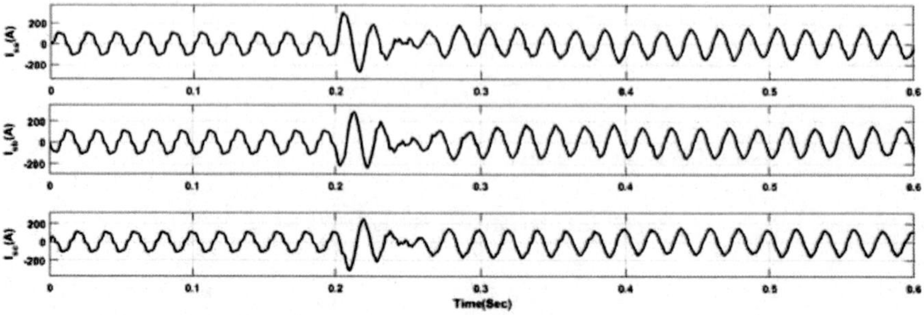

Fig.4 Source current waveform

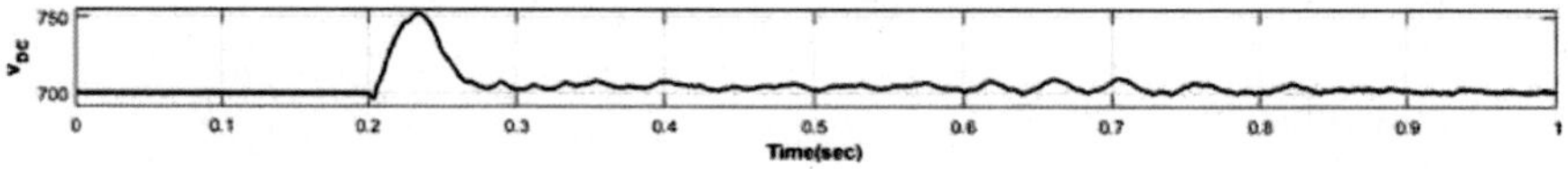

Fig. 5 DC capacitor voltage waveform

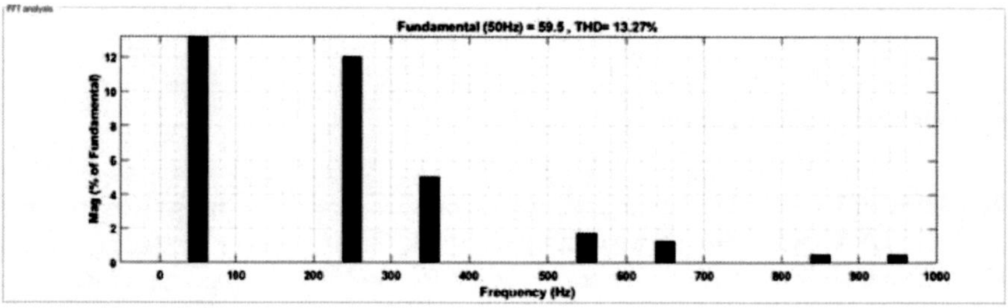

Fig. 6 Supply current THD in off state

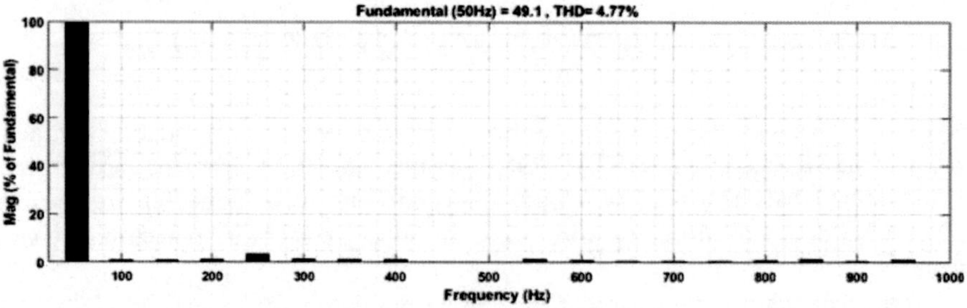

Fig.7 Supply current THD in on state

in source current after switching on SRF-based DSTATCOM. Now, the THD value is 4.77% which is under the boundary of harmonic standard.

4.2 *Performances of DSTATCOM Under UVT Technique*

Figure 8 represents the waveform of source current at each phase, i.e., phases a, b, c up to 0.6 s. Till 0.1 s, the DSTATCOM is in off condition or in other words, the system is running without DSTATCOM. So from 0 to 0.1 s, source currents are distorted and contain harmonics. After 0.1 s, the breaker is closed and DSTATCOM remains in the system. Here, we observe a transient period for very little time, when DSTATCOM is connected. Supply current waveforms are approximately sine waves and have fewer harmonics. Figure 9 illustrates the waveform of DC voltage regulation. A constant value of 700 V value is obtained till 0.1 s. At 0.1 s (during DSTATCOM is ON), there is a transient period that causes a sudden drop off in the capacitor DC voltage for a very small time. From 0.5 to 1 s, there is steady state and DC voltage will stabilize to 699 V. This indicates DC voltage regulation which is very close to 700 V.

Figure 10 depicts THD in the supply current without DSTATCOM switching. THD value of the supply current is 20.97% which crosses the limit of IEEE standards.

Figure 11 demonstrates the THD in the supply current with DSTATCOM switching. The supply current THD is 4.63%, which is under the criteria of IEEE-519 and IEC 61,000–3 harmonic standard.

Table 1 demonstrates the THD comparison of the supply current under the SRF and UVT control scheme of DSTATCOM. In this table, it has been explored that UVT-based DSTATCOM represents better THD concert than SRF-based DSTATCOM as source current THD in UVT-based DSTATCOM is less, i.e., 4.63%. Figure 12

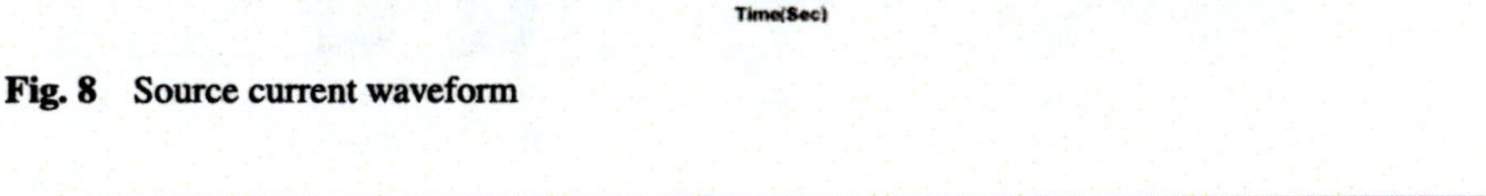

Fig. 8 Source current waveform

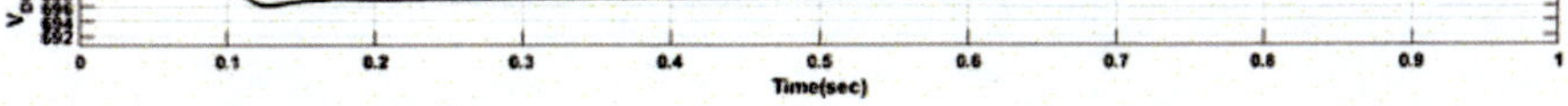

Fig. 9 DC capacitor voltage waveform

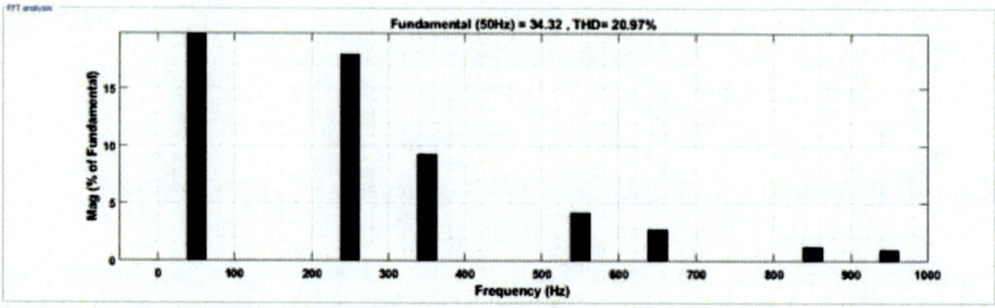

Fig.10 THD spectrum of the supply current in off state

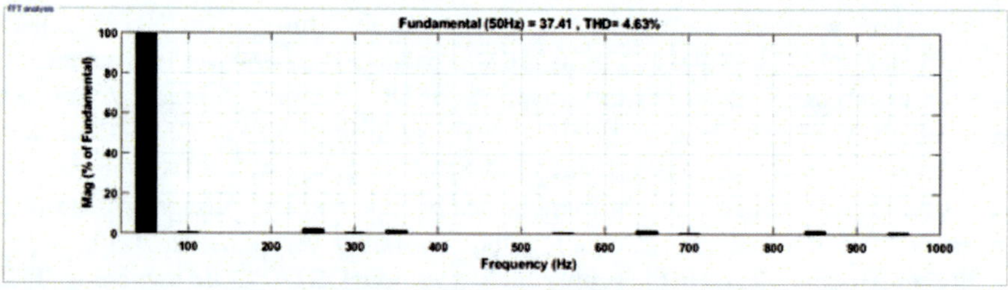

Fig.11 THD spectrum of the supply current in on state

Table1 THD comparison of source current

Control scheme	Before compensation	After compensation
SRF	13.27%	4.77%
UVT	20.97%	4.63%

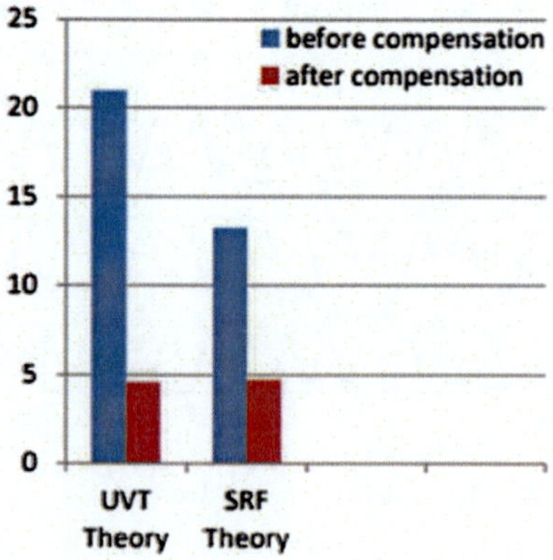

Fig.12 Bar representation of Source current THD comparison

depicts the bar representation of source current THD's comparison under SRF and UVT control scheme of DSTATCOM before and after compensation mode.

5 Conclusion

The DSTATCOM with SRF and UVT control is modelled in MATLAB/Simulink background. This DSTATCOM system is helpful to study power quality improvement tasks such as DC voltage regulation and current harmonic mitigation in the distribution system. From the outcomes, it has been found that DSTATCOM with UVT represents better THD performance than DSTATCOM with SRF. The THD analysis of the supply current in both control structures successfully fulfill the criteria of IEEE-519 and IEC 61,000–3 harmonic standard.

References

1. Chattopadhyay S, Mitra M, Sengupta S (2011) Electric power quality. Springer Netherlands, New Delhi, India
2. Fang J (2021) More-electronics power systems: power quality and stability. Springer, Singapore
3. Çiçek A, Erenoğlu AK, Erdinç O, Bozkurt A, Taşcıkaraoğlu A, Catalão JPS (2021) Implementing a demand side management strategy for harmonics mitigation in a smart home using real measurements of household appliances. IJEPES 125:1–10
4. Brinkel NBG, Gerritsma MK, AlSkaif TA, Lampropoulos I, van Voorden AM, Fidder HA, van Sark WGJHM (2020) Impact of rapid PV fluctuations on PQ in the L.V. grid and mitigation strategies using EVs. IJEPES 118:105741
5. Casaleiro Â, e Silva RA, Teixeira B, Serra JM (2021) Experimental assessment and model validation of PQ parameters for V2G systems. EPSR 191:106891
6. Rodriguez MA, Sotomonte JF, Cifuentes J (2021) A classification method for PQ disturbances using hilbert-huang transform and LSTM recurrent NN. JEET 16:249–266
7. Jiao S, Potti K, Ramachandran R, Rajashekara K, Pramanick SK (2020) A novel DROGI-based detection scheme for PQ improvement using 4-leg converter under unbalanced loads. IEEE Trans Ind Appl 56(1):815–825
8. Elmetwaly AH, Eldesouky AA, Sallam AA (2020) An Adaptive D-FACTS for PQ enhancement in an isolated microgrid. IEEE Access 8:57923–57942
9. Langella R, Testa A, Alii E (2014) IEEE recommended practice and requirements for harmonic control in EPS. IEEE Std 519–2014 1–29
10. Singh B, Solanki J (2009) A comparison of control algorithms for DSTATCOM. IEEE Tran Ind Electr 56(7):2738–2745
11. Benhabib MC, Saadate S (2005) New control approach for 4-wire APF based on the use of SRF. EPSR 73:353–362
12. Rahmani S, Hamadi A, Al-Haddad K (2012) A Lyapunov-function-based control for a 3- ph. shunt HAF. IEEE Trans Ind Electr 59:1418–1429
13. Senthilnathan K, Iyswarya Annapoorani K (2016) ANN control strategy for MCUPQC for PQ improvements in 3-feeder system. AI and Evol Comput Engg Syst Conf 1105–1111
14. Munoz AM (2007) Power quality: mitigation technologies in a distributed environment. U.K., Springer-Verlag, London
15. Singh B, Chandra A, Al-Haddad K (2015) Power quality: problems and mitigation techniques. Wiley, U.K.

Performance Analysis of Solar PV Array Fed Induction Motor Drive for Irrigation Applications

R. L. Josephine, Y. Ganga Prasad Reddy, and B. P. Rachaputi

Abstract Performance of single-stage and two-stage PV fed V/F-controlled induction motor (IM) drive for irrigation applications are analyzed in this paper. The single-stage and two-stage configurations comprise a photovoltaic system and an inverter-fed motor-pump setup in common. The latter stage comprises an additional step-up converter, which is not present in the former stage. In a two-stage configuration, the step-up converter is used to pull out peak power from the PV system by power slope method and the V/F speed control technique is utilized to drive inverter fed IM. In single-stage configuration, the converter is not used to pull out peak power. MPPT control and speed control are fed to the inverter itself. The system employing single-stage and two-stage PV array fed V/F-controlled IMD is modeled and simulated in MATLAB at different insolation conditions.

Keywords Photo Voltaic (PV) array · Step-up converter · 3-ϕ VSI · Induction motor (IM) · Pump

1 Introduction

In India, there is no electricity supply for many rural areas because of the non-existing grid system. In some areas, even if there is a grid system, electricity is supplied only for a limited number of hours. Hence, farmers cannot rely on electricity for irrigation applications. Moreover, rains are also unpredictable. So, they should have an alternative to supply water for fields. Hence utilization of solar photovoltaic system [1] for irrigation applications has been increased in many areas, where there is bright sunlight throughout the day. Usually, there are two configurations for irrigation applications that are popular. They are

R. L. Josephine (✉) · Y. Ganga Prasad Reddy · B. P. Rachaputi
Department of Electrical and Electronics Engineering, National Institute of Technology, Tiruchirappalli, India
e-mail: josephinerl@nitt.edu

A. R. Gupta et al. (eds.), *Power Electronics and High Voltage in Smart Grid*, Lecture Notes in Electrical Engineering 817,
https://doi.org/10.1007/978-981-16-7393-1_5

1. Single-stage configuration
2. Double-stage configuration

Both single-stage [2, 3] and two-stage [4, 5] configurations are analysed in this paper. The objective of this paper is to operate the photovoltaic system at peak power point and feed it to inverter fed motor by V/F control. In single-stage configuration, "Maximum Power Point Tracking" and speed control techniques are supplemented in 3-φ DC-AC converter whereas in double-stage system a step-up converter is used to pull out peak power. It has two motives, to pull out peak power and to maintain the input voltage of the inverter [5]. DC voltage regulation and speed control are supplemented in a 3-ϕ DC-AC converter. These two configurations differ in control complexity, efficiency, viability of grid integration and cost. The main drawback of single-stage configuration is that there is no DC voltage regulation which results in the reduction of DC-link voltage. Another limitation of single-stage configuration is more complexity in its control technique as both control techniques are done by the single converter. Its efficiency is less when compared to the two-stage configuration because of DC-DC converter losses in a two-stage configuration. Its cost is also less.

Power from the solar PV system mainly depends on irradiation levels and temperature. Moreover, the PV array supplies maximum power only at one point. Hence, it is to be ensured that the PV system is at its peak power point. Several MPPT techniques have been introduced in the literature [6–13]. Which differ in accuracy, speed of response, etc. Here Power Slope method (InC method) is used to pull out peak power. Regarding speed control, several speed control techniques are introduced [14, 15]. Despite vector control having many advantages over scalar control, the latter is easy to implement and its control complexity is also less. So V/F control technique is used to control speed in this work. The limitation with V/F control is its poor dynamics in terms of speed when compared to vector control. Hence, by assuming constant water availability and pump conditions, simple V/F control is sufficient for speed control and does not require a complex vector control as in [2–4]. In this paper, IM is used to pump water because it is more reliable and robust and has many applications [16–18].

2 Model Configuration and Design of Single-Stage Configuration

2.1 Design of PV System

Solar PV design is application dependent. It depends on the rating of the motor that is selected to drive the pump under various insolation levels. The power pulled out from the photovoltaic system ought to be higher than motor power to meet the losses that occur at intermediate stages so that the performance of the motor does not get affected [2]. In the proposed model, the power ratings of the PV array and induction

Table 1 Motor specifications and design of PV array

	Parameters of the proposed system	Rating
PV module	Current at short circuit, I_{sc}	4.6 A
	Voltage at open circuit, V_{oc}	36 V
	Current, Imp (at maximum power)	3.9 A
	Voltage, V_{mp} (at maximum power)	28.33 V
	Maximum power Pm at standard insolation	110.4 W
Induction motor	Nominal power	2000 W
	Nominal speed	1430 rpm
	Nominal frequency	50 Hz
	Nominal voltage (L-L)	400 V

motor are chosen as 2.2 and 2 KW, respectively, to drive the centrifugal pump. The design considerations of the proposed model are mentioned in Table 1.

2.2 *Computation of DC-Link Voltage*

The input voltage of the 3-ϕ DC-AC converter ought to be more than that of the maximum line-line voltage to control the motor current [4].

$$V_{dc} = \sqrt{2}^{*} V_{L-L} = 1.414{*}400 = 566V \tag{1}$$

Hence, input (DC) voltage of 3-ϕ DC-AC converter is chosen as 580 V. Here V_{dc} is the input voltage of VSI (Fig. 1).

The specifications of IM and PV are mentioned in Table 1. The ratings of PV mentioned in the table are for a single module. In order to obtain the required voltage and power rating from PV, 20 modules are connected in series.

2.3 *Computation of DC-Link Capacitance*

The capacitance is computed as follows [19]

$$\omega_{rated} = 2^{*}\pi^{*} f_{rated} = 314\ rad/s \tag{2}$$

$$1/2^{*}C_{dc}{}^{*}\left(V_{dc}{}^{2} - V_{dc1}{}^{2}\right) = 3^{*}a^{*}V_{p}{}^{*}I^{*}t \tag{3}$$

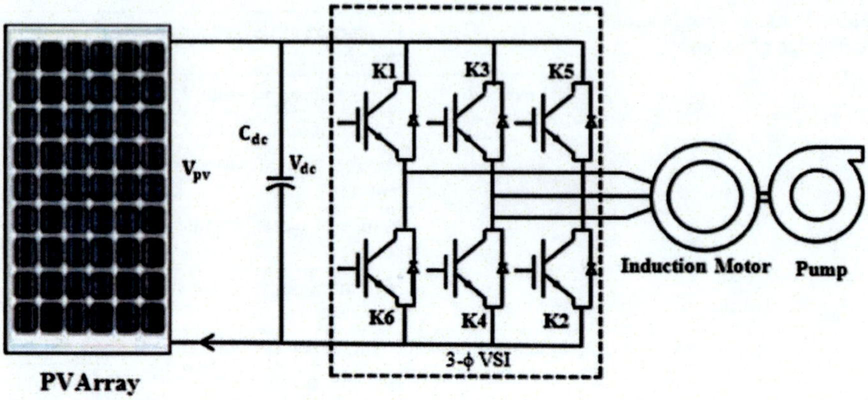

Fig. 1 Configuration of single-stage PV-fed IMD

$$0.5^{*}C_{d}c^{*}(580^{2}-566^{2})=3^{*}1.2^{*}230.9^{*}3.61^{*}0.05 \tag{4}$$

$$C_{dc}=18.7\text{ mF} \tag{5}$$

Here, V_{dc} is the input voltage of 3-ϕ VSI. V_{dc1} is the attained DC voltage at the input of the voltage source inverter, t is the time required for V_{dc1} to reach V_{dc}, V_p is the phase voltage and I is the phase current.

2.4 Design of Water Pump

Centrifugal pump has square law relationship motor speed and load torque [19]

$$T_L=K_p{}^{*}\omega_m{}^{2} \tag{6}$$

Here K_P is pump constant, T_L is load torque ωm is the speed of IM.

2.5 Control Scheme of Single-Stage PV Fed IMD

See Fig. 2.

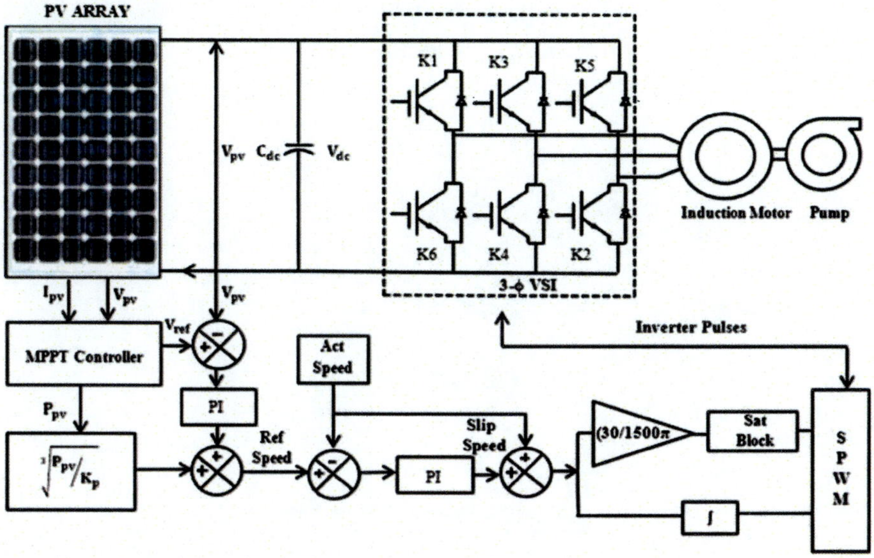

Fig. 2 Configuration and control strategy of single-stage PV-fed IMD

2.6 MPPT Control

$$p = V_t{}^*I_t$$

$$\frac{dP}{dV_t} = I_t + V_t\frac{dI_t}{dV_t} \tag{7}$$

at max power operating point

Fig. 3 PV characteristics

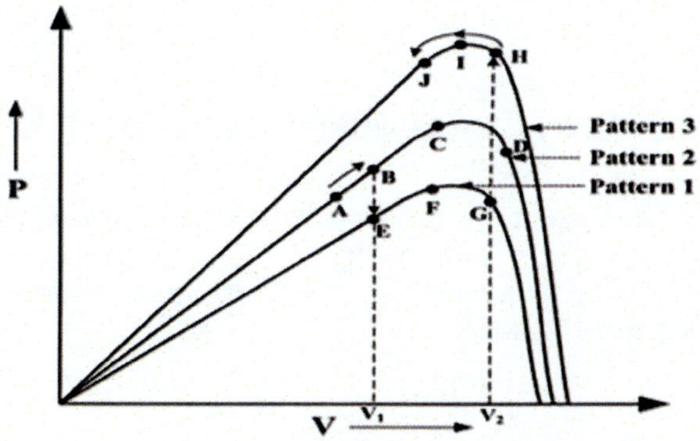

$$\frac{dP}{dV_t} = 0, \ \frac{dI_t}{dV_t} = \frac{-I_t}{V_t}$$

$$\frac{dI_t}{dV_t} = \frac{\frac{dP}{dV_t} - I_t}{V_t} \tag{8}$$

$$\frac{dP}{dV_t} > 0 \text{ from } 0 \text{ to } V_{mp}$$

$$\frac{dP}{dV_t} > 0 \text{ at } V_{mp}$$

$$V_{mp} < 0 \text{ from } V_{mp} \text{ to } V_{oc}$$

Hence

$$\frac{dI_t}{dV_t} + \frac{I_t}{V_t} > 0 \text{ from } 0 \text{ to } V_{mp}$$

$$\frac{dI_t}{dV_t} + \frac{I_t}{V_t} = 0 \text{ at } V_{mp}$$

$$\frac{dI_t}{dV_t} + \frac{I_t}{V_t} < 0 \text{ from } V_{mp} \text{ to } V_{oc}$$

The control strategy of single-stage configuration is shown in Fig. 2. Inputs of PV array are observed and are given to MPPT controller whose output is Vref. Vref is then contrasted with PV array voltage and the difference is fed to the PI controller. The output of PI plus the gain obtained from PV array power is taken as reference speed, which is then contrasted with actual speed and the difference is fed to the PI controller. The output obtained from the PI controller is slip speed. Now actual speed is added with slip speed to get synchronous speed. Modulation index and ɵ are obtained as shown in Fig. 2. SPWM is used to obtain pulses that are fed to the 3-φ VSI (Figs. 3 and 4).

3 Model Configuration and Design of the Two-Stage System

A two-stage configuration consisting of a PV system, a DC-DC converter and a motor pump setup is shown in Fig. 5. The step-up converter interfaced with PV is used to pull out peak power from PV and this is fed to the inverter whose job is to control the speed of IM.

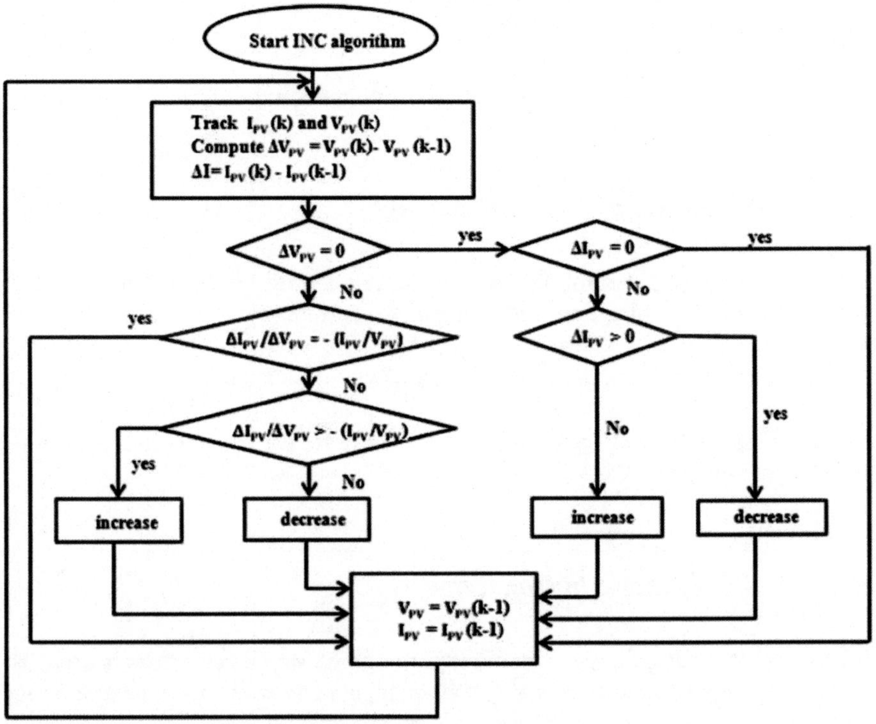

Fig. 4 Flowchart for MPPT algorithm

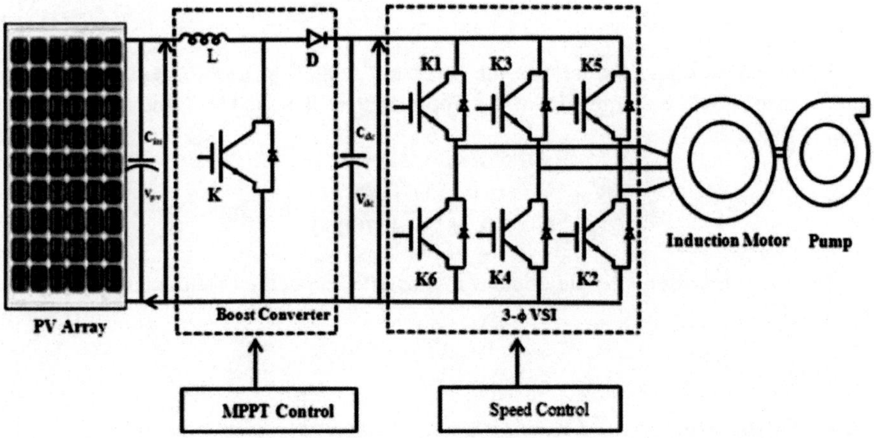

Fig. 5 Configuration of Double-stage PV array fed IMD

3.1 Design of the PV System

PV design is the same as that of single-stage configuration.

3.2 Computation of DC-Link Voltage

The input voltage of the 3-ϕ DC-AC converter ought to be more than that of the maximum line-line voltage to control the motor current [4].

$$V_{dc} = \sqrt{2}^{*} V_{L-L} = 1.414^{*}400 = 566V \tag{9}$$

Hence input (DC) voltage of 3-ϕ DC-AC converter is chosen as 580 V. Here V_{dc} is the input voltage of VSI.

3.3 Design of Step-Up Converter

To regulate the voltage at the maximum power point, step-up converter is used. The input of the step-up converter is V_{mp} and the input of voltage source inverter is V_{dc}. The duty cycle of the step-up converter is estimated as

$$D = 1 - \frac{V_{in}}{V_o} = 1 - \frac{567}{667} = 0.85 \tag{10}$$

Ripple current and inductance are inversely related to each other. The value of inductance should be large so that the ripple content is less. The value of inductance is calculated as [4]

$$L = \frac{V_{mp}{}^{*}D}{\Delta_{IL}{}^{*}f_s} = \frac{567^{*}0.85}{0.25^{*}3.61^{*}10000} = 5.3mF \tag{11}$$

Here Δ_{IL} represents ripple content in current. The value of inductance is chosen as 6 mH.

3.4 Computation of DC–Link Capacitance

The DC-link capacitance is calculated as follows [19]

$$\omega_{rated} = 2 * \pi * f_{rated} = 314 \text{ rad/s} \tag{12}$$

Table 2 Motor specifications and design of PV array

	Parameters of the proposed system	Rating
PV array	Current at short circuit, I_{sc}	4.6 A
	Voltage at open circuit, V_{oc}	700 V
	Current, Imp (at maximum power)	3.9 A
	Voltage, V_{mp} (at maximum power)	570 V
	Maximum Power Pm at standard insolation	2223 W
Induction motor	Nominal power	2000 W
	Nominal speed	1430 rpm
	Nominal frequency	50 Hz
	Nominal voltage (L-L)	400 V

$$1/2 * C_{dc} * \left(V_{dc}^2 - V_{dc1}^2\right) = 3 * a^* V_p * I^* t \tag{13}$$

$$0.5 * C_{dc} * \left(580^2 - 566^2\right) = 3 * 1.2 * 230.9 * 3.61 * 0.05 \tag{14}$$

$$C_{dc} = 18.7\ \text{mF} \tag{15}$$

Here, V_{dc} is the input voltage of 3-ϕ VSI. V_{dc1} is the attained DC voltage at the input of the voltage source inverter, t is the time required for V_{dc1} to reach V_{dc}, V_p is the phase voltage and I is the phase current.

The specifications of IM and PV are mentioned in Table 2. The ratings of PV mentioned in the table are for a single module. In order to obtain the required voltage and power rating from PV, 20 modules are connected in series.

3.5 Design of Water Pump

Centrifugal pump has square law relationship with motor speed and load torque [19]

$$T_L = K_p * \omega_m^2 \tag{16}$$

Here K_P is pump constant, T_L is load torque ωm is the speed of IM.

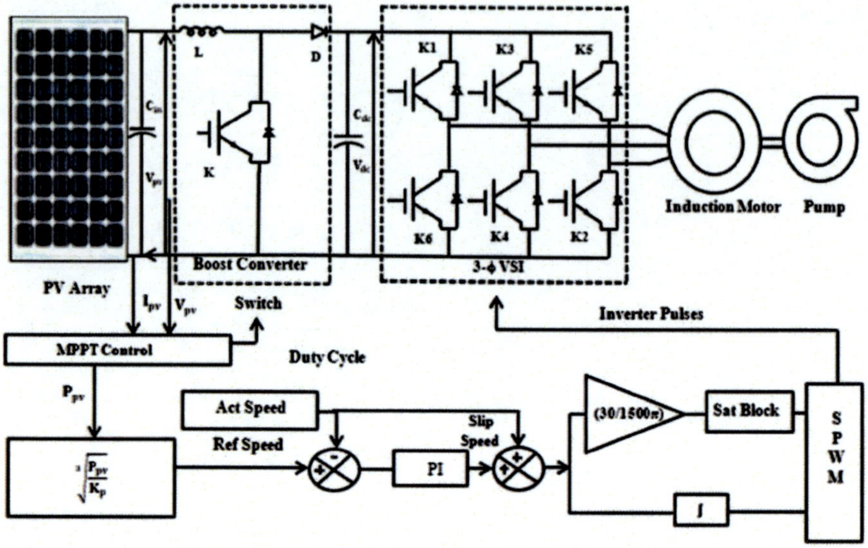

Fig. 6 Configuration and control scheme of two-stage PV fed IMD

3.6 Control Scheme of Double-Stage PV Fed IMD.

In the single-stage control scheme shown in Fig. 2, the MPPT extraction and speed control are integrated with the inverter whereas in the double-stage configuration they are controlled independently using a step-up converter and inverter, respectively, as shown in Fig. 6. MPPT algorithm used in the double-stage configuration is the same as that of single-stage system. Inputs from the PV array are observed and are given to the MPPT controller. The output of it is Vref, which is compared with PV array voltage and the difference is fed to the PI controller. The output of is duty ratio, which is compared with repeating sequence and the obtained pulses are fed to the step-up converter switch to pull out peak power from the PV array. The gain obtained from the PV array power is taken as reference speed, which is compared with actual speed and the error is fed to the PI controller. The output thus obtained is slip speed. Now actual speed is summed with slip speed to get synchronous speed. Modulation index and ɵ are obtained as shown in Fig. 6. SPWM is used to obtain pulses that are fed to the 3-φ VSI.

4 Simulation Results

Thus, the propounded single-stage and double-stage systems are simulated in MATLAB and the results are shown. Insolation is varied from 1000 to 700 W/m^2 at t = 1 s and the analysis of PV array, step-up converter, 3-ϕ VSI, and an induction motor

(IM) is given in below figures. Figure 7 shows the variation of PV current for a step change of irradiation at t = 1 s. Figures 8 and 9 show the variation of line-line voltage and current of the stator. Figures 10, 11, 12 and 13 show the variation of motor speed, electromagnetic torque, PV power and motor power, respectively. When irradiation is reduced at 1 s, the power delivered by PV is reduced. Consequently, power fed to the IM, electromagnetic torque and speed will be reduced. The output waveforms shown in Figs. 14, 15, 16, 17, 18, 19 and 20 of double-stage configuration are also varied in the same manner as that of single-stage configuration.

RESULTS OF SINGLE-STAGE SYSTEM

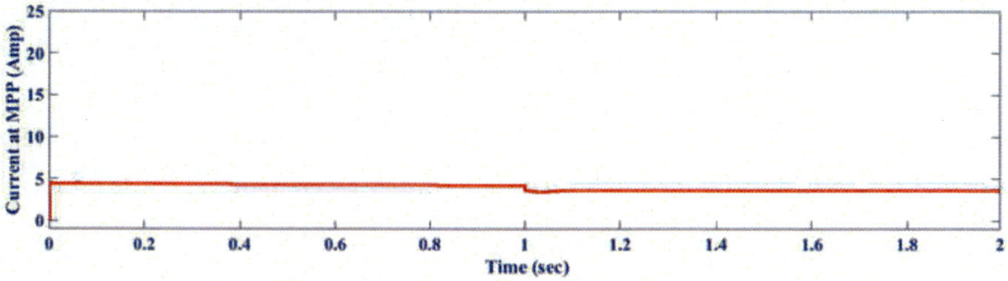

Fig. 7 PV current (A)

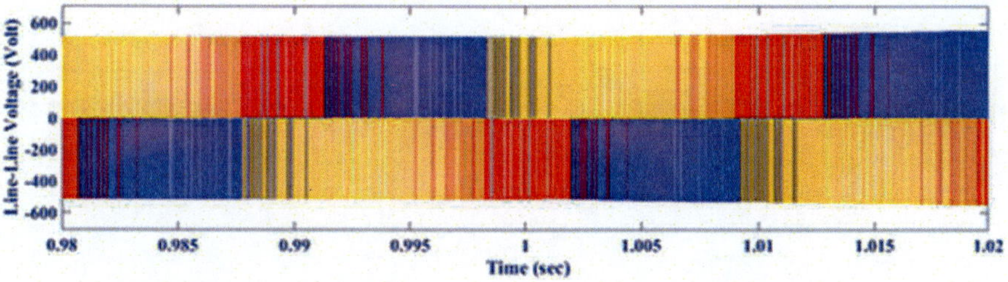

Fig. 8 Line-line voltages of inverter (V)

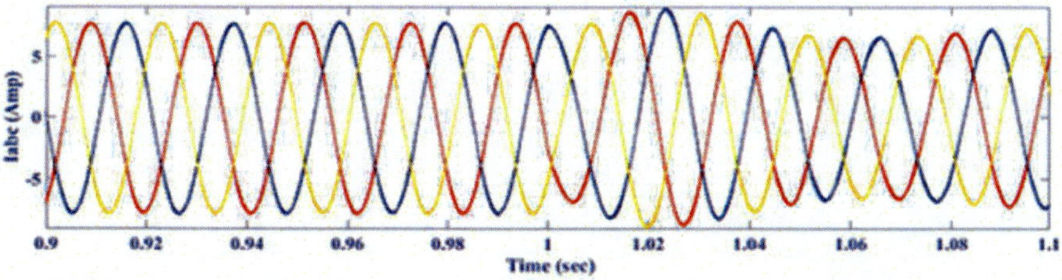

Fig. 9 Output currents of inverter (A)

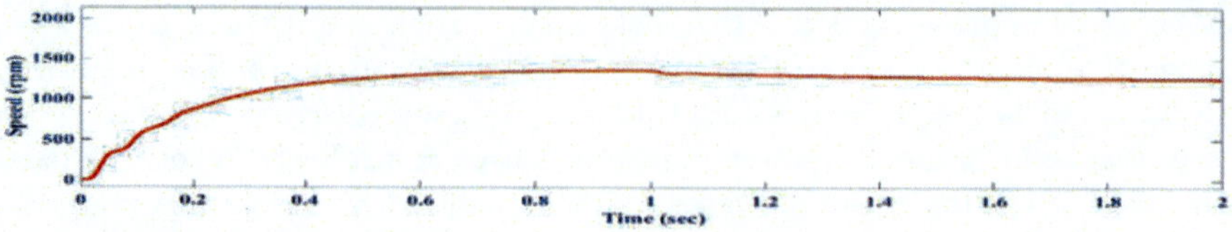

Fig. 10 Motor speed (rpm)

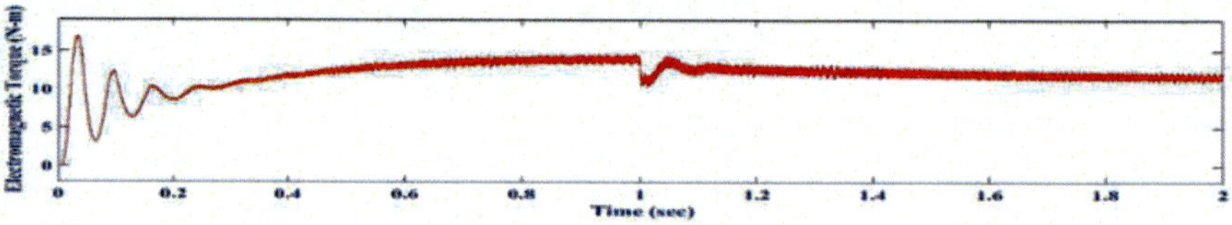

Fig. 11 Electromagnetic torque (N-m)

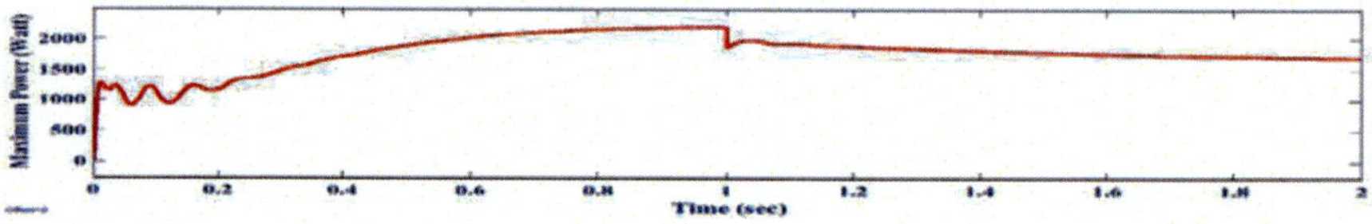

Fig. 12 PV power (W)

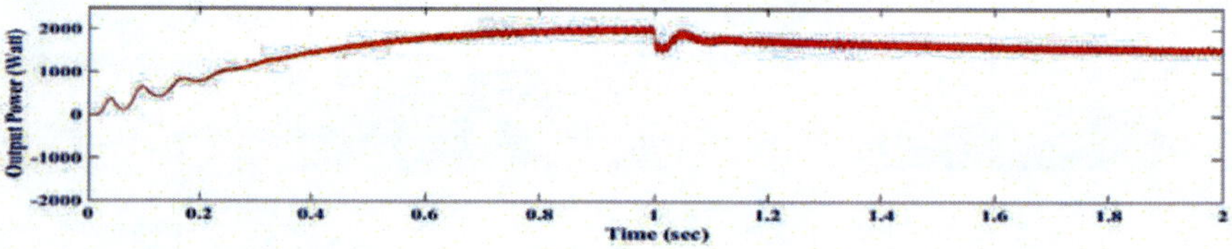

Fig. 13 Motor power (W)

RESULTS OF TWO-STAGE SYSTEM

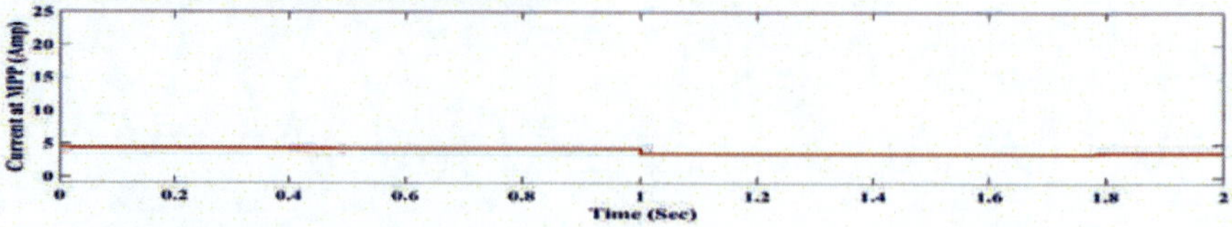

Fig. 14 PV current (A)

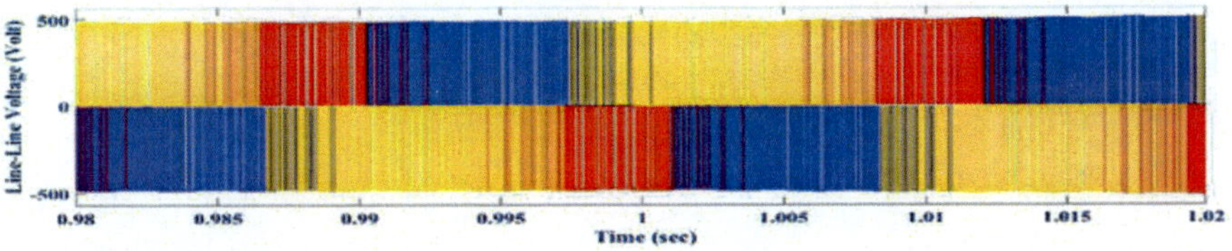

Fig. 15 Line-line voltage of inverter (V)

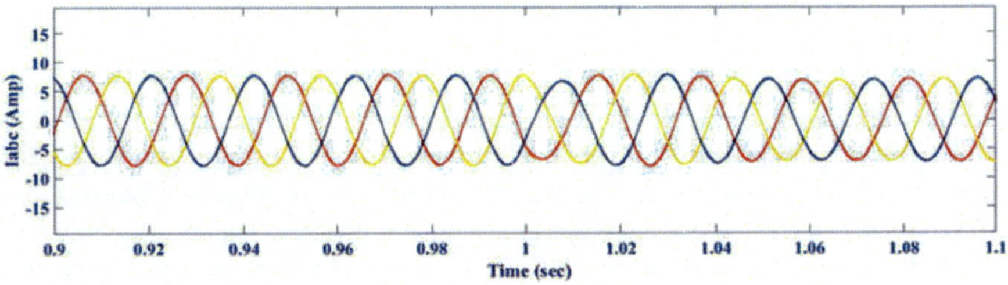

Fig. 16 Output currents of inverter (A)

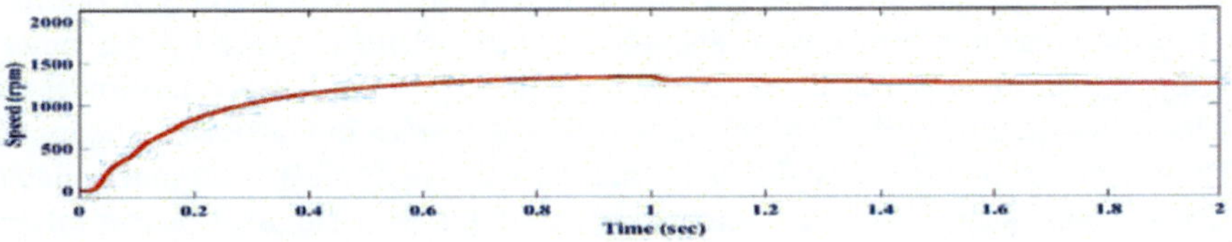

Fig. 17 Motor speed (rpm)

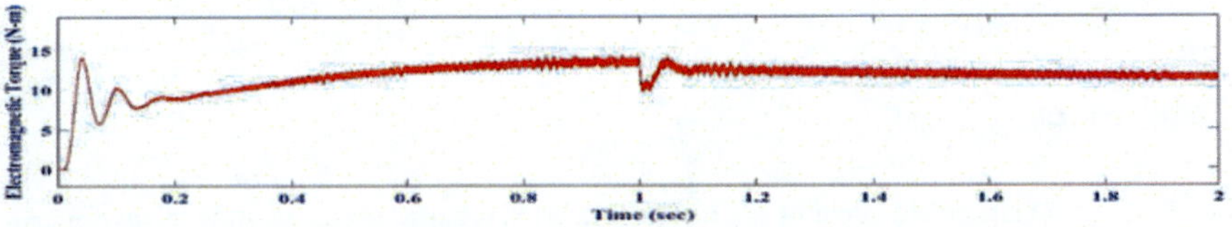

Fig. 18 Electromagnetic torque (N-m)

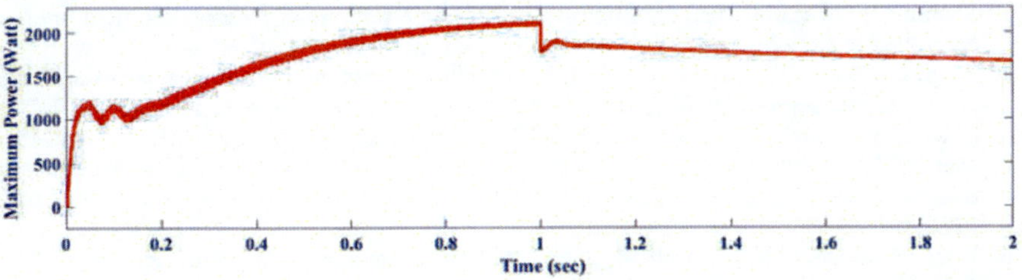

Fig. 19 PV power (W)

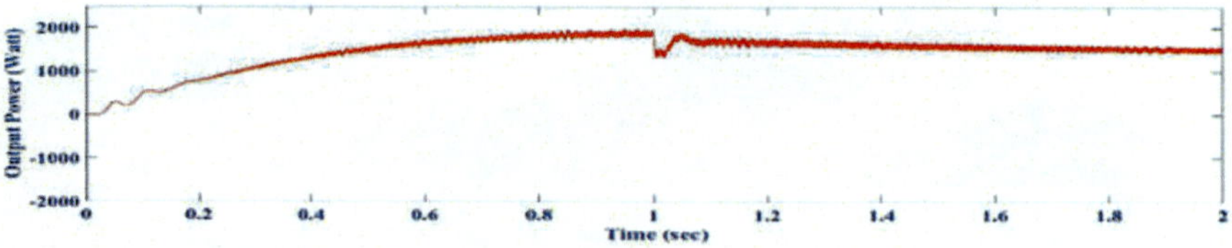

Fig. 20 Motor power (W)

5 Conclusion

The simulation of single-stage and double-stage PV fed induction motor drive (IMD) for irrigation applications is carried out. In a single-stage system, MPPT and V/F speed control are imparted in the inverter. The two-stage system is employing a step-up converter and a conventional 3-ϕ voltage source inverter. The step-up converter is used to pull out the peak power of 2.2 KW from the PV array. The V/F control method is used to control the motor and is utilized to provide peak power to the motor-pump setup. It is observed that the rated speed and the rated power is obtained for the induction motor drive to perform its operation. The extraction of maximum power for change in insolation is also observed as a change in speed at the motor-pump set. The drawback is that the motor is supplied with rated power only under the standard insolation level. If the insolation level is reduced motor power is reduced. The other limitation in employing V/F control is the high ripple in the torque content. Hence, single-stage and double-stage configurations with simple V/F control are implemented in this paper for irrigation applications.Please provide respective details for the sects. 17 and 18 as they are mandatory.

References

1. Foster R, Ghassemi M, Cota M (2010) Solar energy: renewable energy and the environment. CRC Press, Boca Raton, FL, USA
2. Shukla S, Singh B (2018) Single-stage PV array fed speed sensorless vector control of induction motor drive for water pumping. IEEE Trans Ind Appl 54(4):3575–3585
3. Sharma U, Dwivedi S, Jain C, Singh B (2015) Single stage solar pv array fed field oriented controlled induction motor drive for water pump. Natl Power Elect Conf
4. Singh B, Shukla S, Chandra A, Al-Haddad K (2016) Loss minimization of two stage solar powered speed sensorless vector controlled induction motor drive for water pumping. In IECON 2016–42nd annual conference of the IEEE industrial electronics society. IEEE, pp 1942–1947
5. Murshid S, Singh B (2016) Double stage solar PV array fed water pump driven by permanent magnet synchronous motor. In: 2016 IEEE 7th Power India International Conference (PIICON). IEEE, pp 1–5
6. Islam H, Mekhilef S, Shah NBM, Soon TK, Seyedmahmousian M, Horan B, Stojcevski A (2018) Performance evaluation of maximum power point tracking approaches and photovoltaic systems. Energies 11(2):365
7. Ali US (2016) Z-source DC-DC converter with fuzzy logic MPPT control for photovoltaic applications. Energy Proc 90:163–170

8. Sera D, Mathe L, Kerekes T, Spataru SV, Teodorescu R (2013) On the perturb-and-observe and incremental conductance MPPT methods for PV systems. IEEE J Photovolt 3(3):1070–1078
9. Esram T, Chapman PL (2007) Comparison of photovoltaic arraymaximum power point technique. IEEE Trans Energy Convers 22(2):439–449
10. Raju AB, Kanik S, Jyoti R (2008) Maximum efficiency operation of a single stage inverter fed induction motor PV water pumping system. In 2008 First International Conference on Emerging Trends in Engineering and Technology. pp 905–910
11. Ramesh T (2018) Solar powered based water pumping system using perturb and observation MPPT technique. In: IEEE International students' conference on electrical, electronics and computer science (SCEECS)
12. Elgendy MA, Atkinson DJ, Zahawi B (2016) Experimental investigation of the incremental conductance maximum power point tracking algorithm at high perturbation rates. IET Renew Power Gener 10(2):133–139
13. Songbai Z, Xu Z, Li Y, Ni Y (2006) Optimization of MPPT step size in stand-alone solar pumping systems. In: Proceedings of the IEEE Power Engineering. Society General Meeting. pp 1–6
14. Tsuj M, Chen S, Hamasaki SI, Zhao X, Yamada E (2008) A novel V/f control of induction motors for wide and precise speed operation. In: 2008 International symposium on power electronics, electrical drives, automation and motion. IEEE, pp 1130–1135
15. Martinez-Hernandez MA, Gutierrez-Villalobos JM, Malagon-Soldara SM, Mendoza-Mondragon F, Rodriguez-Resendiz J (2016) A speed performance comparative of field oriented control and scalar control for induction motors. In: 2016 IEEE conference on mechatronics, adaptive and intelligent systems (MAIS). IEEE, pp 1–7
16. Jones WV (2013) Motor selection made easy: Choosing the right motor forcentrifugal pump applications. IEEE Ind Appl Mag 19(6):36–45
17. Bose BK (2002) Modern power electronics and AC drives, vol 123. Prentice hall, Upper Saddle River, NJ
18. Parikh AP, Tekwani PN, Patel V (2017) "Design and implementation of solar pumping system with induction motor and submersible pump. In: Nirma University International Conference on Engineering (NUiCONE)
19. Sharma A, Parakh A (2018) Design of solar powered induction motor drive for pumping application. Int J Latest Trends Eng Technol 10(1):228–237

Diode-Based Passive Li-Ion Battery Balancer

Bharat Singh, Gaurav Pratap Singh, Yash Lehri, Lakshay Bhatia, and Yogesh Sehgal

Abstract Li-ion batteries are being employed in various industries for various applications. They have around 99% Coulombic efficiency, high energy density and no memory effect, and this makes them one of the best energy storage devices of these times. In order to increase the lifetime and overall performance of Li-ion batteries, a technique called balancing is employed while charging to prevent overcharging and balance each and every cell in a battery pack to extract maximum energy from them, meanwhile preventing their charging capacity from degrading quickly. There are two types of Balancing—Passive and Active. Passive Balancing is simple and cheap while Active Balancing is complicated and expensive.

Keywords Li-ion · Battery · Balancing · Active balancing · Passive balancing · Balancer · Diode

1 Introduction

Lithium-ion Batteries have energy density around 250–693 Wh/L (0.90–2.43 MJ/L), while their cost is getting reduced every day due to increasing production quantitates across the globe. They also have high Coulombic efficiency which peaks around 99%, high energy density, low self-discharge rates and no memory effect. These qualities make it one of the most desirable energy storage devices present these days. Li-Ion cells have found their use into many modern technologies like e-Mobility, Electronics and Smart Grids. But all this comes with a flaw, i.e. Li-Ion Cells are highly inflammable and can easily cause a fire, if not handled well. They must be prevented from overcharging, operated within specified voltages and the cell must not encounter overloading or damage [1].

They also have a major flaw when cells are combined to form a battery pack. In order to extract the maximum power from battery pack, all the cells must have equal

B. Singh (✉) · G. P. Singh · Y. Lehri · L. Bhatia · Y. Sehgal
Department of Electrical and Electronics Engineering, Bharati Vidyapeeth's College of Engineering, New Delhi, India
e-mail: coedelhi@bharatividyapeeth.edu

A. R. Gupta et al. (eds.), *Power Electronics and High Voltage in Smart Grid*,
Lecture Notes in Electrical Engineering 817,
https://doi.org/10.1007/978-981-16-7393-1_6

Fig. 1 Lithium ion cells

State of Charge (SOC) to perform efficiently. To achieve equal SOC, "Balancing" is employed. Balancing is a technique which makes it possible to achieve almost equal levels of SOC, meanwhile also prevents overcharging of each cell. The maximum energy that can be extracted from the battery pack depends upon the cell, connected in series, which has lowest SOC in the series connected cells of the pack. "Balancer" are the circuits which make it possible to maintain the SOC and also prevents overcharging by simply directing the incoming current to either undercharged cells or to a passive element such as resistor or diode [2, 3] (see Fig. 1).

2 Battery Balancing

Battery balancing, also called battery redistribution, refers to a technique that is employed to improve the available energy in the battery pack to increase the life of the battery pack. In balancing, the major aim is to balance out the SOC of all the cells in series, and prevent overcharging of any cell.

There are two major balancing techniques, (A) Passive Balancing and (B) Active Balancing.

2.1 Passive Balancing

Energy is extracted from the most charged cell or cell with highest SOC and is dissipated as heat, through a passive element (see Fig. 2).

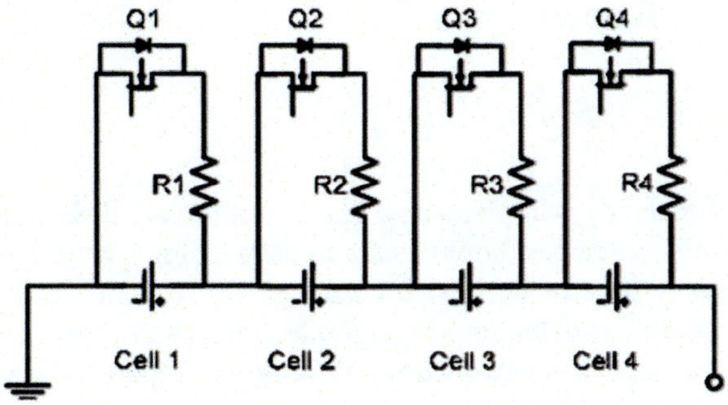

Fig. 2 Passive balancer circuit

Energy is lost during passive balancing, but the circuit is simple and cost-efficient.

2.2 *Active Balancing*

Energy is extracted from the most charged cell or cell with highest SOC and is supplied to the least charged cell or cell with lowest SOC (see Fig. 3).

Energy is not lost in active balancing, but the circuit is complicated and expensive [4].

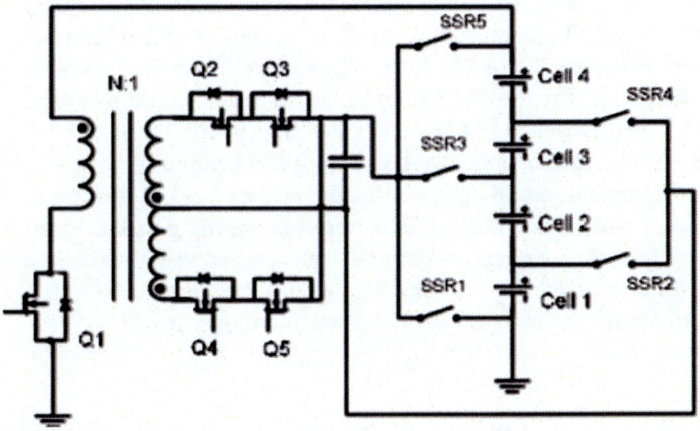

Fig. 3 Active balancer circuit

3 Battery Balancer

3.1 Introduction

Balancing is implemented using circuits called "Balancer." They are generally parallelly connected components to each cell in the battery pack, controlled by a switch to activate or deactivate balancing for that individual cell. Active balancing being complicated requires regular maintenance to ensure proper functioning of the circuit, but passive balancing has simple circuit design and implementation is quick and cost-effective, and it can be implanted using simple electronic components and placing them parallelly while the battery is being charged [5].

3.2 Need of Battery Balancer

Lithium-ion batteries are sensitive to overcharging, overheating and physical mishandling, which may lead to chemical damage in the cells even if the overcharging is within the range of few millivolts or milliamperes and may lead to explosion. Therefore, not only preventing overcharging and overheating, but continuous monitoring of each cell is also necessary in certain applications as a safety measure.

Some of the notable incidents include recalling of Nokia's 46 million BL-5C batteries in 2007 across the globe, when one of the cellphone's battery overheated and exploded in Philippines. Other incidents include recall of batteries of several smartphones since late 2000s and early 2010s as well as grounding of Boeing 787 in 2013 due to thermal runaway of lithium-ion batteries being used in Auxiliary Power Unit (APU) which resulted in emergency landing [9, 10].

Hence, battery management systems are also required to monitor the voltage, current and temperatures of each cell of the battery pack if the application of the device being powered is critical, like batteries powering personal electronics, electric vehicles or sensitive electronic systems, etc. and prevent overcharging and overheating. Moreover, battery balancer contributes in keeping the batteries healthy and efficient, which results in better performance and longer life.

3.3 Applications of Battery Balancer

In order to avoid overcharging of Li-ion cells, battery balancers are employed in Battery Management Systems (BMS) to automatically dissipate or reuse the energy coming from the charger. Balancing could either be "Active" or "Passive," depending upon the complexity of battery management system and application. In addition to balancing, modern battery management systems come with cell temperature and

SOC monitoring capabilities, which ensure safe and efficient operation of Li-ion cells and can trigger warnings in case of abnormal behavior of cells.

4 Simulation of Passive Battery Balancer

4.1 Simulation Using Simulink

Simulation of an ideal battery balancer can be realized in Simulink. Battery block in Simulink provides detailed output regarding SOC, voltage and current of the cell. It is also capable of simulating constant charge—constant voltage charging algorithm of Lithium-ion cells (see Fig. 4).

In the simulation, the cell was connected parallelly to a resistor and a control logic that turns ON or OFF the balancer switch when the SOC reaches 100%. This logic was implemented using Simulink's switch block, in which the input was cell's SOC which was controlling the switching logic and preventing overcharging by activating the balancer.

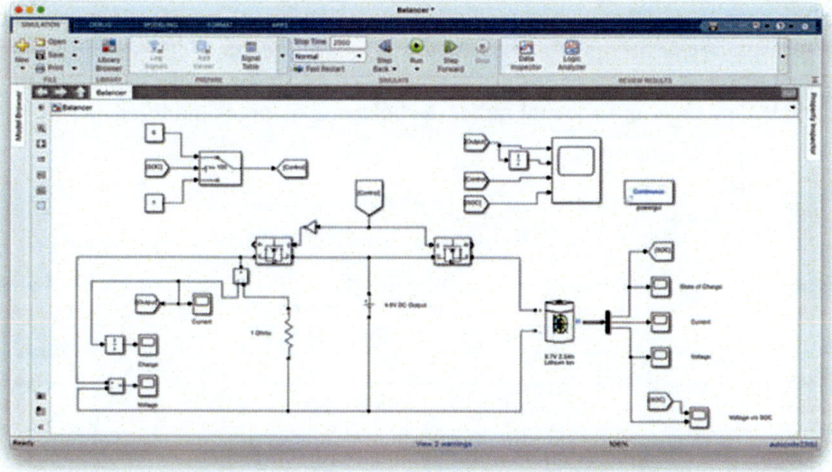

Fig. 4 Modelling of passive battery balancer

4.2 Simulation Results

When the SOC reaches 100%, the switch changes the flow of current from the charging source to the resistor. This dissipates the energy being supplied to the cell, and prevents overcharging (see Fig. 5).

The results were plotted on a graph using Simulink's scope block, and a slight dip in cell's voltage was observed. This is known as diffusion voltage. Switch makes sure to pass the current through resistor for heat generation and keeps the cell safe from overcharging (see Fig. 6).

5 Hardware Implementation of Diode-Based Passive Battery Balancer

Passive battery balancer dissipates the energy provided to the charged cell in the form of heat and this is generally done using a resistor but for small battery pack dissipation can also be implemented using multiple series connected diodes. Since, each diode has some internal resistance and a voltage drop this leads to dissipation of heat when current is passed through it (see Fig. 7).

This implementation works for 3S battery packs due to small charging current, and has a modular design which enables it to balance each cell individually and the number of balancers can be reduced or increased as per the needs. Efficiency of the circuit is not taken into account rather the balanced output is modified.

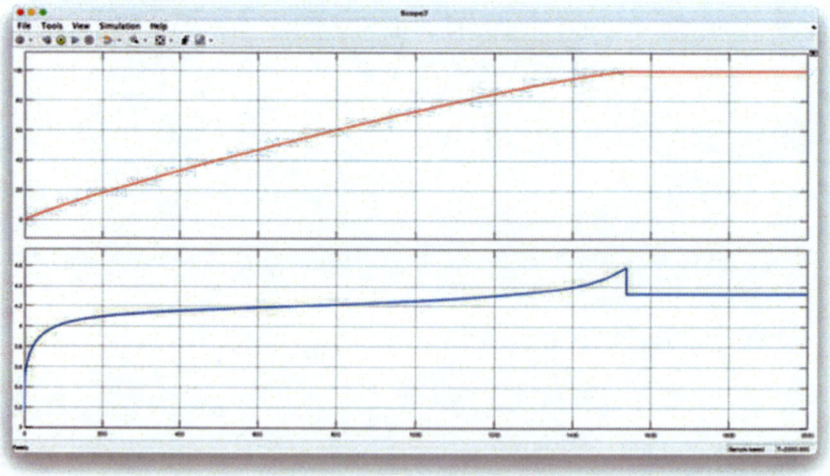

Fig. 5 SOC versus time (red) and cell voltage versus time (blue)

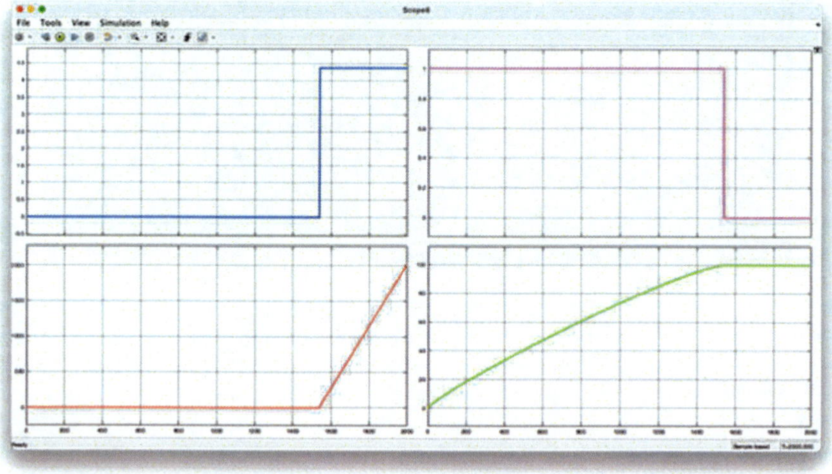

Fig. 6 Input voltage (blue), control signal (pink), charge (red) and SOC (green)

Fig. 7 Single cell balancer circuit

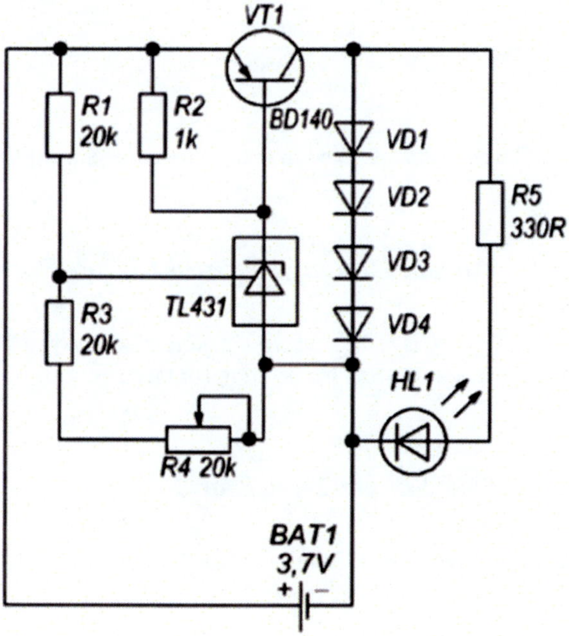

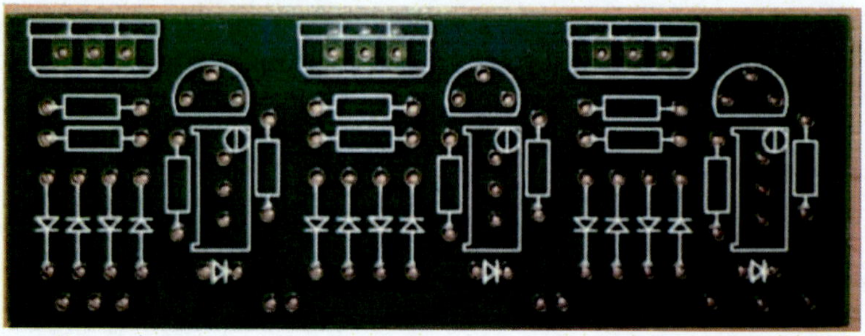

Fig. 8 PCB

Fig. 9 BD140

The components used in battery balancer are explained below.

5.1 Custom Designed Printed Circuit Board

A Custom PCB Schematic was taken from Kasyan TV, and was modified to meet the size requirements for printing [6] (see Fig. 8).

5.2 PNP Silicon Transistor

BD140, PNP Power Transistor (see Fig. 9).

5.3 Bipolar Transistor Switch

TL431, Adjustable Precision Shunt Regulator (see Fig. 10).

Fig. 10 TL431

Fig. 11 Trimpot

5.4 Trimpot

20 KΩ Trimpot Variable Resistor (3296 Package) (see Fig. 11).

5.5 Resistors, Diode and LED

20 and 1 KΩ Resistors, 1N4007 Diode and 5 mm LED (see Fig. 12).

6 Working of Diode-Based Passive Battery Balancer

When the cell reaches around 4.2 V, the power transistor's (BD140) collector gets activated and the incoming current flows through the diodes and LED. This activates the LED, which notifies that the charging is complete and cell can be disconnected.

The cutoff voltage can be adjusted using Trimpot, which is basically a variable resistant and controls the switching voltage magnitude for Power Transistor (BD140). This feature can be useful when cells from different manufacturers or

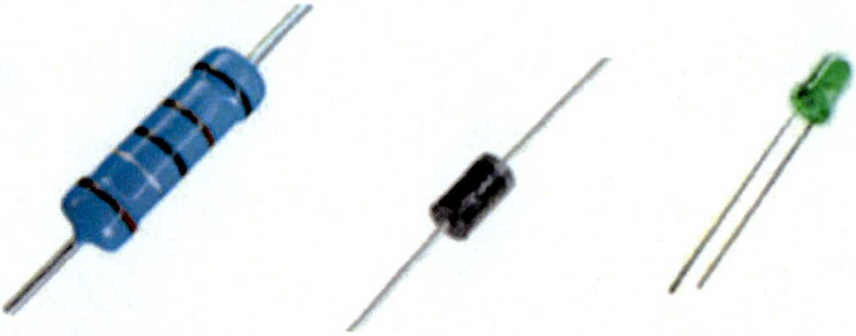

Fig. 12 Resistor, diode and LED

different chemistries are being charged. Bipolar Transistor Switch (TL431) makes it sure to keep the voltage across the cell constant even when there is change in the input voltage during CC-CV Charging Algorithm [7] (see Fig. 13).

The current from the charger gets directed toward the diodes, and the energy is dissipated in the form of heat. Overcharging is prevented using this technique and the cell remains safe even when its SOC level reaches 100%.

Since the circuit does not have CC-CV Charging ability, the cutoff voltage here should be selected with respect to 80% SOC of the cell [8] (see Fig. 14).

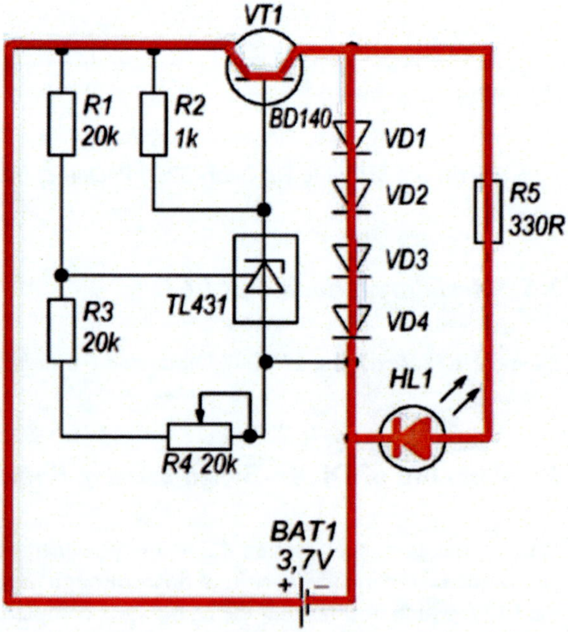

Fig. 13 Diode and LED activated when cutoff voltage is reached

Fig. 14 Prototype of diode-based passive battery balancer

7 Results

The following results were obtained when two Li-ion cells were charged using battery balancer circuit (see Tables 1 and 2).

It can be noticed even after the cell reaches 4.2 V, the voltage remains constant even after 7 min of charging. The circuit prevents overcharging and energy is dissipated as heat.

Table 1 Cell 1 results

Cell voltage (V)	Time (min)	LED status (on/off)
3.2	0	Off
3.7	15	Off
4.2	23	On
4.2	30	On

Table 2 Cell 2 results

Cell voltage (V)	Time (min)	LED status (on/off)
3.4	0	Off
3.6	5	Off
3.9	18	Off
4.2	24	On

8 Conclusion

It was noticed that the designed circuit prevents overcharging after a certain cutoff voltage is reached; however, this cutoff voltage can be altered by adjusting the trimpot as the cutoff voltage of cells varies due to various chemistries and manufacturers.

The extra energy is dissipated in the form of heat through parallelly placed diodes, which get activated using a power transistor when a certain voltage level is reached across the cells. This voltage level is controlled by a trimpot and a bipolar transistor switch.

When the cutoff voltage is reached, the LED glows and diodes actively dissipate heat to prevent overcharging and keep the cells efficient and safe.

References

1. Deng D (2015) Li-ion batteries: basics, progress and challenges. In: Energy science and engineering
2. Ohsaki K, Kuboki T (2005) Overcharge reaction of lithium ion batteries. J Power Sour 146(1):97–100
3. Barnett O, Sriramulu S (2012) Lithium ion batteries lithium ion battery, safety lithium ion battery safety. In: Encyclopedia of sustainability science and technology, pp 6097–6122
4. Arnaudov K (2020) An algorithm and circuits for active balancing systems. In: 21st international symposium on electrical apparatus & technologies
5. Xiong S (2018) Advanced battery management technologies for electric vehicles. Wiley
6. Kasyan TV, https://www.youtube.com/watch?v=v_RfcPI3WKc
7. Zhang Z, Xu C (2020) Advanced electrolytes for fast-charging high-voltage lithium-ion batteries in wide-temperature range. In: Advanced energy materials, Wiley
8. Wang H, Zhang Z (2020) On-line charging and discharging control of lithium ion battery. J Phys Conf Ser
9. BBC News UK, http://news.bbc.co.uk/2/hi/6945593.stm
10. Tampa Bay Times, https://www.tampabay.com/news/business/airlines/boeing-787-dreamliner-investigation-focuses-on-batteries/1271675/

Comparison of Used Conductor for Corona, Radio Interference Voltage (RIV) Characteristics and Audible Noise Values with Unused Conductor

K. Devender Rao, G. Venkateshwara Rao, G. Ramesh, and K. Govardhanachari

Abstract EHV/UHV transmission lines will be in the service for more than 20 years. During service these lines will be exposed to different environmental conditions and experience various electrical and mechanical stresses. It is important to understand that these transmission lines whose performance parameters are within acceptable limits during establishment remains same after usage for couple of years in the field. Due to ageing, it is evident that the lines deteriorate in its performance both in electrical and mechanical characteristics for various obvious reasons. Few of the important parameters of the conductor are Corona inception and extinction voltages, radio interference characteristics and audible noise. As per Indian/International standards and utility specifications, the acceptable limits for these performance parameters are defined. In this paper, an attempt has been made to evaluate the corona inception and extinction voltages, radio interference characteristics and audible noise of a moose conductor used in 400 kV line which was in service for about 15 years and compared with the values of an unused conductor.

Keywords Corona inception and extinction voltages · Radio interference characteristics · Audible noise · Used and unused conductor

1 Introduction

The important characteristic of transmission line is its power handling capacity, which means the bulk power that can be transferred on the line. During handling the higher power or higher voltages, it is essential that the technical parameters do not exceed specified limits. Few of the parameters such as visible corona, radio interference voltage and audible noise are to be maintained within the limits [1] even after using the conductor for couple of years in power transmission. During operation of the transmission line, the conductor is exposed to various environment conditions and the morphology of the conductor changes. Due to this change in the conductor,

K. D. Rao (✉) · G. V. Rao · G. Ramesh · K. Govardhanachari
Ultra-High Voltage Research Laboratory, Central Power Research Institute, Hyderabad, India
e-mail: kdrao@cpri.in

A. R. Gupta et al. (eds.), *Power Electronics and High Voltage in Smart Grid*, Lecture Notes in Electrical Engineering 817,
https://doi.org/10.1007/978-981-16-7393-1_7

corona discharge phenomena which are common around HVAC and HVDC transmission lines, and the electromagnetic disturbance in the vicinity results in radio interference (RI), audible noise (AN), etc. all of which are both technically and environmentally undesirable [2, 3]. As per the requirement of Indian/international standards and customer/utilities, the nominal values for corona inception voltage shall be ≥330 kV (minimum 3% higher than the corona extinction voltage) and corona extinction voltage 320 kV. Similarly radio interference values shall be less than 1000 μvolts at the test voltage of 267 kV for a 400 kV conductor. The noise values in decibels shall be as lower as possible and there is no commercial test for assessing audible noise from a conductor but it should be less than 80 dB as per the standard ISO 3740:2019. It is expected that these nominal values when tested in laboratory remains same after usage of couple of years in the service. It is known that these parameters (corona inception and extinction values, radio interference values and audible noise values) constitute corona and any deviation from these values will have impact of increase in corona losses, interference in the radio signal frequencies and audible noise disturbance for nearby residences. Also, if the problem is not attended/replaced, and left the conductor to be in service, there is a possibility that these values will further deteriorate. As of now, there is no data available for comparing the values that after usage of conductor in service for said years, any kind of tolerances may be allowed. Hence, an attempt is made to evaluate the values of used conductor for above-mentioned parameters. This paper describes the corona, RIV characteristics and audible noise values of conductor used in the field for 15 years and compared with the values of unused conductor.

2 Conductor Configuration Used in the Field

For a 400 kV transmission line, twin conductor configuration is generally used with Aluminum Conductor Steel Reinforced (ACSR) moose or ACSR Bersimis as the main conductor. In this case ACSR moose conductor is selected which is 31.77 mm overall diameter, 54 number of aluminum strands, 7 number of steel strands, 3.53 mm nominal diameter of each strand (aluminum or steel). The cross-sectional area of the aluminum conductor is 597.0 mm^2.

The above conductor configuration is usually used in the power utilities in India for 400 kV transmission line. The conductor which is in operation will be replaced for damages found, if any or change in the configuration of the transmission line.

In this case study, the ACSR moose conductor which was being used in the ambient conditions as shown below are removed due to replacement, as the old conductor was damaged in few of the locations. The environmental conditions where the conductor was in service is shown Table 1.

The conductor was removed from location for the reasons already mentioned. The conductor was used in the transmission line for about 15 years in the transmission line at 400 kV voltage level. The locations where conductor replaced is 400 kV line from

Table 1 Ambient conditions

Sl. no.	Parameters	Readings
1	Average temperature	29 °C
2	Average rainfall	300–650 mm
3	Ambient temperature	29 °C
4	Min. temperature	18 °C
5	Max. temperature	44 °C
6	Wind speed	10.0 miles per hour
7	Atmospheric pressure	718 mmHg

one of the Indian utility which was erected in year 2000 and the line was removed for a certain part of line during the year 2015.

The new (Unused) available ACSR moose conductor was used for measurements and compared with the results of used conductor. The used conductor details such as period of usage, place of conductor, dimensions, etc. have been already mentioned. The used conductor was selected in such a way where there are no damages on the conductor. 10 m length of conductor was selected for both used and unused conductors for carrying out measurements. Photographs of unused conductor and used conductor for 15 years are shown below (see Fig. 1).

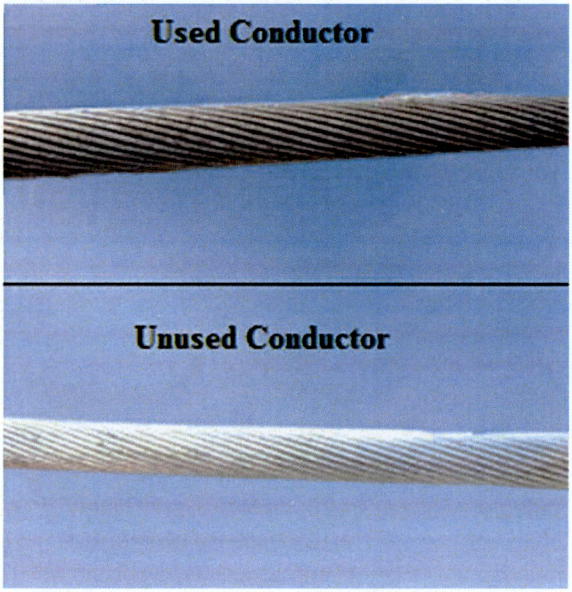

Fig. 1 Photographs of used and unused conductor

3 Measurements on Used and Unused Conductor

Following measurements were carried out on Aluminum Conductor Steel Reinforced (ACSR) unused conductor and used conductor in the transmission line.

- Visible corona discharges
- Radio interference voltages (RIV)
- Audible noise under the experimental conductor.

3.1 *Equipment Used for the Measurements*

Visible Corona Discharges such as Corona Inception Voltage (CIV) and Corona Extinction Voltage (CEV) are verified with a naked eye.

Radio Interference Voltage values are measured using Electro-Magnetic Interference (EMI) test receiver. EMI test receiver which is designed as per CISPR 16 is used to record the RIV values in dB. The voltage is fed to the EMI test receiver from an impedance box of about 40 ohms which is coupled to a coupling capacitor of 2nF. The recorded values are then converted to micro-volts.

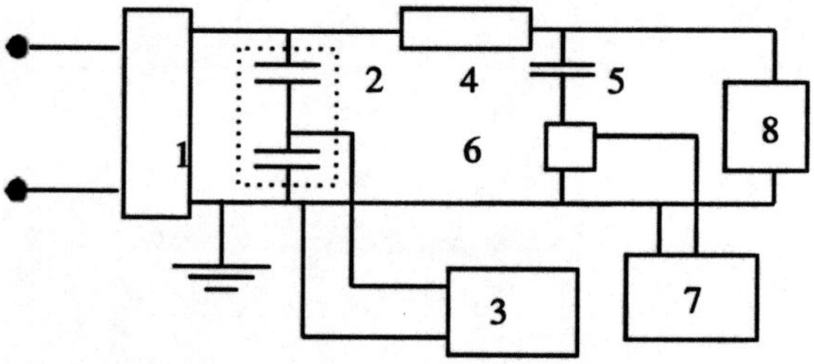

Legend:

1: HV AC Source	2: Voltage Divider
3: RMS Voltmeter	4: RF Rejection Coil
5: Coupling Capacitor	6: Impedance Box
7: EMI Test receiver	8: Conductor

Audible Noise (AN) measurements are carried out using an AN meter which can measure values in dB. The AN meter is a precision impulse sound level meter. This meter consists of a microphone, weighting networks, an rms detector and a digital display that shows the weighted electrical signal from the microphone. The microphone is an electric-acoustic transducer which responds to the sound waves and converts to the electrical signals. This meter has frequency weighting networks, namely, A B C and D. A-weighting represents to human perception and used for AN measurements. Frequency spectra of AN measurements from lines is obtained using octave band filters (see Fig. 2).

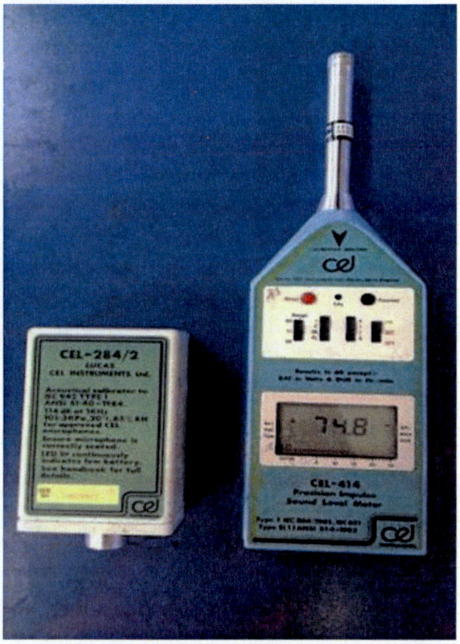

Fig. 2 Audio noise meter

4 Test Arrangement

Test is carried out on a single Aluminum Conductor Steel Reinforced (ACSR) moose conductor of 10 m length and twin ACSR moose conductor of 10 m length where conductor spacing is 457 mm during twin moose. Since conductor is 10 m length, the conductor is tightened at both ends with negligible sag. Necessary hardware was used for single and twin conductor configuration. The ends of the conductor termination were shielded with corona rings to suppress the Radio Interference Voltage (RIV) emitting from end terminating hardware. The conductor was erected at a height of about 10 m from the ground. The variation in the end hardware has influenced any of the Corona Inception Voltage (CIV), Corona Extinction Voltage (CEV), RIV and Audible Noise (AN) measurements. The conductor test arrangement and measurement setup and equipment remains identical for both single and twin bundle conductor. Test arrangement is as shown below (see Fig. 3).

Fig. 3 Twin conductor arrangement

5 Test Procedures for Conducting Evaluating CIV, CEV, RIV and Audible Noise

5.1 *Visible Corona Discharges*

Corona Inception Voltage (CIV) and Corona Extinction Voltage (CEV) are verified with the naked eye with complete darkness around the conductor. The voltage is slowly raised to CIV value. After the inception of corona, the voltage is slowly reduced to CEV value. This procedure is repeated for 3 times. The third repetition values [or] lower values out of 3 repetitions are taken as final values (as per customer requirement) of CIV and CEV.

5.2 *Radio Interference Voltages*

The RIV measurements at 1 MHz. The voltage is raised to 130% of the test voltage. This voltage is maintained for 5 min to precondition the conductor. Then the voltage is reduced to a value of 30% of test voltage in steps of 10% of test voltage. The rate of raise of test voltage is as per IEC 60,060-1 [4]. At each voltage, RIV value is noted. This procedure [5, 6] is repeated for 3 times. The third repetition values are taken as final values of RIV.

5.3 *Audible Noise*

The audible noise in dB is measured in the middle under the experimental conductor, and at 2 and 4 m away from the middle of conductor. The audible noise is measured at each voltage level as measured for RIV. This procedure is repeated 3 times. The third repetition is taken as final values of audible noise.

6 Results and Discussion

6.1 *Visible Corona Discharges for Twin Moose Conductor*

Discussion: Corona Inception Voltage (CIV) and Corona Extinction Voltage (CEV) values are found to be higher for unused conductor for a twin moose conductor configuration. As the usage of conductor occurs in the field, the CIV and CEV values are reduced. The CIV values are reduced from 348 to 308 kV and CEV values are reduced from 341 to 297 kV for an unused conductor to used conductor. This means that the initiation of corona occurs early in used conductors. The differences in voltages are found to be about 40 kV for both CIV and CEV values. The increases in these values are due to looseness in the conductor strands which occur due to sag in the self-weight of the transmission line in due course of time, environmental influences, etc. Corona loss (Watts/meter) of a particular conductor or conductor bundle operating at a specific surface voltage gradient will be affected by changing the corona on set gradient. Since, the CIV value is very high in case of used conductors (though it is a twin bundled conductor), the corona power loss in case of used conductor will be obviously high. The used conductor was selected with no physical damages on it, so parameters such as surface irregularity factor *m*, relative air density δ, etc. of the conductor are not discussed here (see Fig. 4).

6.2 *Visible Corona Discharges for Single Moose Conductor*

Discussion: Corona Inception Voltage (CIV) and Corona Extinction Voltage (CEV) values are found to be higher for unused conductor for a single moose conductor configuration. A similar trend is found with single conductor as seen in twin bundled

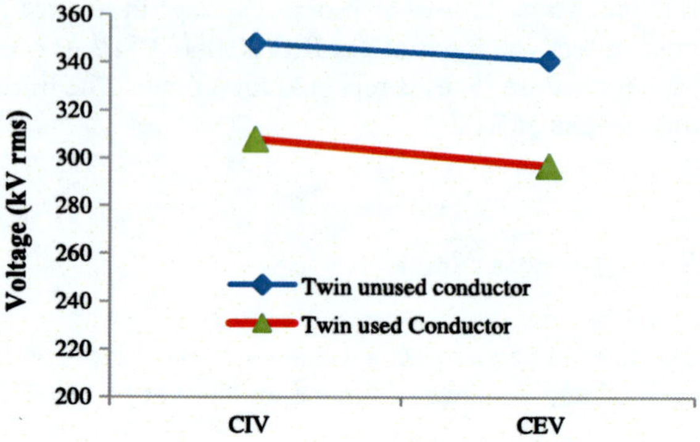

Fig. 4 Visible corona discharges for Twin moose conductor

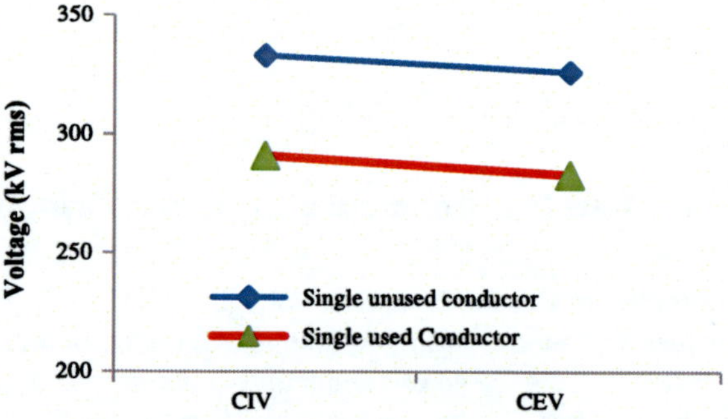

Fig. 5 Visible corona discharges for Single moose conductor

conductor where both CIV and CEV values are reduced for used conductor. The CIV values are reduced from 333 to 291 kV and CEV values are reduced from 326 to 283 kV for an unused conductor to used conductor. The differences in voltages are again found to be about 40 kV for both CIV and CEV for the same reasons as mentioned for twin bundled conductor (see Fig. 5).

6.3 Radio Interference Voltage (RIV) Characteristics for Twin Moose Conductor

Discussion: RIV values are found to be higher for used conductor for a twin moose conductor configuration. The RIV values show similar trend till 40% of the test voltage for both used and unused conductor and thereafter, i.e. from 50% of the test voltage level, the RIV level increases for unused conductor. The value of RIV level at 130% of the test voltage is below 1000 micro-volts for unused conductor, whereas it is 2300 micro-volts for used conductor. The acceptable value for RIV is 1000 micro-volts by power distribution utilities and relevant standards. RIV produced by line depends on many factors that includes line configuration for multi-conductors

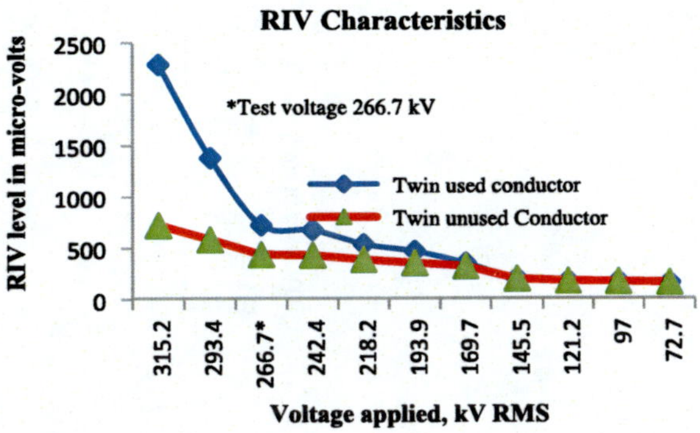

Fig. 6 RIV for Twin moose conductor

(stranded conductor). If the stands are loosened in the conductor after usage for couple of years, the corona current generation resulting to electric and magnetic field occurrence on line generates early Electromagnetic Interference (EMI) or early RIV. This is one of the reasons where higher RIV values are found in used conductor. The increases in these values are due to looseness in the conductor strands which occur due to sag in the self-weight of the transmission line in due course of time, environmental influences, etc. The higher RIV values are not acceptable to use in high voltage lines for utilities and as per standards (see Fig. 6).

6.4 *Radio Interference Voltage (RIV) Characteristics for Single Moose Conductor*

Discussion: RIV values are found to be higher for used conductor for a single conductor configuration. The RIV values show similar trend till 50% of the test voltage for both used and unused conductor and thereafter, i.e. from 60% of the test voltage level, the RIV level increases for unused conductor. The value of RIV level at 130% of the test voltage is below 1000 micro-volts for unused conductor whereas it is 5800 micro-volts for used conductor. The acceptable value for RIV is 1000 micro-volts by power distribution utilities and relevant standards. A similar trend is found with single conductor as seen in twin bundled conductor where RIV values are much higher than required limits for used conductor for the same reasons as mentioned for twin bundled conductor (see Fig. 7).

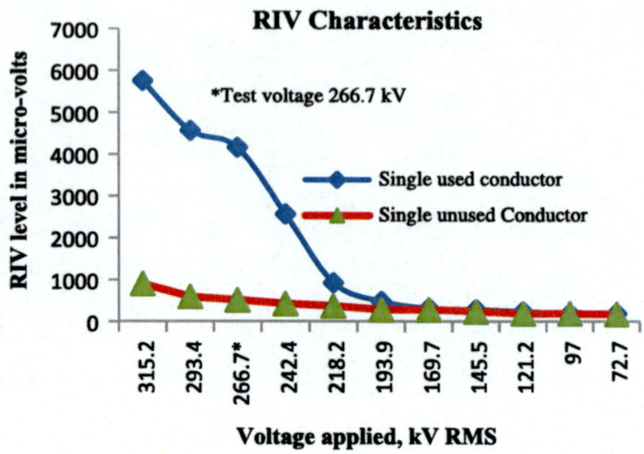

Fig. 7 RIV for Single moose conductor

6.5 Audible Noise Under the Experimental Conductor Result for Single and Twin Moose Conductor

Discussion: The Audible noise values are measured at the height of 1 m from the ground level and in the middle of the conductor (0 m lateral distance). Also the measurements were carried out at 2 and 4 m lateral distances from the middle of the conductor. The audible noise values are shown in Figs. 8, 9 and 10. The sound pressure level of AN in dB (A) of HV line is a function of conductor surface gradient, number of sub-conductors, sub-conductor diameter, the radial distance between the conductor and the point of measurement. Of these factors, the conductor diameter and the multi-conductor/strands tightness in a conductor is showing the variation in the sound level as per the experimentation data. As shown in Fig. 8, single conductor noise levels of an unused conductor are lesser than used conductor values. Similarly,

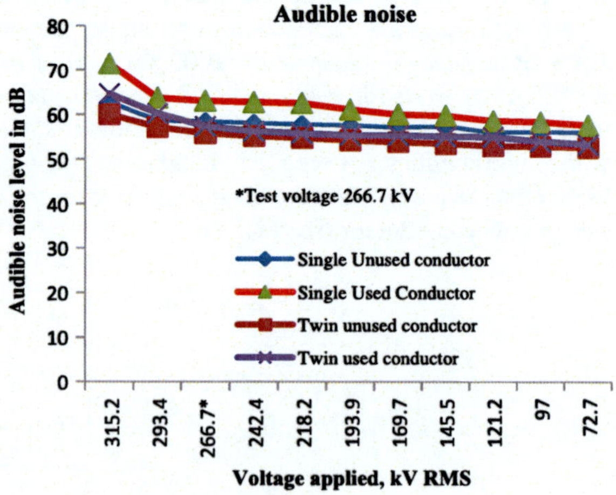

Fig. 8 Audible noise under the experimental conductor for Single and twin moose conductor (at middle of the conductor)

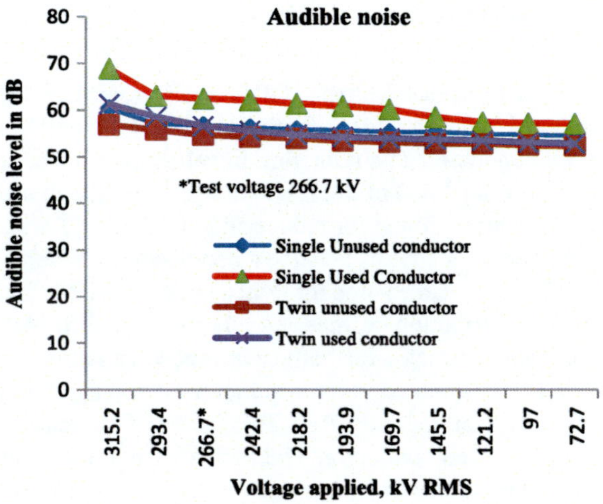

Fig. 9 Audible noise under the experimental conductor for Single and twin moose conductor (at 2 m from middle of the conductor)

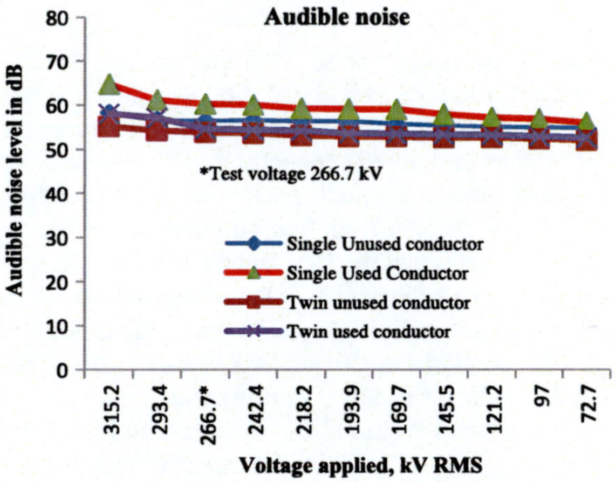

Fig. 10 Audible noise under the experimental conductor for Single and twin moose conductor (at 4 m from middle of the conductor)

the twin bundle conductor noise levels of an unused conductor are lesser than used conductor values. However, an observation made is that the twin bundled conductor values are lesser than single conductor irrespective of unused or used conductor. Similar trends were found during experiments as shown in Figs. 9 and 10. If a human is exposed to hear the sound level of around 90 dB continuously for 8 h, there is a possibility of hearing damage. Thus, the acceptable limits of hearing noise levels are below 80 dB. However, in the experiments carried out, the sound levels were less than 80 dB for all configurations and used or unused conductor. ISO 3740:2019(en) dealt with Acoustics—Determination of sound power levels of noise sources—Guidelines for the use of basic standards.

7 Conclusions

Corona Inception Voltage (CIV) and Corona Extinction Voltage (CEV) values are reduced (deteriorated) for both conductor configurations, i.e. single and twin bundled used conductor. The reduction in values have direct influence in increasing corona power loss (W/m) of the high voltage lines due to early corona onset gradient. One of the reasons found for reduction in CIV and CEV values for used conductor is that the conductor strands gets loosened due to self-weight of the transmission lines.

The acceptable value for Radio Interference Voltage (RIV) is 1000 micro-volts by power distribution utilities and relevant standards. RIV values are found to be higher for used conductor for both twin moose and single conductor configuration. If the stands are loosened in the conductor after usage for couple of years, the occurrence on line generates early RIV. This is one of the reasons where higher RIV values are found in used conductor. The increases in these values are due to looseness in the conductor strands which occur due to sag in the self-weight of the transmission line in due course of time, environmental influences, etc.

The acceptable limits of hearing noise levels are below 80 dB noise levels. The experiments carried out on two different configurations for both conductor configurations show that the noise levels are less than 80 dB. The measurements carried out at lateral distances of 0, 2 and 4 m also show the sound levels are less than 80 dB. In Audible Noise (AN) measurements too, the noise values of used conductor are higher than the unused conductor for all measurement setups. The variation of values is due to conductor strands get loosened due to self-weight of the transmission lines.

The comparison from unused conductor to the used conductor in the field for 15 years shows that the CIV and CEV values, RIV values and audible noise values deteriorate with the ageing of the conductor used in the transmission lines. A similar trend is found with single conductor and twin bundled conductor.

This trend of the CIV and CEV, RIV and audible noise remains same for both single conductor and twin moose conductor.

One of the reason for the deterioration of values is that conductor strands gets loosened due to self-weight of the transmission lines during usage in the field/service.

In the future, it is necessary to understand the deterioration of conductor for different service usage of years, i.e. 5, 10, 15, 20 years soon. These studies have to be experimentally carried out in the laboratories and compared with theoretical evaluations.

Acknowledgements Authors thank CPRI Management for permitting to publish the paper. Authors thank Dr. PradeepNirgude, Additional Director, CPRI for providing guidance and technical support during the course of experiments. The assistance rendered by Mr. K Srinivas, Mr. L Balakrishna, Mr. G Jangaiah, Mr. Nawaz shareif and during conducting experiments is highly acknowledged.

References

1. Maruvada PS (2000) Corona performance of high voltage lines. Taylor & Francis Group, London UK
2. Bian X, Yu D et al (June 2011) Influence of Aged conductor surface conditions AC corona discharge with a corona cage. IEEE Trans Dielectr Electr Insul 18(3)
3. Bian X, Chen L, Yu D et al (December 2011) Influence of Aged conductor surface conditions AC corona generated Audible noise with a corona cage. IEEE Trans Dielectr Electr Insul 19(6)
4. IEC 60060–1: 2010 (2010) High voltage test techniques–general definitions and test requirements
5. IEC 60284: 1997 (1997) Radio interference test on high voltage insulators
6. IEC60437: 1997 (1997) Overhead lines–requirement sand tests for fittings

A Switched-Capacitor Based Generalized Topology for Multilevel Inverter with Cross-Switched Structure

Avinash Maurya and Ambarisha Mishra

Abstract In recent years, various new topologies for multilevel inverter (MLI) have been proposed to reduce circuit elements. But most of these topologies require high voltage switches and thus limit the usage for high-voltage applications. In this paper, a switched-capacitor (SC) based generalized MLI with cross-switched structure is proposed, which requires low voltage switches for any high output voltages. The proposed topology comprises a smaller number of switches, and also it has voltage boosting ability. It does not require any external voltage balancing circuit for the capacitors. The proposed topology eliminates the requirement of end side H-bridge converter to generate negative voltage levels because switches of the H-bridge converter (for polarity reversal) have to withstand total operating voltage. Thus, the proposed topology can be easily adopted for high-voltage applications using low-voltage switches. The structure of MLI facilitates equal load sharing among the sources using redundant switching path. To examine the generalized MLI, nine-level inverter is simulated in MATLAB/SIMULINK software. Comparisons with some latest topologies are also presented to show the effectiveness of the proposed topology.

Keywords Cross-switched structure · Switched-capacitor unit · Voltage boosting ability · Total harmonic distortion

1 Introduction

Multilevel inverters are power electronics converter that employ low-voltage switches, diodes, DC sources and gate driver circuits. The basic principle of MLI is to distribute maximum required operating voltage among low-rated power electronics switches [1]. For the same output voltage, as the number of voltage levels

A. Maurya (✉) · A. Mishra
National Institute of Technology Patna, Bihar 800005, India

A. Mishra
e-mail: ambrish.mishra@nitp.ac.in

A. R. Gupta et al. (eds.), *Power Electronics and High Voltage in Smart Grid*,
Lecture Notes in Electrical Engineering 817,
https://doi.org/10.1007/978-981-16-7393-1_8

increases, total harmonic distortion (THD) decreases and output becomes more sinusoidal. Some important features of MLIs are reduced dv/dt, improved power quality, reduced power loss and less EMI, etc.

Conventional topologies such as neutral point clamped (NPC) and flying capacitor (FC) inverter requires large number of diodes and capacitors, respectively. Cascaded H-bridge (CHB) inverter requires large number of separate isolated DC sources and switches. The conventional topologies employ external circuit to balance the capacitor's voltages. This makes the system complex and costly [2].

To harness renewable energy, MLI is one of the best options for high power applications. For low voltage sources in case of renewable energy sources, inverters having inherent boosting capabilities are required [3].

A five-level inverter is introduced in [4]. The inverter does not require any voltage balancing circuit for the capacitors. Also, it has voltage boosting ability. But it requires seven switches, one voltage source and two capacitors for five voltage levels. In [5], a self-balancing five-level inverter is proposed. This inverter also does not employ any voltage balancing circuit for the capacitors. For the five level and nine level, it requires nine switches and nineteen switches, respectively. Large number of switches are the only drawback of this inverter. A high-gain switched-capacitor MLI is introduced in [6]. But it needs a H-bridge converter for the polarity reversal of voltage across load. Seven-level inverter with twelve switches and two capacitors is presented in [7]. In [8], six-level inverter is proposed that employs four capacitor, eight switches and one voltage source. But it lacks voltage boosting ability. Reference [9] requires a few switches, but it also does not have any inherent voltage boosting ability for the capacitors. Single stage five-level boost inverter is presented in [10]. It requires two voltage sources, two capacitors and eight switches. It uses a controller for balancing of capacitor's voltage. A cross connected source based MLI is introduced in [14], similar to the proposed MLI. But it has no voltage boosting ability. Mostly, in traditional SC-based MLI H-bridge is employed for polarity reversal of output voltage of the converters [6, 11, 14].

Keeping all the good features and removing the demerits of above-mentioned MLIs, a generalized switched-capacitor based multilevel inverter with cross-switched structure is presented. Unlike other topologies, this topology does not employ high voltage switches to lessen the number of switches, H-bridge converter to get negative voltage levels and complicated circuit or any intricate logic scheme to keep the capacitors' voltages in a permissible limit. Also, the proposed topology has voltage boosting ability. These factors make it suitable for high-voltage applications. The topology is suitable for symmetrical as well as asymmetrical mode of MLI. Further sections describe the proposed MLI in detail.

2 Proposed Generalized SC-Based MLI

2.1 Topology Description

The proposed generalized topology is shown in Fig. 1. This proposed structure employs a combination of SC units and low-rated power electronic switches. The SC unit is the basic fundamental unit of the proposed structure. Each basic unit (SC unit) includes a DC source, a capacitor, two switches and an active diode. During charging and discharging mode, the capacitor is connected parallel and series with DC source, respectively. The proposed structure may also be extended for a higher number of phases as per requirement. Since the basic working principle of single phase and n-phase are similar, so further discussion would be on single phase.

Proposed topology mainly generates an odd number of voltage levels. This structure generates minimum five voltage levels, and each SC unit adds four levels. It is explained by the following equations.

$$L = 4\mathrm{N_{SC}} + 1 \tag{1}$$

$$\mathrm{N_{DC}} = \frac{L - 1}{4} \tag{2}$$

$$L = 4\mathrm{N_{DC}} + 1 \tag{3}$$

$$\mathrm{N_S} = 2(2\mathrm{N_{DC}} + 1) \tag{4}$$

where $\mathrm{N_{SC}}$, $\mathrm{N_{DC}}$, L and $\mathrm{N_S}$ are number of SC units, number of DC sources, number of voltage levels and number of power electronics switches, respectively.

Using Eqs. (3) and (4), relationship between the number of switches and number of voltage levels can be written as follows:

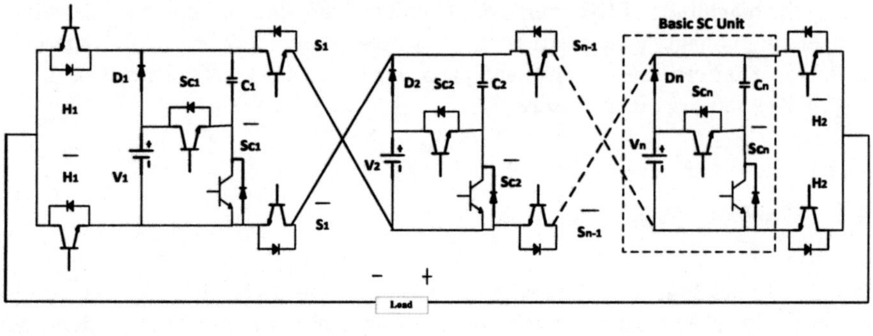

Fig.1 Proposed generalized topology for MLI

$$N_S = L + 1 \tag{5}$$

2.2 Switches Ratings

It is seen from Fig. 1 that outermost switches H_1, $\overline{H}_1$, H_2, and $\overline{H}_2$ differ from other switches. Maximum open voltage stress for these four switches will be equal to 2 V_{DC}. But for S_n and $\overline{S}_n$, maximum open voltage stress will be equal to 4 V_{DC}. Sc_n and $\overline{Sc}_n$ that suffer minimum voltage stress V_{DC}.

Assuming $V_{o,\max}$ as maximum operating voltage of proposed MLI. V_s and V_T are termed as maximum voltage stress across outer switches (H_1, $\overline{H}_1$, H_2, and $\overline{H}_2$) and inner switches (S_n and $\overline{S}_n$), respectively.

$$V_s = 2V_{DC} \tag{6}$$

$$V_T = 4V_{DC} \tag{7}$$

$$V_{o,\max} = 2N_{DC}V_{DC} \tag{8}$$

$$V_{o,\max} = \frac{L-1}{2}V_{DC} \tag{9}$$

$$V_{DC} = \frac{2}{L-1}V_{o,\max} \tag{10}$$

$$V_T = 4\frac{2}{L-1}V_{o,\max} \tag{11}$$

It is clear from the equations that for the fixed maximum operating voltage ($V_{o,\max}$), as the number of voltage levels increases, voltage stress across the switches decreases. But, if the magnitude of DC source of SC unit is fixed, then for any high magnitude of operating voltage, voltage stress of the switches does not alter, i.e., for outermost switches, inner crossed switches and capacitor's switches, voltage stress will be 2 V_{DC}, 4 V_{DC} and V_{DC}, respectively.

2.3 Working Principle

Working principle is described with the help of single phase nine-level inverter, as shown in Fig. 2. There are five pairs of complementary switches. Using redundancy of switching modes, equal utilization of the sources is feasible. Sequence of

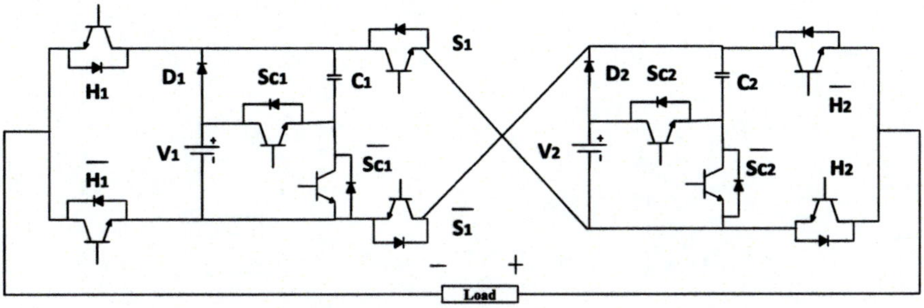

Fig. 2 Proposed nine-level inverter

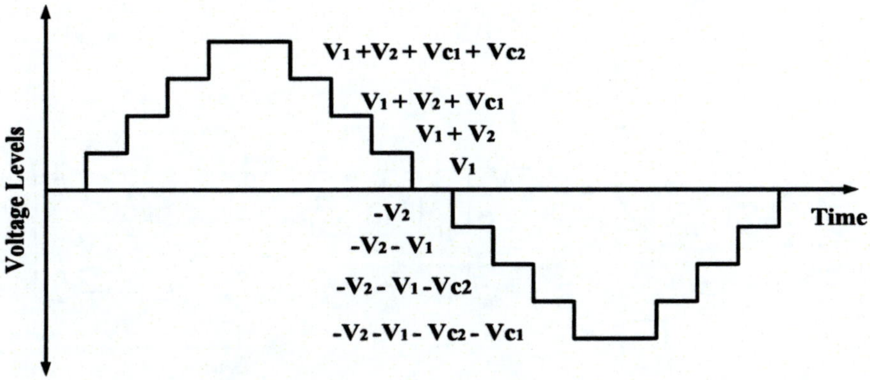

Fig. 3 Sequence of switching modes for equal utilization of the sources

switching modes for equal utilization of the sources for the nine-level MLI is shown in Fig. 3. All the prime operating (working) modes are shown in Fig. 4 to get nine output voltage levels and complete switching status of the switches are presented in Table 1. S_1 remains ON for all possible positive levels and first zero (mode fifth) level, whereas $\overline{S}_1$ remains ON for all possible negative levels and another zero (mode sixth) level. Thus, S_1 and $\overline{S}_1$ operate on fundamental frequency.

2.4 Switching Scheme

In the proposed topology alternate phase opposition disposition pulse width modulation (APOD-PWM), a type of level-shifted PWM, is used, as shown in Fig. 5. For the nine level, eight triangular carrier signals are required. A single 50 Hz sinusoidal signal is taken as modulating (reference) signal. Each carrier signal is compared with the modulating signal. If the magnitude of the modulating signal becomes more than positive carrier signal C_x^+ ($x = 1, 2, 3 \ldots$), then comparator output will be 1, otherwise

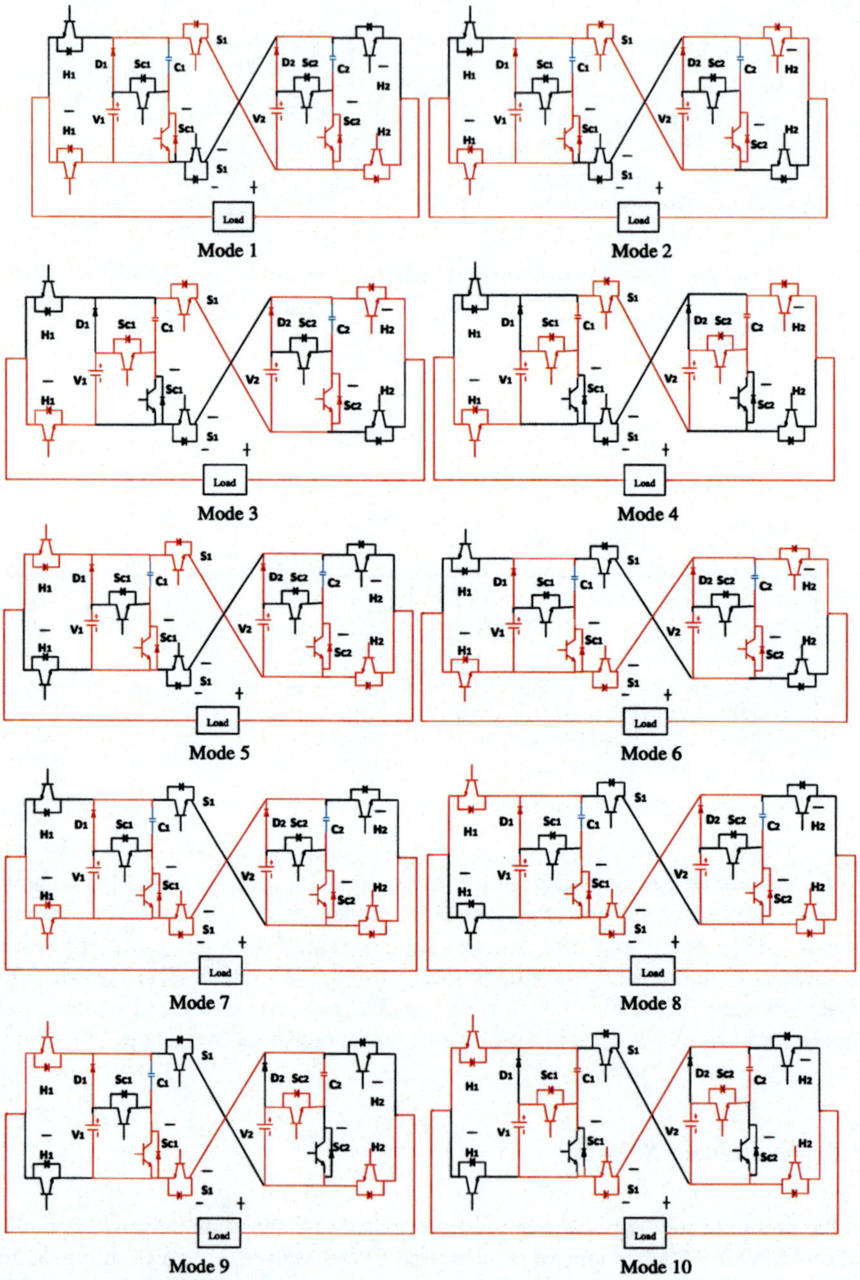

Fig. 4 Working modes for the nine-level inverter

Table 1 Switching status for the nine-level MLI

Mode	H_1	$\overline{H}_1$	Sc_1	$\overline{S}c_1$	S_1	S_1	Sc_2	Sc_2	H_2	H_2	V_{outpu}	Status of C_1	Status of C_2
I	0	1	0	1	1	0	0	1	1	0	V_{dc}	Charging	Charging
II	0	1	0	1	1	0	0	1	0	1	$2V_{dc}$	Charging	Charging
III	0	1	1	0	1	0	0	1	0	1	$3V_{dc}$	Discharging	Charging
IV	0	1	1	0	1	0	1	0	0	1	$4V_{dc}$	Discharging	Discharging
V	1	0	0	1	1	0	0	1	1	0	0	Charging	Charging
VI	0	1	0	1	0	1	0	1	0	1	0	Charging	Charging
VII	0	1	0	1	0	1	0	1	1	0	$-V_{dc}$	Charging	Charging
VIII	1	0	0	1	0	1	0	1	1	0	$-2V_{dc}$	Charging	Charging
IX	1	0	0	1	0	1	1	0	1	0	$-3V_{dc}$	Charging	Discharging
X	1	0	1	0	0	1	1	0	1	0	$-4V_{dc}$	Discharging	Discharging

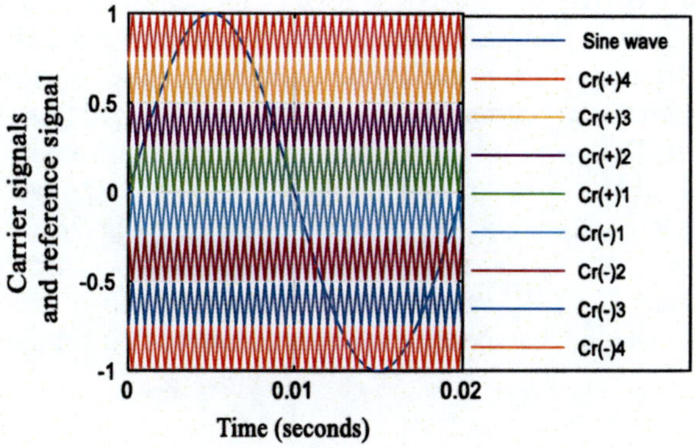

Fig. 5 APOD pulse width modulation

0. Similarly, if the magnitude of the modulating signal becomes more than negative carrier signal C_x^- ($x = 1, 2, 3\ldots$), then comparator output will be 0, otherwise 1.

Generally, output signals of all the comparators are combined to acquire an aggregated signal. But in this proposed topology, the aggregated signal is modified to use the redundancy of switching status. It is seen from the wave shape of the modified aggregated signal that there are two zero levels (mode fifth and mode sixth), as

Fig. 6 Modified aggregated signal

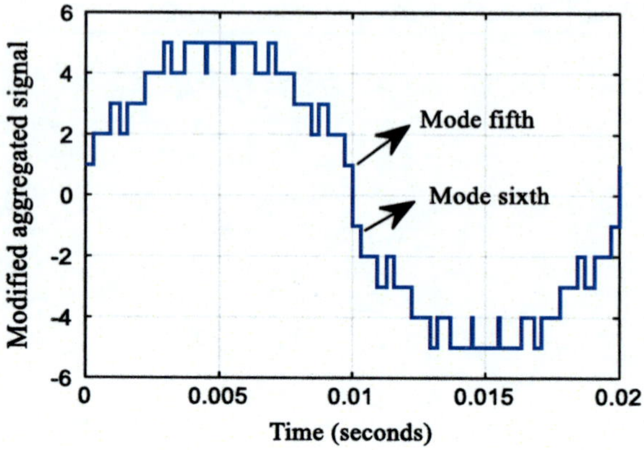

shown in Fig. 6. This modified aggregated signal is given to multiport switch, which generates pulses sequentially according to switching status of the switches for the particular voltage level.

3 Simulation Results and Discussion

To examine the proposed generalized topology, nine-level inverter is simulated in MATLAB/SIMULINK software. Two equal DC voltage sources $V_1 = V_2$ of 50 V are taken, reference signal of 50 Hz is taken and carrier signals of 2 kHz frequency are chosen for better THD. 80 mf value is selected for both the capacitors. Output voltage and current waveforms for resistive load ($R = 10\ \Omega$) and inductive load ($R = 10\ \Omega$, $L = 10$ mH) with its harmonic spectrum are shown in Fig. 7a–f, respectively. Modulation index is taken one. For R load, THD of voltage and current waveform are same and are equal to 14.01%, as shown in Fig. 7b, c. Current waveform of R-L load is nearly similar to ideal sinusoidal signal, because the load behaves as a low-pass filter. So, THD value of current waveform comes to 3.18%, as shown in Fig. 7f. It is examined that there is no effect of the load on voltage harmonics.

4 Comparison

Proposed topology is compared with some latest MLIs to show its efficacy. Comparisons among various terms are shown through Table 2. The table comprises number of voltage levels, number of switches, number of DC sources, number of capacitors and voltage boosting ability. In [4], for five-level voltage output seven switches are needed, unlike the proposed topology that requires only six switches. For five-level and nine-level voltage output [5] needs nine and nineteen switches, respectively.

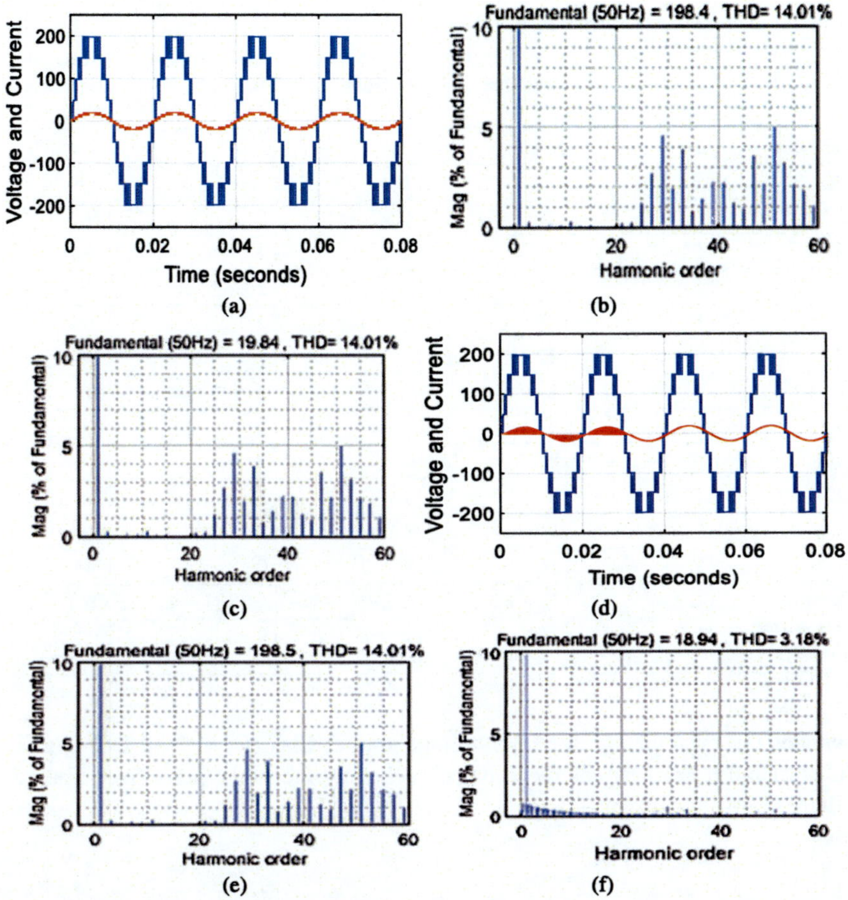

Fig. 7 Simulated results **a** Output voltage and current waveform for R load, **b** Voltage THD for R load, **c** Current THD for R load, **d** Output voltage and current waveform for R-L load, **e** Voltage THD for R-L load, **f** Current THD for R-L load

References [8, 9, 12] do not have any voltage boosting ability. Reference [10] needs external voltage balancing circuit and it employs eight switches, two voltage sources and two capacitors for five voltage levels. For seven-level, [13] requires nine switches with one boosting capacitor and two DC link capacitors. It can boost only 1.5 times of the magnitude of input voltage. Reference [14] requires ten switches to acquire nine level of output voltage, like the proposed topology. But it does not have any voltage boosting ability.

Overall, it is concluded that all the good factors are contained in the proposed topology. It can boost twice the applied DC voltage, and also it doesn't use any end side H-bridge converter to get negative voltage levels. Voltage stress across the

Table 2 Comparison with recent MLIs (for single phase)

Inverter type	Number of levels	Number of switches	Number of sources	Capacitor (as a source)	Voltage boosting ability
NPC	L	2(L − 1)	1	L − 1	No
FC	L	2(L − 1)	1	L − 1	No
CHB	L	2(L − 1)	0.5(L − 1)	0	No
[4]	5	7	1	2	Yes
[5]	5	9	1	1	Yes
[5]	9	19	1	3	Yes
[7]	7	12	1	2	Yes
[8]	6	8	1	4	No
[9]	9	8	1	2	No
[10]	5	8	2	2	Yes
[12]	7	8	1	3	No
[13]	7	9	1	3	Yes
[14]	9	10	4	0	No
Proposed generalized topology	L	L + 1	0.25(N − 1)	0.25(N − 1)	Yes
	9	10	2	2	

switches does not change for high-voltage applications (for a fixed DC source of SC unit) and it decreases as the voltage level count increases (for a fixed operating voltage).

5 Conclusions

A switched-capacitor based MLI with voltage boosting ability has been presented in this paper. Mathematical relations between the number of switches, DC sources, SC units and voltage levels have been developed. Using redundancy of switches, voltage sources are equally utilized in a full alternating cycle of the voltage. Inner crossed switches work at fundamental frequency. Unlike other topologies, proposed topology does not employ any H-bridge converter to acquire negative voltage levels. This merit makes it suitable for high-power applications. Nine-level inverter has been simulated in MATLAB/SIMULINK software, and the simulated results are shown for resistive and inductive load. Voltage THD and current THD are 14.01% and 3.18%, respectively. There is no requirement of any external arrangements to balance the capacitor's voltage. Since all the switches are complementary, the number of gate driver circuits would be half of the number of total switches. Overall, the proposed topology may be an agreeable choice in the race of advancement of topologies for MLIs.

References

1. Gupta KK, Ranjan A, Bhatnagar P, Sahu LK, Jain S (2016) Multilevel inverter topologies with reduced device count: a review. IEEE Trans Power Electron 31:135–151
2. Raushan R, Mahato B, Jana KC (2016) Comprehensive analysis of a novel three-phase multilevel inverter with minimum number of switches. IET Power Electron 9:1600–1607
3. Bana PR, Panda KP, Naayagi RT, Siano P, Panda G (2019) Recently developed reduced switch multilevel inverter for renewable energy integration and drives application: topologies, comprehensive analysis and comparative evaluation. IEEE Access 7:54888–54909
4. Saeedian M, Hosseini SM, Adabi J (2018) A five-level step-up module for multilevel inverters: topology, modulation strategy, and implementation. IEEE J Emerg Sel Top Power Electron 6:2215–2226
5. Sandeep N, Ali JSM, Yaragatti UR, Vijayakumar K (2019) A self-balancing five-level boosting inverter with reduced components. IEEE Trans Power Electron 34:6020–6024
6. Panda KP, Bana PR, Sanjeevikumar P, Panda G, Leonowicz Z, Mitolo M (2020) A single-source high-gain switched-capacitor multilevel inverter with inherent voltage balancing. In: Proceeding of the 2020 IEEE international conference on environment and electrical engineering and 2020 IEEE industrial and commercial power systems Europe EEEIC/ICPS Europe 2020, pp 8–13
7. Lee SS (2018) A single-phase single-source 7-level inverter with triple voltage boosting gain. IEEE Access 6:30005–30011
8. Le QA, Lee DC (2016) A novel six-level inverter topology for medium-voltage applications. IEEE Trans Ind Electron 63:7195–7203
9. Sharifzadeh M, Al-Haddad K (2019) Packed E-Cell (PEC) converter topology operation and experimental validation. IEEE Access 7:93049–93061
10. Nguyen MK, Tran TT (2017) Quasi cascaded h-bridge five-level boost inverter. IEEE Trans Ind Electron 64:8525–8533
11. Barzegarkhoo R, Kojabadi HM, Zamiry E, Vosoughi N, Chang L (2016) Generalized structure for a single phase switched-capacitor multilevel inverter using a new multiple DC link producer with reduced number of switches. IEEE Trans Power Electron 31:5604–5617
12. Liu J, Wu J, Zeng J (2018) Symmetric/asymmetric hybrid multilevel inverters integrating switched-capacitor techniques. IEEE J Emerg Sel Top Power Electron 6:1616–1626
13. Zeng J, Lin W, Liu J (2019) Switched-capacitor-based active-neutral-point-clamped seven-level inverter with natural balance and boost ability. IEEE Access 7:126889–126896
14. Sulake NR, Pranupa S, Indira I (2019) Cross connected source based reduced switch count multilevel inverter topology with fault tolerance. In: 2019 global conference for advancement in technology, GCAT 2019, pp 1–7

Coordinated Charging of Electric Vehicles in Smart Grids to Minimize Distribution Loss

Sandipan Roy, Arjun Visakh, and M. P. Selvan

Abstract In the last few years, the popularity of Electric Vehicles has seen steep growth all over the world. Large-scale penetration of these vehicles can produce negative impacts like increased power loss and an overall reduction in power quality, especially at the distribution level. These effects can be mitigated by coordinating and scheduling the charging process using two-way communication between the utility and its customers. The search for the best charging schedule can be formulated as a mathematical optimization problem. In this paper, a coordinated charging strategy that minimizes the power loss in distribution lines and improves the voltage profile is simulated using MATLAB. Quadratic programming is applied to solve the optimization problem. A smart grid environment in which the chargers can be remotely controlled to operate at the optimal charging power is crucial for the implementation of this charge scheduling scheme.

Keywords Electric vehicles · Distribution network · Smart grid · Coordinated charging · Optimization · Loss minimization · Quadratic programming

1 Introduction

The first hybrid plug-in vehicles were introduced in the markets at the beginning of the previous decade, and thereafter the Electric Vehicles (EVs) have gained immense popularity throughout the world. Since they are cleaner with no harmful emissions and less dependent on fossil fuels, EVs are gaining ground on the traditional combustion engine-based vehicles and could play a significant part in future transportation systems. In India, the diverse consumer base creates a huge potential for EVs, and its penetration is expected to reach around 44% by the end of 2030 [1].

S. Roy (✉) · A. Visakh · M. P. Selvan
National Institute of Technology Tiruchirappalli, Tiruchirappalli, Tamil Nadu, India

M. P. Selvan
e-mail: selvanmp@nitt.edu

A. R. Gupta et al. (eds.), *Power Electronics and High Voltage in Smart Grid*,
Lecture Notes in Electrical Engineering 817,
https://doi.org/10.1007/978-981-16-7393-1_9

As the population of EVs continues to grow further, the operation and control of power systems become more challenging. There will be a substantial change in the patterns of power flows, grid losses, and voltage profiles along the distribution network. Furthermore, with the estimated levels of driving and coincidence of charging patterns with peak load hours, reinforcement of the grid may be required, which could cost up to 19% of the original network costs [2].

If the EVs are allowed to charge in a haphazard manner, there will be a negative impact on grid performance. According to a study [3], every 10% increase in EV penetration creates a 17.9% increase in peak demand when uncontrolled charging takes place. Furthermore, it leads to a rise in power loss, and voltage deviations that leads to deterioration of the power quality in the local distribution system. It may also lead to overloading of distribution transformers and cables, reduced efficiency, and grid reliability [4].

In order to combat these challenges, various coordinated and smart charging techniques have been developed over the years. A demand response strategy that accommodates EV charging without increasing the peak load was introduced in [5]. The system load profile can be flattened by minimizing the load variance using sequential quadratic programming [6], convex optimization [7], or grey wolf optimization [8]. Load leveling was also achieved by a game-theoretic approach that formulated the optimal charging profile by computing the Nash equilibrium [9]. A scheduling problem that maximizes the energy delivered to EVs without violating the network limits was solved using linear programming in [10]. The network efficiency can be increased by minimizing the power loss in the distribution lines using quadratic and dynamic programming [11], convex optimization [12], or meta-heuristic methods [13]. A real-time smart load management system was able to improve voltage profile and minimize power losses through maximum sensitivities selection optimization [14].

The voltage variations that accompany EV charging can be minimized using dynamic programming [15]. EV charging strategies that reduce frequency fluctuations using mixed integer linear programming [16] and hierarchical game theory [17] are also present in the literature. The satisfaction of EV owners mainly depends upon the cost of charging and the well-being of the vehicle. Continuous charge–discharge cycles are known to have a degrading effect on the performance and life of EV batteries [18]. Hence several research works have approached the optimal charging problem by considering both the state of health of batteries and the total energy cost [18–20]. Charge scheduling techniques based on whale optimization algorithm [21] and augmented epsilon-constraint method [22] combine multiple objectives to provide benefits at the customer level and system level simultaneously.

In this paper, domestic charging of EVs is coordinated with the goal of minimizing the power loss occurring in distribution lines. According to the correlation between the variation of load, load factor, and feeder losses [23], minimization of line loss is accompanied by a minimization in load variance. Thus, attempting to minimize the power loss will lead to load leveling and an improved load factor. Furthermore, the reduction of peaks and valleys in system loading will then lead to a flatter and improved voltage profile. The search for the optimal charging schedule is formulated

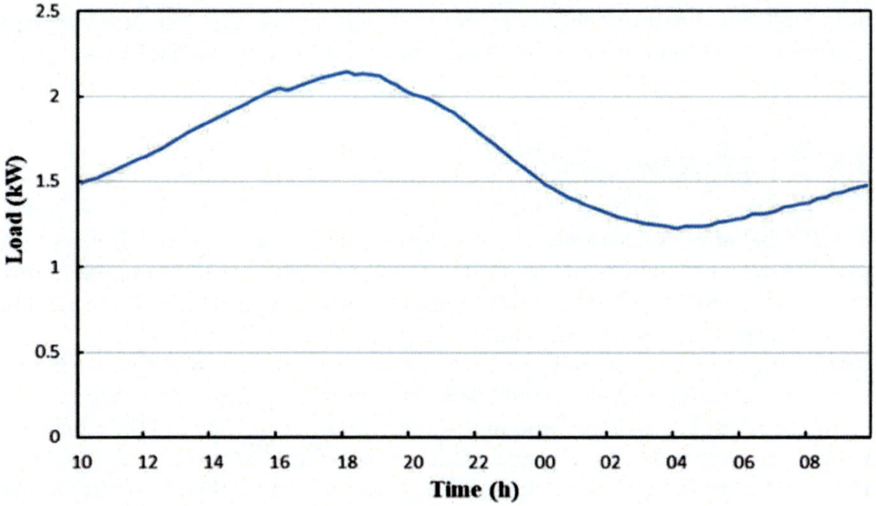

Fig. 1 Household load profile

as a quadratic programming problem here. This optimization algorithm can also ensure overload protection of equipment like distribution transformers and overhead lines by virtue of constraints on their loading. The impact of this coordinated charging on peak load, power loss, and maximum voltage deviation is analyzed on a radial distribution system in MATLAB.

2 Assumptions and Modelling

2.1 Load Scenarios

A residential load profile is selected randomly from a large pool of load data [24]. It has been normalized to the Indian scenario by downscaling in accordance with the ratio of per capita electricity consumption between the two countries [25]. The load profile provides power consumption data for every 15-min interval over the course of a day. The load profile is depicted in Fig. 1.

2.2 EV Specifications

The EV specifications considered in this paper are based on those of the TATA Tigor EV, launched in October 2019. It is equipped with a 21.5 kWh battery that provides a full charge range in excess of 200 km. The EV battery can be recharged by connecting the 3.3 kW-rated onboard charger to any 15 A plug socket [26]. The flow of energy is assumed to be unidirectional, or in other words, vehicle-to-grid is not considered.

2.3 EV Charging Load

The EV batteries are assumed to be recharged overnight at home. The EVs are modeled as constant power loads [12]. The majority of the EV users plug in their cars for charging upon returning home from work around 6 p.m., and starting time for charging is between 1 p.m. and 11 p.m. for around 90% of EV users [27]. This trend is implemented here with the help of a normal distribution or Gaussian curve with its mean at 6 p.m. and a standard deviation of 5 h, as displayed in Fig. 2.

Randomness is also introduced in the battery SOC at the plug-in time employing a Gaussian probability distribution model with 0.5 p.u. as mean and 0.2 p.u. as standard deviation [27]. All the EVs are expected to be fully charged by the hour of departure. It is also assumed that not more than one vehicle is charged at a time in each household.

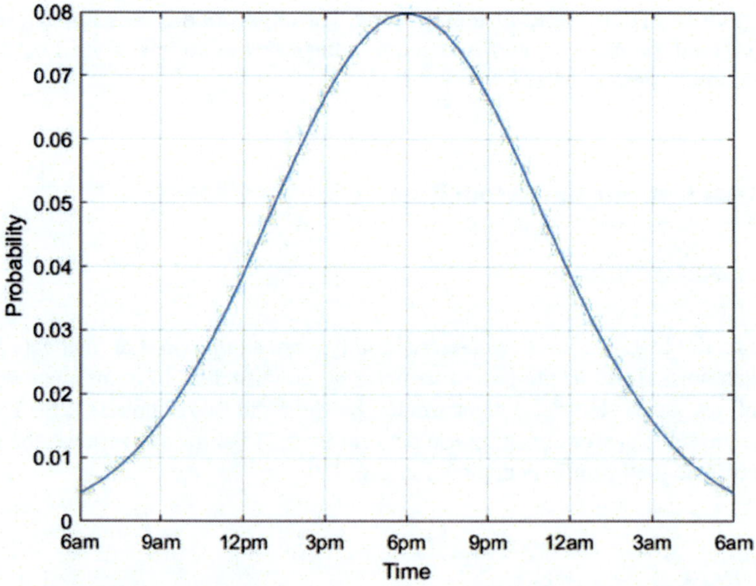

Fig. 2 Starting time of charging [27]

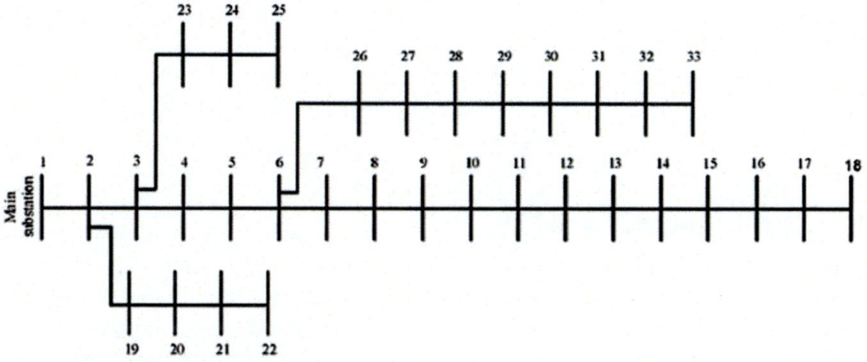

Fig. 3 IEEE 33-node test feeder [28]

2.4 Grid Topology

The IEEE 33 bus radial distribution system [28] shown in Fig. 3 is used as the test system for the studies performed in this paper. The network was downscaled from 12.66 kV to 400 V line voltage to characterize India's standard residential distribution system. Each node represents the connection between a domestic consumer and the feeder. The nodes with charging EVs are chosen in a random manner.

The consumers in this network are assumed to be equipped with smart meters. These advanced meters will enable two-way communication and permit remote load control. In this smart grid environment, the distribution system operator (DSO) will be capable of controlling the charging power of individual EVs.

2.5 Load Flow Analysis

Load flow analysis is required to calculate the voltage deviations and the power loss in the network. As the test system is radial in structure, the node voltages, line currents, and node currents were determined by implementing the backward–forward sweep power flow method [29]. Initially, a flat voltage profile is considered for the node voltages. All the loads in the system are represented by the constant power load model.

In the backward step, the node currents are calculated based on the previous iteration's node voltages. Thereafter, the node voltages are recomputed based on the root node voltage, and the voltage drops across the lines between the nodes in the forward step. The voltages and currents are updated iteratively until the node voltages-based stopping criterion is reached.

Table 1 Impact of uncoordinated EV charging

Penetration	0% (no EVs)	40%
Maximum voltage deviation (%)	5.13	9.21
Ratio of power loss to total power (%)	1.53	4.74
Peak load (kVA)	68.53	101.93

3 Uncoordinated Charging

Uncoordinated charging represents a situation where the batteries start charging as soon as the owners plug in their EVs. The power output remains fixed throughout the charging period, without any kind of control by the distributor and irrespective of the starting time of charging.

At the beginning of the 24-h cycle, the daily load profile selected earlier is assigned to all the nodes. In order to visualize the effect of EV penetration, a reference case without EVs is considered as a benchmark for comparisons. The penetration of EVs, i.e., the proportion of nodes with EV present, is assumed to be 40% since it is the predicted penetration by 2030 [1]. The EVs are placed randomly in the network. The load flow analysis to calculate voltage deviation, line loss, and system loading is performed at the beginning of each 15-min interval.

The impact of uncoordinated charging on the performance of the distribution system can be seen in Table 1. The maximum voltage deviation from the nominal value has risen to 9.21%, which is precariously close to the maximum permissible value of 10% in distribution networks [30]. The power loss, expressed as the ratio of total loss and total load, has tripled. The rise of distribution loss will ultimately be borne by the end customer in the form of a higher tariff. Furthermore, the system's peak load also shows a significant rise with the addition of EV loads. The increase in peak demand on the grid may lead to equipment damage and faults owing to overloading. Thus, there is a need for a coordinated charging algorithm to reduce such high losses and voltage deviations when the EV penetration increases.

4 Coordinated Charging

In the previous section, the charging of EV batteries was random and up to the owners' discretion, which creates some critical grid stability problems. Therefore, this section aims to formulate an optimization problem that would minimize the line loss in the network by finding the optimal charging profile for EVs. The optimization problem is solved by the quadratic programming (QP) technique. This method can optimize a quadratic function made of several variables. In this case, the total line loss resembles a quadratic function of line currents, which depends on the currents drawn by EV chargers.

4.1 Optimization Problem

The objective is to minimize the power loss in the distribution network using sequential quadratic optimization [31]. The loss during each interval is obtained by solving a system of nonlinear power flow equations. $P_{n,t}$ represents the charging power of EV n at time t obtained by quadratic optimization, and it should never exceed the maximum power of the charger $P_{\max}$. At the end of the charging period, the batteries must be fully charged. Hence, the energy transferred into the batteries should be equal to $(1 - s_i)$ times $C_{\max}$, where s_i depicts the initial SoC of battery with a maximum capacity $C_{\max}$. x_n denotes whether node n has an EV or not, i.e., x_n is one if node n has an EV and is zero if it does not have EV. The line currents are constrained to not exceed their rated capacity. These constraints along with the objective function can be summarized as follows:

$$\text{Minimize} \sum_{t=1}^{t\max} \sum_{l=1}^{\text{lines}} R_l I_{l,t}^2$$

such that

$$\forall n \epsilon \{\text{nodes}\} : \sum_{t=1}^{t\max} P_{n,t}.\Delta t.x_n = (1 - s_n).C_{\max}$$

$$\forall t, \forall n \epsilon \{\text{nodes}\} : 0 \leq P_{n,t} \leq P_{\max}$$

$$\forall n \epsilon \{\text{nodes}\} : x_n \epsilon \{0, 1\}$$

$$\forall t, \forall l \epsilon \{\text{lines}\} : I_{l,t} \leq I_l^{\max}$$

4.2 Methodology

In coordinated charging, the EV owners do not have control over the EV charging profile. However, the EV owners can decide the time by when the batteries need to be fully charged. As most owners would go to work in the morning, the end of the charging period is considered to be 9 a.m. The charging power is no longer fixed but can be varied between zero and maximum.

The placement of EVs within the test system is retained from the uncoordinated case. The EVs are placed randomly after the load profile is assigned to each node. The initial SoC is assigned randomly to each of the connected EVs. The node voltages and line currents are calculated by employing the backward–forward sweep technique considering no EVs in the network and a flat voltage profile initially. The optimal charging profile is determined using quadratic optimization.

5 Results and Analysis

The power loss, maximum voltage deviation, and system peak load under uncoordinated and coordinated charging have been compared in Table 2. The maximum voltage deviation at 7.84% is now well within the 10% limit. The loss and peak load on the system have been reduced by around 38% and 17%, respectively, compared to the uncoordinated case.

Figure 4 compares the charging profiles of the EV connected at node 5 during uncoordinated and coordinated charging. It is clear that during coordinated EV charging, the charger is controlled to operate below its rated power and to avoid

Table 2 Comparison between Uncoordinated and Coordinated charging

Parameters	Uncoordinated charging	Coordinated charging
Maximum voltage deviation (%)	9.21	7.84
Ratio of power loss to total power (%)	4.74	2.95
Peak load (in kVA)	101.93	84.87

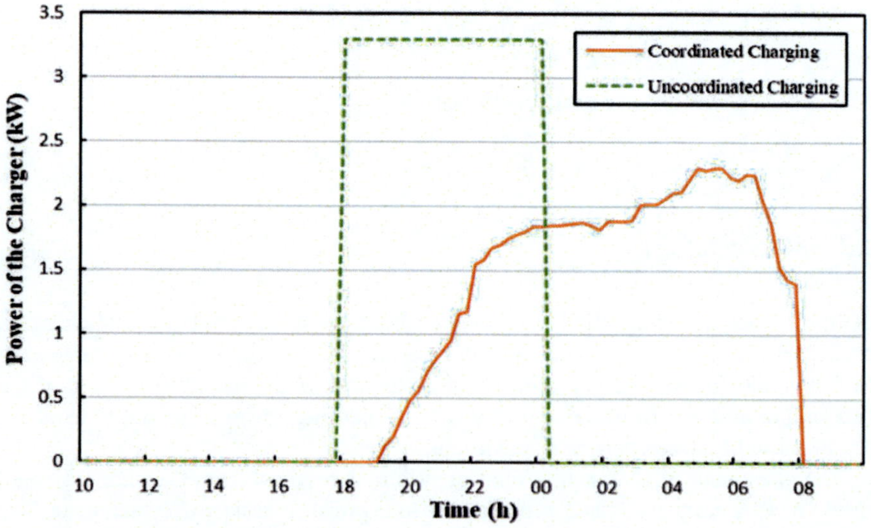

Fig. 4 Charging profile of EV charger at node 5

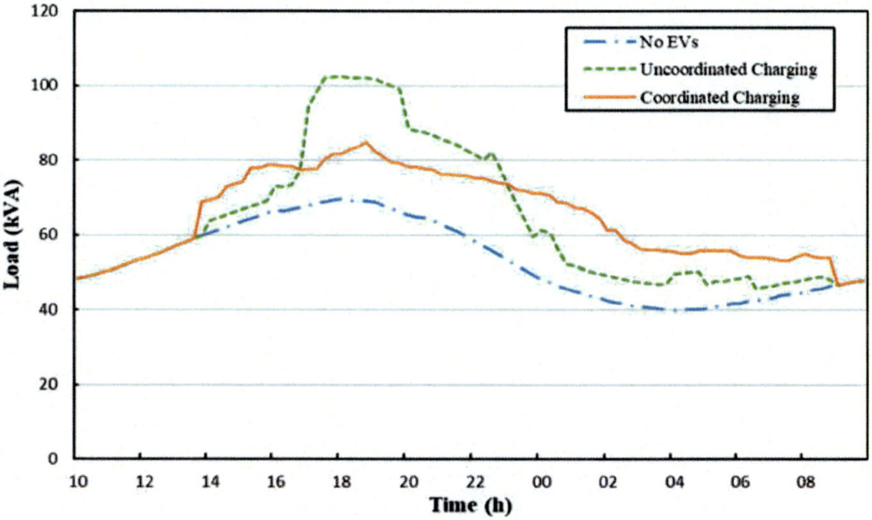

Fig. 5 Total system demand

coincidence with the peak load hours, thereby reducing the total demand and the resulting losses in the system.

Thus, with coordinated charging, the number of EVs charging during peak hours is minimized, leading to peak shaving. By maximizing the EVs charged during off-peak hours, valley filling is also achieved. Together, they achieve load leveling, which can be seen in the total system demand as depicted in Fig. 5.

Figure 6 illustrates the voltage profile at node 18, which is farthest from the main substation and is most susceptible to voltage sags. During the peak load hours, its voltage dips to 0.922 p.u. which corresponds to the maximum voltage deviation of 7.84%, as seen in Table 2. The troughs and peaks in voltage associated with the turning on and off of individual EV chargers have also been smoothened.

6 Conclusion

The rapid rise in the penetration of EVs brings in various challenges in the operation of the distribution network. Therefore, there is a need for coordinated charging techniques that reduces power loss and voltage deviations by flattening the peak load. It prevents overloading of distribution equipment like transformers and overhead lines, which avoids the need for reinforcement of the network. Although coordinated charging requires additional costs, it is more economical in the long run.

In this paper, a scheduling problem that optimizes the EV charging profile in a radial distribution system was implemented using the Quadratic Programming

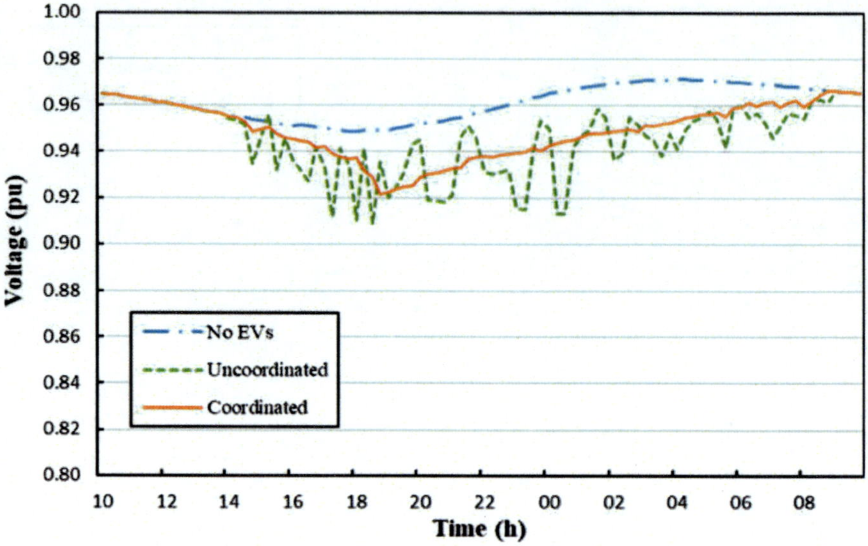

Fig. 6 Voltage profile at node 18

technique. Coordinated EV charging shows significant improvement in the operation of the test feeder in terms of power loss, voltage deviations, and peak load. The methodology can be extended to larger distribution networks, along with including other objective functions to improve parameters such as frequency fluctuations and load power factor.

Such coordinated charging schemes cannot be effectively implemented with the available infrastructure in today's distribution systems. However, with the advent of smart grids that combine smart metering with two-way communication, DSOs will be able to monitor, communicate with, and control individual EV loads for achieving the results presented here. Smart grids will also present opportunities for EV owners to provide Vehicle-to-Grid services.

References

1. Phadke A, Gopal A, Sheppard R, Abhyankar N, Park WY (January 2017) All electric passenger vehicle sales in India by 2030: value proposition to electric utilities, government, and vehicle owners. Ernest Orlando Lawrence Berkeley National Laboratory, Energy Analysis and Environmental Impacts Division
2. Fernandez LP, San Roman TG, Cossent R, Domingo CM, Frias P (2010) Assessment of the impact of plug-in electric vehicles on distribution networks. IEEE Trans Power Syst 26(1):206–213
3. Putrus GA, Suwanapingkarl P, Johnston D, Bentley EC, Narayana M (September 2009) Impact of electric vehicles on power distribution networks. In: 2009 IEEE vehicle power and propulsion conference. IEEE, pp 827–831

4. Masoum MA, Moses PS, Hajforoosh S (January 2012) Distribution transformer stress in smart grid with coordinated charging of plug-in electric vehicles. In: 2012 IEEE PES innovative smart grid technologies (ISGT). IEEE, pp 1–8
5. Shao S, Pipattanasomporn M, Rahman S (2012) Grid integration of electric vehicles and demand response with customer choice. IEEE Trans Smart Grid 3(1):543–550
6. Zhou B, Yang D, Bu S, Du G, Littler T (2017) The impact of electric vehicle un- certainties on load levelling in the UK. In: 2017 IEEE power and energy society general meeting. IEEE, pp 1–5
7. Kisacikoglu MC, Erden F, Erdogan N (2017) Distributed control of pev charging based on energy demand forecast. IEEE Trans Industr Inf 14(1):332–341
8. Singh J, Tiwari R (December 2018) Multi-objective optimal scheduling of electric vehicles in distribution system. In: 2018 20th national power systems conference (NPSC). IEEE, pp 1–6
9. Ma Z, Callaway DS, Hiskens IA (2011) Decentralized charging control of large populations of plug-in electric vehicles. IEEE Trans Control Syst Technol 21(1):67–78
10. Richardson P, Flynn D, Keane A (2011) Optimal charging of electric vehicles in low- voltage distribution systems. IEEE Trans Power Syst 27(1):268–279
11. Clement-Nyns K, Haesen E, Driesen J (2009) The impact of charging plug-in hybrid electric vehicles on a residential distribution grid. IEEE Trans Power Syst 25(1):371–380
12. Sortomme E, El-Sharkawi MA (2010) Optimal charging strategies for unidirectional vehicle-to-grid. IEEE Trans Smart Grid 2(1):131–138
13. Suyono H, Rahman MT, Mokhlis H, Othman M, Illias HA, Mohamad H (2019) Optimal scheduling of plug-in electric vehicle charging including time-of-use tariff to minimize cost and system stress. Energies 12(8):1500
14. Deilami S, Masoum AS, Moses PS, Masoum MA (2011) Real-time coordination of plug-in electric vehicle charging in smart grids to minimize power losses and improve voltage profile. IEEE Trans Smart Grid 2(3):456–467
15. Dubey A, Santoso S (2015) Electric vehicle charging on residential distribution systems: impacts and mitigations. IEEE Access 3:1871–1893
16. Kaur K, Kumar N, Singh M (2018) Coordinated power control of electric vehicles for grid frequency support: MILP-based hierarchical control design. IEEE Trans Smart Grid 10(3):3364–3373
17. Tan J, Wang L (2016) A game-theoretic framework for vehicle-to-grid frequency regulation considering smart charging mechanism. IEEE Trans Smart Grid 8(5):2358–2369
18. He Y, Venkatesh B, Guan L (2012) Optimal scheduling for charging and discharging of electric vehicles. IEEE Trans Smart Grid 3(3):1095–1105
19. Wei Z, Li Y, Cai L (2018) Electric vehicle charging scheme for a park-and-charge system considering battery degradation costs. IEEE Trans Intell Veh 3(3):361–373
20. Ma Z, Zou S, Liu X (2015) A distributed charging coordination for large-scale plug-in electric vehicles considering battery degradation cost. IEEE Trans Control Syst Technol 23(5):2044–2052
21. Adetunji K, Hofsajer I, Cheng L (July 2020) A coordinated charging model for electric vehicles in a smart grid using whale optimization algorithm. In: IEEE 23rd international conference on information fusion (FUSION), pp 1–7
22. Crow ML (June 2017) Multi-objective electric vehicle scheduling considering customer and system objectives. In: IEEE Manchester PowerTech, pp 1–6
23. Sortomme E, Hindi MM, MacPherson SJ, Venkata SS (2010) Coordinated charging of plug-in hybrid electric vehicles to minimize distribution system losses. IEEE Trans Smart Grid 2(1):198–205
24. California ISO Homepage (2020) http://www.caiso.com/Pages/default.aspx. Accessed 2020/11/26
25. World Bank Electric Power Consumption (kWh per capita) (2020) https://data.worldbank.org/indicator/EG.USE.ELEC.KH.PC. Accessed 2020/11/26
26. TATA Tigor EV Specifications (2020) https://tigor.tatamotors.com/electric/specification. Accessed 2020/11/26

27. Cao Y, Tang S, Li C, Zhang P, Tan Y, Zhang Z, Li J (2011) An optimized EV charging model considering TOU price and SOC curve. IEEE Trans Smart Grid 3(1):388–393
28. Ponoćko J (2020) Data analytics-based demand profiling and advanced demand side management for flexible operation of sustainable power networks, 1st edn. Springer International Publishing, Switzerland
29. Kersting W (2002) Distribution system modeling and analysis. CRC Press, Boca Raton
30. Central Electricity Authority (Grid Standard) Regulations (2010) https://www.cea.nic.in/reports/regulation/grid_standards_reg.pdf
31. Haesen E, Driesen J, Belmans R (2007) Robust planning methodology for integration of stochastic generators in distribution grids. IET Renew Power Gener 1(1):25–32

Solar-Powered PMSM-Driven Battery-Supported Water Pumping System

Apoorva Sharma and Tripurari Nath Gupta

Abstract The intent behind the presented paper is to propose an efficient solar-powered water pumping system which is driven by permanent magnet synchronous motor, and it is supported by a backup storage such as battery. The battery is utilized to maintain PMSM parameters unaffected during varying solar irradiance. This system deploys a Solar PV array, a DC–DC boost converter, a Voltage Source Inverter (3-phase, six switches), a Permanent magnet synchronous motor, a lead-acid battery, and in final a centrifugal pump. Electrical energy generated by solar PV array supplies to the system, so it behaves like a source of electrical energy. The DC–DC boost converter is situated between solar pv array and VSI and that is connected to PMS motor. Battery is connected to dc link to bidirectional converter. The incremental conductance technique comes into the frame to extract the optimized power output from solar pv array and it is best suited for the system because it responds better than others under varying irradiance scenario. The system is designed and tested using MATLAB/Simulink under fixed and varying solar PV irradiance.

Index Terms Solar PV array · Permanent magnet synchronous motor (PMS motor) · Voltage source inverter (VS Inverter) · Maximum power point tracking (MPPT)

1 Introduction

Solar Energy, with its essentially infinite potential and free accessibility, declining establishment cost, and zero working cost, solar photovoltaic (PV)-based power age is picking up more extensive agreeableness, and presents to a nonpolluting energy [1, 2]. Albeit solar-based PV incorporated water pumping system presents a possible arrangement; nonetheless, irregular nature of solar energy restricts its utilization for active hours. For a powerful use of WPS, this drawback needs to be resolved [3, 4].

A. Sharma (✉) · T. N. Gupta
National Institute of Technology, Srinagar, Uttarakhand, India

A. R. Gupta et al. (eds.), *Power Electronics and High Voltage in Smart Grid*,
Lecture Notes in Electrical Engineering 817,
https://doi.org/10.1007/978-981-16-7393-1_10

The proposed system comprises a solar PV array, a DC–DC boost converter, a Voltage source inverter (VSI), a battery with buck-boost converter, and a centrifugal pump. The electrical energy is supplied by solar PV array. Boost converter is utilized for tracking the MPP and to boost the input of Solar PV array which is used to supply to three phase, six switches-based VS Inverter. Basically, the DC–DC conversion is a DC–DC converter for which duty cycle of DC–DC converter is adjusted in a manner that solar PV array run operation is at its maximum power. The MPPT operations that are conspicuous in tracking MPP are open-circuit voltage method, short-circuit current method, perturb and observe method, incremental conductance method (INC), and neural network and fuzzy techniques. Here INC (Incremental conductance) is opted. The VSI mainly works as to convert DC into AC power and drives the PMS motor. Stationary reference frame control is used to control Solar PV water pumping system. The PMS motor which is connected to a centrifugal pump and that pump is used to rotate by the motor.

In the proposed system, battery is connected to the DC link in a bidirectional flow to resolve the intermittent nature of solar energy and tested the system for different irradiance which also validates whether system is efficient or not. When the supplied power by solar PV array is over to load power then remaining power is ingested by the battery and it will get charged, on other hand in the situation of unavailability of the sun or supplied power by solar PV array is not sufficient, then battery will supply the power and attain the load demand.

When solar irradiance varies, the power supplied by the Solar PV array also varies and that affects the PMSM parameters and surely efficiency and accuracy of the system too. To resolve these problems, the system is proposed and designed using MATLAB/Simulink and tested for various application.

2 A Brief Review on Literature

According to literature review, solar-energized water pumping system can be separated out as mainly two parts, the first one, in which we have any storage element attached to the system and in second one the system is based on the motor. A brief discussion about these given below:

2.1 Storage Based

(a) In battery-linked solar PV array pumping frameworks, the produced power charges a battery. When the sunlight is not able to generate appropriate power for supply, overcast days or around night hours, attached battery can be utilized to give supply to the water pump. Advantage of this framework is battery backup for night and drawbacks are significant expense and low efficient [5, 6].

(b) For direct determined PV pumping schema, the power produced is utilized to operate the whole water pumping system. Subsequently, when the sunlight is unavailable system will not work because there is no backup power available. Preferred position of this framework is simple and more affordable and detriment is no battery backup [7, 8].

2.2 *Motor Based*

(a) Pumping systems which are motor based can't be straightforwardly associated with the source PV array. To feed the motors such as induction motor which is an AC motor, there is an inverter system which should be connected with the converter to convert output voltage into variable AC voltage. Benefits of this framework are ease, robustness, and negative marks are low efficiency and heating issues [9].

(b) The system with DC motor comprises a solar PV array and a pump. It could possibly have a converter connected to the system. This system is more efficient rather than the system with induction motor. Favorable position is more efficient than others and disservice is that it requires high maintenance [10].

3 Essentials for Designing of the System

These are the essentials for constructing the solar-energized water pumping system [11].

1. The Solar PV Array
2. The MPPT technique
3. Boost Converter
4. Battery Storage
5. VS Inverter
6. PMS motor
7. Pumping system—(centrifugal pump).

3.1 *Solar PV Array*

1. To make a high voltage and high current solar PV array, solar cells are configured in series and parallel manner according to our load requirement.
2. The power extracted from the solar PV array should be at its maximum point ($P_{pv} = P_{\mathrm{mpp}}$).
3. Required modules calculated such as

- No. of the required solar PV module for the system:

$$N = \frac{P_{\text{mpp}}}{P_{mp}} \tag{1}$$

- No. of the required solar PV modules in series:

$$N_s = \frac{V_{\text{mpp}}}{V_{mp}} \tag{2}$$

- No. of required solar PV modules in parallel, and are attached in the system:

$$N_p = \frac{P_{\text{mpp}}}{(N_s)(V_{mp})(I_{mp})} \tag{3}$$

where P_{mpp} is the power of PV array and P_{mp} is the power of a single module.

For this system, the number of required modules is 12 in which 4 modules are arranged in series and 3 are in parallel.

3.2 MPP Tracking Technique

To extract the maximum power from PV array MPP tracking is required. Among numerous MPP tracking methods in the system INC method is used because it provides a good performance in every condition (steady-state and varying environmental conditions).

For this system evaluated MPP power is 3.661 KW.

3.3 PMS Motor

PMS motor has an advantage of higher efficiency, long life span, high reliability, and large torque to weight ratio and low maintenance also.

3.4 DC–DC Converters

The converter duty cycle can be calculated as

$$d = 1 - \frac{V_{pv}^{*}}{V_{dc}^{*}} \tag{4}$$

where $V_{pv}{}^{*}$ is the reference PV voltage estimated by MPP technique.

3.5 VS Inverter

A three-phase voltage source inverter is used. The power which is input to the inverter is provided by boost converter and processed to the PMS motor.

3.6 Centrifugal Pump

Pump is mechanically connected to the PMS motor. The torque speed characteristic for the pump is given by equation:

$$T_1 = K_p * \omega_r{}^2 \tag{5}$$

$$K_p = \frac{T_1}{\omega_r{}^2} \tag{6}$$

where K_p is the pump constant.

3.7 Battery Storage

- A Ni-MH battery is coupled to the DC link via a bidirectional converter which maintains the DC link voltage.
- Battery is used for bidirectional power flow.
- Attaching battery makes the system more efficient and works as a storage backup also.

4 System Under Study

Figure 1 shows the system under study in this paper. According to the above-calculated structure and logical arrangement a solar-energized water pumping system is designed. The maximum power is extracted from solar PV array by MPP tracking INC mechanism. The PMS motor is driven by utilizing a VS Inverter which is controlled by utilizing stationary reference frame theory. A current controller (Hysteresis) is utilized which powers the motor currents to remain within a particular band. This band is called hysteresis band (HB). In the system PI (Proportional

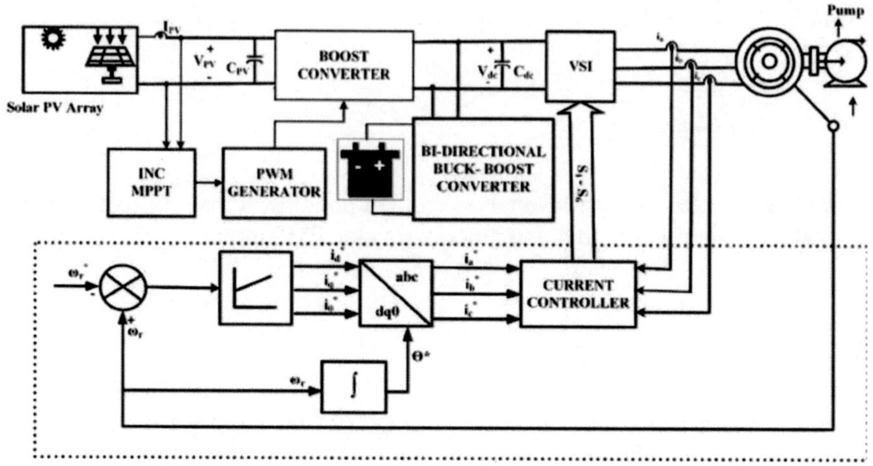

Fig. 1 Layout of the solar PV water pumping system

Integral) controllers are utilized in this methodology. Purpose of the PI controller is to diminish the error between the reference and set value.

5 Controlling for the System

The controlling mechanism of the proposed methodology is presented in Fig. 1, and is separated out in two major categories for better understanding. First we will discuss about the MPPT operation and after which we will discuss about the speed control of PMS motor [12].

5.1 MPP Tracking

An INC MPP tracking mechanism can perform under rapidly varying atmospheric conditions. The duty ratio of the converter is adjusted through MPP tracking in small step size for good MPP tracking.

- At MPP:

$$\frac{\Delta I_{pv}}{\Delta V_{pv}} = -\frac{I_{pv}}{V_{pv}} \tag{7}$$

- At left of MPP:

$$\frac{\Delta I_{pv}}{\Delta V_{pv}} > -\frac{I_{pv}}{V_{pv}} \tag{8}$$

- At right of MPP:

$$\frac{\Delta I_{pv}}{\Delta V_{pv}} < -\frac{I_{pv}}{V_{pv}} \tag{9}$$

where ΔI_{pv}, ΔV_{pv} are changes in solar PV array current and voltages, respectively.

5.2 *Vector Control of PMS Motor*

Vector controlling for PMS motor is chosen here because its advantages such as it gives a superior response for speed by parting out the stator current into direct axis ($I_d{}^*$) and quadrature axis ($I_q{}^*$) segments, and it controls them individually as field current and armature current. The (Id^*) is kept zero. The ($I_q{}^*$) can be determined as follows:

$$I^*{}_{q(k)} = I^*{}_{q(k-1)} + I^*{}_{q(k)} + K_p\left(\omega_{\text{error}(k)} - \omega_{\text{error}(k-1)}\right) + K_i\left(\omega_{\text{error}(k)}\right) \tag{10}$$

6 Simulation Results and Applicability During Varying Solar Irradiance

The propounded framework of PMS motor-driven Solar PV array fed water pumping system has been validated through MATLAB/Simulink. Also, it has been demonstrated under consistent and varying insolation levels. For getting a decent acknowledgment of framework execution, the initial, steady state, dynamic executions are concentrated in detail such as.

6.1 *Simulated Results at Initial and Steady-State Condition at a Consistent Solar Insolation of 1000 W/m²*

In Fig. 2a the solar insolation is fixed at 1000 W/m^2 for time period (0–3 s). In Fig. 2b, the power which is extracted from PV array is also almost constant at a value

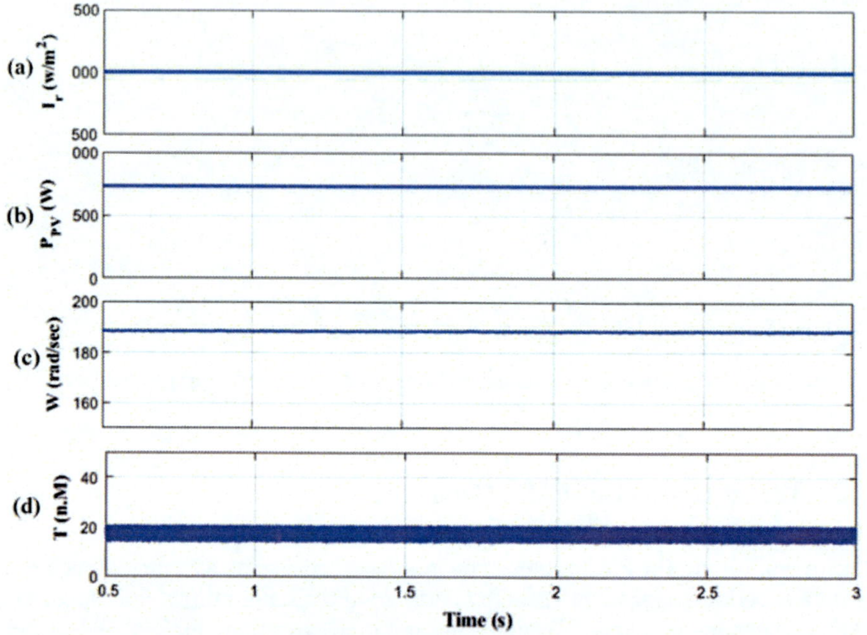

Fig. 2 Performance of the system **a** Solar insolation 1000 w/m^2, **b** PV power at 1000 W/m^2 solar insolation, **c** Rotar speed (ω_m) rad/s, **d** Load torque(T) n*m

of 3.661 kw for 0–3 s. It is a slightly higher than because of consideration of losses in between. Figure 2c, d manifest the initial execution in formation with steady-state execution for Solar PV array. The steady-state parameters of PMS motor such as rotor speed (ω_m), load torque (T) are shown in following Fig. 2.

As there is no change that has occurred in the insolation or we can see there is no fluctuation present in the PV power, therefore PMS motor parameters remain at a constant value for the time period for which the system is tested. The initial and steady-state performance express the motor speed at consistent state estimation of 188 rad/s. The load torque changes in relation with PMSM speed and reaches at an evaluated estimation of 18 Nm. The Solar PV array power (P_{pv}) is tracked at MPP power (P_{mpp}) of 3.661 kW and settles at this value. The motor power additionally reaches at steady-state appraised estimation of 3.384 kw.

6.2 System Demonstration at Dynamic Condition Under Varying Solar Insolation (1000-600-800 W/m^2)

In the dynamic execution of proposed methodology under fluctuating solar PV insolation, the solar PV insolation has been brought down from 1000 to 600 W/m^2. In

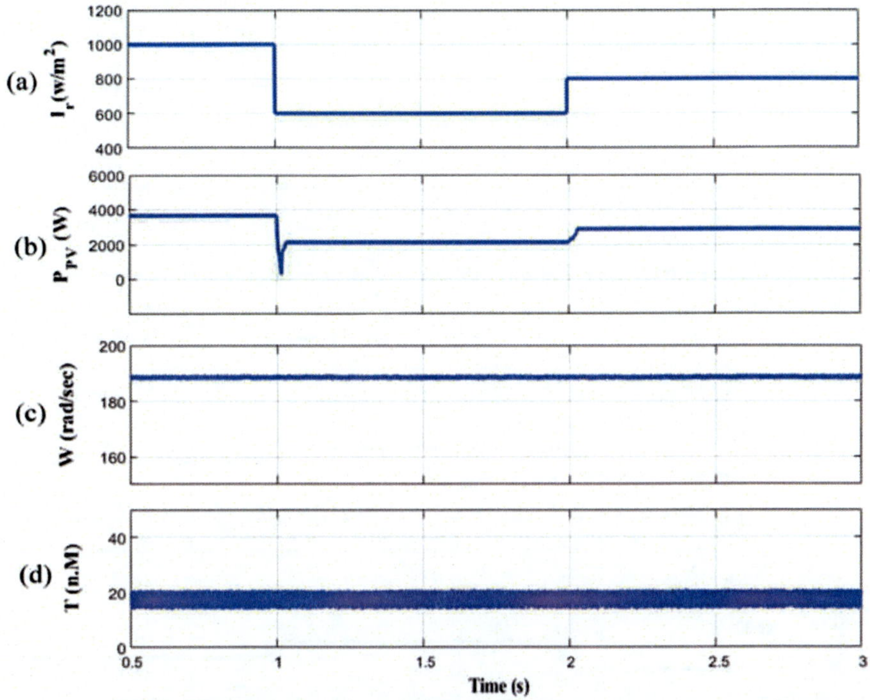

Fig. 3 Performance of the system **a** Solar insolation 1000-600-800 w/m^2, **b** PV power under varying solar insolation, **c** Rotar speed (ω_m) rad/s, **d** Load Torque(T) N*m

Fig. 3a, the solar insolation is at 1000 w/m^2 for 0–1 s and afterward it is diminished to 600 W/m^2 for 1–2 s, and then again raised to 800 w/m^2 for 2–3 s. The system is demonstrated under such varying conditions.

In Fig. 3b, at the point when solar PV insolation gets changed then MPP power also gets affected by that and decreased, respectively. However, in Fig. 3c, d PMS motor parameters stay constant which means they didn't get affected by the fluctuation of solar PV insolation. So, if maximum extracted power from solar PV array decreases the PMS motor runs at constant speed and pump will also drive constant power. We can manifest them by seeing the results. Figure 4 shows the performance of battery when solar irradiance gets changed.

Figure 4b–d shows the voltage, current, and power for the battery for different irradiance. When PV supplied power is low battery supply the motor to maintain constant speed is as shown in Fig. 4d, where at 1000 W/m^2 battery gets charged and takes 0.374 KW, and when irradiance decreased to 600 W/m^2 battery feed the motor with around 1.178 KW and when irradiance goes up to 800 W/m^2 battery provides 0.399 KW power to the motor.

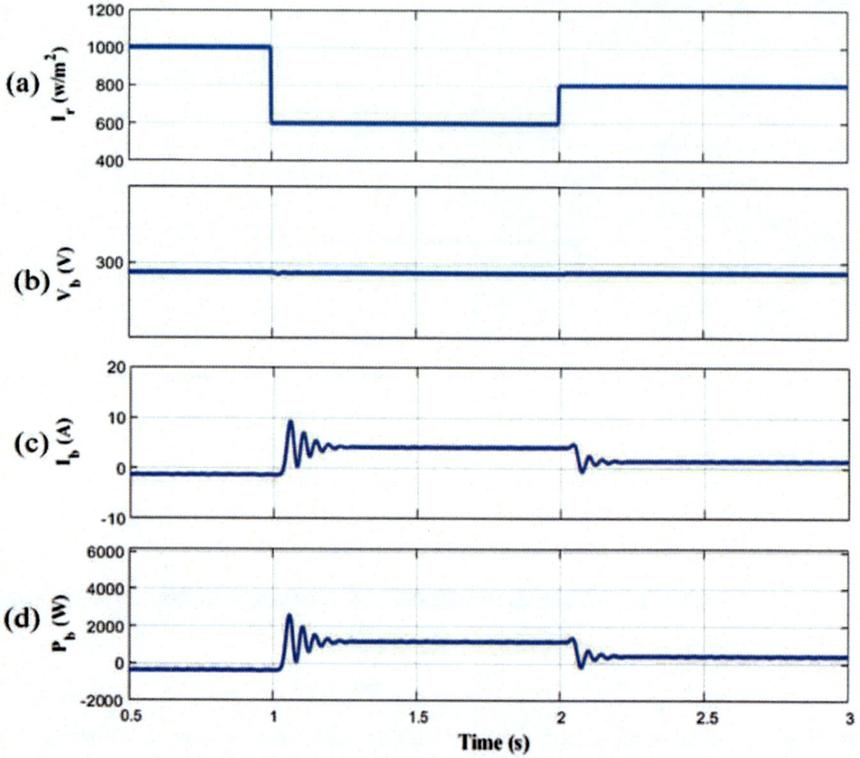

Fig. 4 Performance of the system **a** Solar insolation 1000-600-800 w/m^2, **b** Voltage of battery (V_b) V under varying solar insolation, **c** Current of battery (I_b) A, **d** Power of battery (P_b) W

7 Discussion

To reliably operate the above system in efficient manner a storage backup is required when the power generated from the solar PV array is not able to fulfill the load demand, and in that situation battery will work as a supply source for the load. Moreover, when extracted power from PV array is higher than load requirement, it will charge the battery. It makes the system more worthy and efficient.

So, the addition of the battery keeps the system intact from the varying solar insolation too and can drive the motor at desired speed and power as shown in Fig. 4. According to Fig. 4d during high solar irradiance, battery power is negative because battery will get charged during this period up to 1 s. From 1 to 2 s solar irradiance decreased to 600 W/m^2, and during that time battery provides the power to the PMS motor. Further, irradiance increased to 800 W/m^2 even during that time some part of the required PMS motor is provided by the battery to maintain the speed constant. So the speed of the PMS motor speed remains constant during irradiance change (see Table 1).

Table 1 Particulars of PMS motor and battery

Description	Nominal value
Power	3.384 KW
Rated Torque	18 Nm
Speed	188 rad/s
No. of phases	3
DC link voltage	230 V
Battery rating	200 V,6.5 Ah

It is observed from the results that system under varying solar PV insolation gives satisfactory performance and the objective of running the pump at constant power is also achieved successfully.

8 Conclusion

From the above outcomes, Solar PV water pumping framework utilizing PMSM has been planned, displayed, and simulated with the assistance of MATLAB/Simulink. The framework introduced in this work has given a good execution under varying conditions. The framework has been investigated for a particular range of sun-powered insolation variety. The steady state and dynamic execution are very satisfactory. Introducing storage backup such as battery makes the system more efficient. An INC MPP following technique has improved the power yield of solar PV array. The nonappearance of halfway stage such as DC–DC converter has made the system robust, straightforward, and, furthermore, efficient. Subsequently the proposed system has given a practicable result for Solar PV water pumping framework.

Appendix

See Table 2

Table 2 Particulars of Solar pv array and Pump constant

Description	Parameter	Nominal value
Maximum power	P_{mp}	305 W
Maximum power point current	I_{mp}	5.58 A
Maximum power point voltage	V_{mp}	54.7 V
MPP power	P_{mpp}	3.661 kw
No. of module required	N	12
No. of module in series	N_s	4
No. of module in parallel	N_p	3
Pump constant in Nm/(rad/s)^2	K	0.000509

References

1. Masters GM (2013) Renewal and efficient electrical. Power system, 2nd edn. Wiley-IEEE Press
2. Khan BH (2010) Non-conventional energy resources, 2nd edn. Mc Graw Hill
3. Dubey M, Sharma SK, Saxena R (2020) Solar energy based ZSI fed induction motor drive for water pumping. In: IEEE international conference on computing, power and communication technologies (GUCON), Greater Noida, India, pp 135–139
4. Dhumal KR, Dhamse SS (2018) Solar PV array based water pumping by using SRM drive: a review. In: 2018 international conference on computation of power, energy, information and communication (ICCPEIC), Chennai, India, pp 140–146
5. Mishra AK, Singh B (2018) Stage solar PV powered water pump with a storage system. In: 2018 8th IEEE India international conference on power electronics (IICPE), JAIPUR, India, pp 1–6
6. Shahajahan TK, Mittal SK (2020) Intelligent grid interfaced solar water pumping system. In: 4th international conference on electronics, communication and aerospace technology (ICECA), Coimbatore, India, pp 524–531
7. Sharma U, Singh B, Kumar S (2017) Intelligent grid interfaced solar water pumping system. IET Renew Power Gener 11(5):614–624
8. Madark M, Ba-razzouk A, Abdelmounim E, El Malah M (2020) Nonlinear CTMV control of a solar photovoltaic water pumping system. In: IEEE 2nd international conference on electronics, control, optimization and computer science (ICECOCS), Kenitra, Morocco, pp 1–6
9. Mufsina AM, Shahin M (2020) A modified SEPIC converter based solar water pumping system using SRM drive. In: International conference on power electronics and renewable energy applications (PEREA), Kannur, India, pp 1–6
10. Zamanlou M, Iqbal MT (2020) Design and analysis of solar water pumping with storage for irrigation in Iran. In: IEEE 17th international conference on smart communities: improving quality of life using ICT, IoT and AI (HONET), Charlotte, NC, USA, pp 118–124
11. Murshid S, Singh B (2019) Implementation of PMSM drive for a solar water pumping system. 2019 IEEE Trans Ind Appl 55(5):4956–4964
12. Prajapati YM, Jaiswal NK (2019) Sensor-less control of permanent magnet synchronous motor. In: 2019 3rd international conference on trends in electronics and informatics (ICOEI), Tirunelveli, India
13. Szychta E, Szychta L (2020) Characteristics of collective losses of squirrel cage induction motors used in pumping units: ELEKTRO, Taormina, Italy, pp 1–6

Assessing the Reusability of 3D Printer Filaments with Reference to Legal Aspects of Sustainability

Parth Patpatiya and Nidhi Arora

Abstract Additive manufacturing (AM) is a tool that has revolutionized the way products are manufactured and assembled. As 3D printing is a rapid prototyping process hence owing to its continuous usage of filament, sustainability has become an important issue to be discussed. After 3D printing of parts it is realized that a lot of material is finally ended up in the garbage. This paper acquaints with the feasibility of recyclable 3D printed filament to create a sustainable technology for 3D printing and the environmental implications of non-biodegradable waste and also highlights its responsible management for sustainable ecological protection with reference to the legal provisions. In the constitution of India, under the directive principle of state policy (Part IV) it clearly states that it is the duty of every citizen "to protect and improve the natural environment including forests, lakes, rivers and wildlife". In the case of Dr. B. L. Wadehra versus Union of India it was held that to live in a clean city is a statutory right of every citizen. Hence converting 3D printed wastes and 3D parts at the end of their life into reusable material is the key factor.

Keywords Recycling · 3D Printing · Indian Constitution · Sustainability

1 Introduction

After several years of research, Additive Manufacturing (AM) can be outlined as a procedure of joining materials to form 3D objects, often layer by layer which is contradicted in case of Subtractive Manufacturing. Allowance of AM for a better saving of material than traditional processes make 3D printing a diverse manufacturing technology to filter sustainability as well as circular economy worldwide [1]. The usage of printing material might reach eleven crores kilogram by 2020, and to date most 3D printer filament is made from raw plastic [2]. In addition to the virgin

P. Patpatiya
School of Automation, Banasthali Vidyapith, Vanasthali, Rajasthan, India

N. Arora (✉)
Department of Legal Studies, Banasthali Vidyapith, Vanasthali, Rajasthan, India

A. R. Gupta et al. (eds.), *Power Electronics and High Voltage in Smart Grid*, Lecture Notes in Electrical Engineering 817,
https://doi.org/10.1007/978-981-16-7393-1_11

plastic source material in 3D printing, there are also 3 billion kilograms of plastic inclined of in the United States each year, only 6.5% of which is reprocessed [3]. Approximately 4% of universal manufacturing of oil and gas is expended as a feedstock for plastics; furthermore 3–4% is utilized to provide energy for the conversion and conversion of polymeric materials [4]. The alternative to bio-plastics, oil-based polymers is educed from reusable sources such as natural fibers as well as sugars. Fused Deposition Modelling that uses an incessant filament of a thermoplastic material is an affordable method among AM technologies [5]. According to the additive manufacturing general principles, FDM process belongs to material extrusion category [6]. FDM printers have a non-complex layout involving a Cartesian assembly with all the three positioned axes, containing a maximum of three extruders and a building platform. Though people print amazing 3D outputs, if the material is not correctly disposed or recycled it will lead to generation of waste to a great extent. These energetic demands are popular with industrial FDM printers that make use of a hot closed printing zone [7]. In an estimation given by Gebler et al. the use of AM for creating new parts might lead to save approximately 5% in energy and reduce carbon-dioxide emissions in the production industries worldwide. The second most important facet of FDM is the waste generated and the extra material from 3D printed specimens (e.g. filament ends, support construction and leftovers). The waste generated can be managed unswervingly dealt by either manufacturers or final users through reusing the material and manufacture of reusable filaments. The aim of this paper is to appraisal the use of reusable filaments as well as enlisting various issues and related research areas.

2 Literature Review

The urbanization and revolutionary technological developments have opened doors to prevalent and pervasive growth of waste products in the process. This challenges us to balance economy, ecology and socio-policy in an efficient manner. Primarily sustainable activities intended to limit use of energy and resources in the production process. Now, the aim has shifted toward product sustainability and sustainability in manufacturing [8]. Moreover, European Union's circular economy initiative invariably highlights the role of recycling wherein "the proposed actions will contribute to close the loop of product lifecycles through greater recycling and re-use, and bring benefits for both the environment and the economy" [9]. The practical significance of the aforementioned aspects can be seen in inclination to additive manufacturing over subtractive manufacturing. The dawn of Industry4.0 and its much evident setting in involves a great use of AM technologies. Additive Manufacturing process involves use of digital 3D design data in order to build a component in layers by depositing material instead of milling a workpiece from solid block (subtractive manufacturing). Series production is increasingly making use of additive manufacturing. 3D printing (a synonym to AM) follows a layer by layer printing approach and can be considered a distributed manufacturing technology for enhanced sustainability and circular economy worldwide [10]. The seven different types of 3D printers, namely, fused deposition modeling (FDM), stereolithography (SLA), digital light processing (DLP), selective laser sintering (SLS), selective laser melting (SLM), laminated object manufacturing (LOM), and digital beam melting (EBM) [11]. Fused Deposition Modelling has been the most popular additive manufacturing and 3D printing process worldwide which is a consequence of Stratasys selling a number of FDM machines equivalent to the number of all other AM machines put together. However, the turning point for FDM technology was when RepRap project was made open source [12]. This low-cost technique of Fused Deposition Modelling uses thermoplastic filament for the layer by layer execution process. 3-D printers or Fused Deposition Modelling systems have a modest design comprising a Cartesian coordinate system with axes positioned in three perpendicular directions, up to all the three warm extruders and a structured printing zone. FDM is more formally referred to as material extrusion, in ASTM and ISO terminology owing to the use of extrusion heads. This is a relatively modest way of manufacturing with an affordable minimum price of 100USD. It can be predicted that the use of 3D printing is expected to increase in near future. This increased use entails an increased and intractable generation of waste if the appropriate disposal and recycling are not taken into stride. The handling of emitted emissions, resources, energy and waste must be accounted and hence the environmental impact of 3D printing, in the view of AM sustainability. Among these the use of a warm packed adiabatic entire volume is of critical importance [13]. A two-step model was used for evaluating the huge impact of it on environment because of all 3D printing technologies across the globe till 2025 [14]. The estimates implied an upscale in production efficiency on applying the intended improvements to 3D printing. Also, 5% savings on energy and CO2 emissions in the production industry is expected. However, on the contrary the organization of material of Fused Deposition Modelling waste needs to be discussed, as ignorance may prove to be malignant in the longer run. This scrap and surplus material from 3D printed end use parts such as support structures, filament ends and scraps, contribute significantly to ecological aspect of Fused Deposition Modelling. The waste management can be approached for either at the beginning of the chain or at the end of it, in the sense, waste could be straightforwardly managed by the makers or the end users. There is no particular way for any re-use and recycling process in general and may vary according to the use. For instance, an undergraduate research suggested development of an extrusion process using Polyethylene Terephthalate Glycol (PETG) Pellets. As Polyethylene Terephthalate (PET) plastics has higher melting temperature hence modifications were incorporated by considering the aspects such as viscosity of the fluid and need for drying of plastics before the process. The study's intent was to highlight the feasibility of recyclable plastics into the filament used in manufacturing [15]. In this paper Shastri et al. [25] has analyzed the legal provisions for environment protection and waste management even highlighted the steps taken by apex court through landmark judgement in respect of the environment protection and sustainable development. In the present study Ray et al. [26] has thrown a light on the various legal framework for waste management and also reviewed the drawbacks and challenges in its implementation. In this article Bharadwaj et al. [27] has discussed the repercussions of non-biodegradable waste on sustainable development.

3 Bio-filaments for FDM

Biodegradable thermo-plastics are used as filaments like polyhydroxyalkanoates (PHA), polyvinyl alcohol (PVA), Polyethylene terephthalate (PET) and High impact polystyrene (HIPS). Table 1 lists the properties of these materials.

4 Impact of 3D Printing Non-biodegradable Waste on Sustainable Development

Businesses are looking for opportunities to simplify their manufacturing processes to minimize energy use and emissions, a tactic known as circular production, as we aim for a prosperous future. 3D printing is also considered as one of the leading environmental technologies mainly because of two advantages: it makes designs more effective and causes less waste. Much of the recycling efforts are made in the polymer 3D printing sector, despite the progress in the recycling of metals. Today, several businesses are manufacturing plastic filaments from recycled plastic. Among them are companies that produce entirely or partly recycled filaments, such as GreenGate3D, Filamentive, NefilaTek, Refil and RePLAy 3D.

5 Environmental Law and Case Related to Non-biodegradable Waste

The free life in a protected environment is primary freedom provided by Article 21 of our Constitution, as well as the privileges which have been recognized and upheld by a vast number of courts, such as the Torts Act, the Indian Penal Code, the 1860 Civil Procedure Code of 1908 and the 1973 Criminal Procedure Code. The first 1950 statute, which is the supreme land rule, Article 51-A (g) of the Constitution of India, imposes a legal responsibility on any Indian person to protect and improve the environment. In compliance with Article 48A, the State is responsible for protecting the environment. Often a fundamental human right is the right to live in a safe climate. According to Article 3 of the UNDHR of 1948, everybody has a right to live. Article 25 says that everyone is entitled to a standard of life suitable for his and his family's health and well-being. In the exercise of the expertise provided under Articles 3, 6 and 25 of the Environmental Safeguards Act 1986, the Ministry of Environment and Forestry has national policies, and has enacted the Urban Waste Management and Handling Rules 2000. All local authorities responsible for storing, segregating, treating, distributing, refining and disposing of urban solid waste are subject to these laws. Waste composting is a mandatory responsibility for all local authorities in the region under Urban Solid Waste Management Regulations 2000. "Composting, vermicomposting, anaerobic digestion or other effective biotechnological processing for waste stabilization treatment is required for the treatment of biodegradable waste" under the MSW Rules 2000. The stated period was December 31, 2003 or earlier for the installations of waste management and disposal. The enforcement of the provisions of such laws in the municipality's geographical area and any collecting, transport, segregation, transport, transportation and waste disposal services shall be the responsibility of each of the municipal authorities. In compliance with Sects. 6, 8 and 25 of the Environmental protection Act of 198, the central government has established a set of laws, directives and guidelines including the 1998 Rules for the Economic Management and Handling of Biomedical Waste.

6 The Environment Protection (Control of Non-Biodegradable Garbage) Act, 2016

Environmental emissions and its harmful consequences have drawn the world's attention, and ways and measures are being considered and introduced to control pollution. In this respect, garbage collection in cities and towns has become a big issue. Solid waste management is the responsibility of the agencies of local self-government. The central portion of solid waste is bio-degradable, i.e. the behavior of living beings and microorganisms will kill it. It is possible to turn such biodegradable waste into compost or to use it as a source of energy or manure. At the same time, the bane of industrial life is a non-biodegradable waste.

The introduction of polyvinyl chloride (PVC), polypropylene and polystyrene plastics and other pollutants causes health-related environmental disasters. This pollution chokes gutters, drains and underwater outfalls, creating a nightmare for the wastewater system. It clogs the soil, inhibits the free flow of water into it and depletes its fertility and water tables. The goal of the "The Environment Protection (Control of Non-Biodegradable Garbage) Bill, 2016" is to control the use and disposal of certain non-biodegradable substances.

7 3D Printed Materials and Their Properties

A restricted assortment of resources is accessible for 3D printing and 3D printed portions which generally have a shoddier mechanical performance if a comparison is made with the same materials processed by extrusion, compression or injection molding. Nevertheless, engineers and designers are authorized to design freely over other traditional processes [16, 17]. Table 2 lists the tractile attributes of Fused Deposition Modelling test specimens mass-produced from composite, plastic and recycled filaments. The mechanical properties of biodegradable plastics or composites are unlike the pure matrix metal. The introduction of filler material has a challenge related to increment in the melt viscosity, and consequently the challenge associated with flowability. Several issues such as porosity, lack of fusion between layers and swelling persuaded by natural fibres are also difficult [22]. Biodegradable filaments' filler content is generally below forty percent by volume whereas layer extrudability as well as adhesion tends to increase the content and hence mechanical properties also increases [22]. Polylactic acid (PLA) filament's processability and the mechanical properties contains five percent by weight of craft pine lignin are investigated by Gkartzou et al. [24]. The reason of brittle behavior of PLA is the accumulation of lignin and the elongation at break gets reduced. Moreover the superficial roughness of the PLA filament is caused by the remarkable growth of filler. Li et al. [16] has reported some strength and weaknesses in bio-composites with different cellulose fibres. Strength, hardness, flexibility and moisture sensitivity are some of the examples. Mixing differently refined cellulose fibres as some of these problems gets transformed through. Markstedt et al. [24] have used a modified 3D printer to print a pure cellulose with a liquid base.

Table 1 Characteristics of bio-degradable filament used for 3D printing

Material	Ultimate tensile strength (MPa)	Elongation (%)	Young modulus (MPa)
Acrylonitrile butadiene styrene (ABS) [18]	19.9–29.1	1.5–8.9	1910–2050
Polycarbonate (PC) [18]	29.5–36.9	3–6.7	1620–2000
Polylactic acid (PLA) [19]	49.1–65.5	1.7–5.0	2800–3600
Polylactic acid (PLA) recycled once	51	1.88	3093 ± 194
Polylactic acid (PLA) recycled 5 times [21]	48.8	1.68	3491 ± 98
PLA/PHA+10–20% fibre [17]	20–30	0.9–1.1	3500–4000
PLA/PHA + 10–20% fibre water saturated [17]	15–20	0.5–0.7	3100–3600
Polylactic acid (PLA) + 5% pine lignin [24]	40.2–43.6	2.31–2.83	2160–2200
TPS/ABS biomass [21]	34.8–46.8	Not Available	Not Available
PLA + graphite 2% [22]	50	8.1	Not Available
PLA + graphite 8% [22]	62	6.1	Not Available
High density polyethylene (HDPE) virgin [23]	25.5	16.1	463.4
High density polyethylene (HDPE) recycled once [23]	25.6	16.1	428.4

Table 2 The fused deposition modelling test specimens mass-produced from composite, plastic and recycled filaments

Material	Ultimate tensile strength (MPa)	Elongation (%)	Youngs modulus (MPa)
Acrylonitrile butadiene styrene (ABS) [18]	19.9–29.1	1.5–8.9	1910–2050
Polycarbonate (PC) [18]	29.5–36.9	3–6.7	1620–2000
Polylactic acid (PLA) [19]	49.1–65.5	1.7–5.0	2800–3600
Polylactic acid (PLA) recycled once	51	1.88	3093 ± 194
Polylactic acid (PLA) recycled 5 times [21]	48.8	1.68	3491 ± 98
PLA/PHA + 10–20% fibre [17]	20.0–30.0	0.90–1.10	3500–4000
PLA/PHA + 10–20% fibre water saturated [17]	15–20	0.5–0.7	3100–3600
Polylactic acid (PLA) + 5% pine lignin [24]	40.2–43.6	2.31–2.83	2160–2200
TPS/ABS biomass [21]	34.80–46.80	Not Available	Not Available
PLA + graphite 2% [22]	50	8.1	Not Available
PLA + graphite 8% [22]	62	6.1	Not Available
High density polyethylene (HDPE) virgin [23]	25.5	16.1	463.4
High density polyethylene (HDPE) recycled once [23]	25.6	16.1	428.4

8 Discussions and Conclusion

Waste parts obtained in 3D printing after the end of their life can be recycled using various techniques. PLA and ABS are the two 3D printer filaments designated as type 7 under the International Resin Identifier Codes (ASTM) which are not processed by municipal programs. Process like grounding biodegradable plastic into small particles using a blender can recycle PLA into useable filament. The process involves a complex number of steps like sieving, blending, grinding, and printing the material for various biomedical uses. The same process is carried at Duquesne University. Their focus is to use the recycled plastic for orthotics and prosthetics research. Research is also focused on removing harmful levels of fluoride in drinking water using a water filter. Unlike with PLA, recycling PET into 3D printable filament requires more work. It requires different additives, fillers, and dyes and experiences different processing conditions even for the same polymer type. The company Filabot has partnered with TerraCycle to provide filament recycling for any type of 3D printable plastic. Their method allows a customer to pay for a box that has prepaid shipping to the recycler. The main drawback to this method is that the boxes and shipping tend to be rather expensive (a small box is $85). HDPE is non-biodegradable and can take centuries to decompose. Hence HDPE is then shredded and melted down to further refine the polymer. The plastic is then cooled into pellets which can be used in manufacturing.

It is estimated that by 2050 the ocean will have more plastic trash than fish in weight. This is more problematic since 3D printing is becoming popular and finding its way into homes and businesses across the world. To combat this problem turning plastic trash back into 3D printable filament is of need of an hour.

References

1. Kohtala C (2015) Addressing sustainability in research on distributed production: an integrated literature review. J Clean Prod 106:654–668
2. Karen L (2016) New filament aims to bring sustainability to 3D printing. Plastics Today
3. Daniel C, Sargeant M, Somers K (2016) 3-D printing takes shape. McKinsey & Company
4. Hopewell J, Drovak R, Kosior E (2009) Plastics recycling: challenges and opportunities. Philos Trans R Soc B 364(1526):2115–2126
5. Calignano F, Manfredi D, Ambrosio EP, Biamino S, Lombardi M, Atzeni E, Salmi A, Minetola P, Iuliano L, Fino P (2017) Overview on additive manufacturing technologies. Proc IEEE 105(4):593–612
6. ISO/ASTM Standard, 52900 (2015) Additive manufacturing, General principles: Terminology
7. Drizo A, Pegna J (2006) Environmental impacts of rapid prototyping: an overview of research to date. Rapid Prototyp J 12(2):64–71
8. Herrmann C, Bergmann L, Thiede S (2009) Methodology for the design of sustainable production systems. Int J Sustain Manuf 1(4):376–395
9. "European Union action on circular economy" European Commission (2017)
10. Lepowsky E, Tasoglu S (2018) Emerging anti-fouling methods "towards reusability of 3D-printed devices for biomedical applications. Micromachines
11. Jasveer S, Jianbin X (2018) Comparison of different types of 3D printing technologies. Int J Sci Res Publi 8(4)
12. Wohlers Talk (2016) "Popularity of FDM" 3D printing ,additive manufacturing, future—Terry Wohler
13. McAlister C, Wood J (2014) The potential of 3D printing to reduce the environmental impacts of Production. Sustainable production design & supply chain initiatives, Eceee Ind. Summer Study Proc
14. Gebler M, Schoot Uiterkamp AJM, Visser C (2014) A global sustainability perspective on 3D printing technologies. Energy Policy 74:158–167
15. Lehrer J, Scanlon M (2017) The development of a sustainable technology for 3D printing using recycled materials, Fall 2017 Mid-Atlantic ASEE Conference
16. Li T, Aspler J, Kingsland A, Cormier LM, Zou X (2016) 3d printing–a review of technologies, markets, and opportunities for the forest industry. J Sci Technol For Prod Process 5(2):30
17. Duigou AL, Castro M, Bevanc R, Martin N (2016) 3D printing of wood fibre biocomposites: from mechanical to actuation functionality. Mater Des 96:106–114
18. Cantrell J, Rohde S, Damiani D, Gurnani R, DiSandro L, Anton J, Young A, Jerez A, Steinbach D, Kroese C, Ifju P (2016) Experimental characterization of the mechanical properties of 3D-printed ABS and polycarbonate parts. Adv Opt Methods Exp Mech 3:89–105
19. Letcher T (2014) Material property testing of 3D printed Specimen in PLA on an Entry level 3D printer. In: Proceedings of the ASME 2014 international mechanical engineering congress & exposition
20. Cruz F, Lanza S, Boudaoud H, Hoppe S, Camargo M (2015) Polymer recycling and additive manufacturing in an open source context: optimization of processes and methods. In: Annual international solid freeform fabrication symposium-an additive manufacturing conference. Austin, Texas (USA), pp 10–12

21. Kuo CC, Liu LC, Teng WF, Chang HY, Chien FM, Liao SJ, Kuo WF, Chen CM (2016) Preparation of starch/acrylonitrile-butadiene-styrene copolymers (ABS) biomass alloys and their feasible evaluation for 3D printing applications. Compos B 86:36–39
22. Zhang D, Chi B, Li B, Gao Z, Du Y, Guo J, Wei J (2016) Fabrication of highly conductive graphene flexible circuits by 3D printing. Synth Met 217:79–86
23. Hamod H (2014) Suitability of recycled HDPE for 3D printing filament" B.Sc Thesis. Arcada University of Applied Science, Helsinki
24. Markstedt K, Sundberg J, Gatenholm P (2014) 3D bioprinting of cellulose structures from an ionic liquid. 3D Print Addit Manuf 1(3):115–121
25. Shastri C Solid waste management–an Indian legal profile
26. Ray M, Rahman M (2016) An overview of legal framework for waste management system in india with special allusion to SWM rules. Int J Interdiscip Multidiscip Stud 4(1):13–19
27. Bharadwaj A, Yadav D, Varshney S (2015) Non-biodegradable waste–its impact & safe disposal. Int J Adv Technol Eng Sci 3

Modelling and State of Charge Estimation of Li-Ion Battery for Electric Vehicle

A. Maheshwari and S. Nageswari

Abstract Lithium-ion (Li-ion) battery necessitates an accurate state of charge (SOC) estimation technique to control charging and discharging for the safety of electric vehicle (EV). SOC is an indicator to display the remaining capacity of the battery. To achieve an accurate SOC, model-based SOC estimation algorithm is developed in this paper. Due to the self-correction nature of Kalman filter (KF), KF family algorithm is most promising for SOC estimation when the system is running. As battery modelling has a direct impact on SOC estimation, an equivalent circuit model (ECM) is preferred due to its balance between accuracy and complexity. In this paper, a closed loop approach with the combination of ECM and extended Kalman filter (EKF) algorithm is implemented on Li-ion battery to estimate an accurate SOC. For ease of implementation, model parameters are considered as univariant of SOC. The obtained discharge characteristic curves at different C-rate are close matches with manufacturer's catalogue. The simulation results show minimum estimation error and good convergence rate of EKF.

Keywords Battery management system · C-rate · Equivalent circuit model · Electric Vehicle

1 Introduction

With the increasing of EV technology, research on battery technology is increased since battery is an essential component for EV. Among several batteries, Li-ion battery is best suited due to its good features like high energy density, long service life, low self-discharge rate, etc. [1]. Because of the harsh operating conditions of EV, battery management system is required to display, control and protect the battery. In numerous key functions of BMS, SOC is an indicator to control the charging and discharging limits for safe use of batteries [2]. In addition, the SOC shows the driving range so as to reduce the range anxiety of driver. Therefore, precise estimation of

A. Maheshwari (✉) · S. Nageswari
Department of EEE, Alagappa Chettiar Government College of Engineering and Technology, Karaikudi, Tamilnadu, India

A. R. Gupta et al. (eds.), *Power Electronics and High Voltage in Smart Grid*, Lecture Notes in Electrical Engineering 817,
https://doi.org/10.1007/978-981-16-7393-1_12

SOC is most important for EV. But it's a challenge as it can't be measured directly with any type of sensor or meter [3], it needs to be estimated using measured variables such as voltage, current and temperature.

Although different estimation methods are used, the method of estimation based on the model is now recognized as the technology of choice [4, 5]. So, making a battery model is the foremost step and model accuracy is a direct influence on the state estimation [6]. But, the complex nature of the battery's electrochemical process creates a barrier to battery modelling [7]. On the other hand, to represent entire dynamic characteristics of battery and attain the desired model accuracy, state precision is essential. Because, the parameters are varied while charging and discharging a battery and these are strongly depending on SOC, temperature and aging of battery [8]. Therefore, parameters are frequently identified and updated in battery model for accurate state estimation. So far, numerous battery models and state estimation methods are developed at various levels of accuracy, configuration effort, ease of implementation and computational complexity [9]. After a general survey, ECM was chosen to depict the electrical behavior of battery and it is widely accepted because of its simplicity and accuracy. In the literature, battery parameters are identified by control system approach like least square algorithm [10, 11] or optimization algorithm like GA [12, 13]. However, battery parameters are not constant; it must be updated because parameters are depending on SOC, C-rate and temperature.

In this proposed work, all parameters are considered as univariate function of SOC so that parameters are updated at every sample. Accordingly, time taken by a complex algorithm for parameters identification is saved. In this article, a simplified closed-loop approach such as ECM-based SOC estimation using EKF is presented with the help of MATLAB/Simulink. Although several non-linear state estimators are available in the literature, EKF algorithm is most promising for state estimation of a non-linear system. The model-based SOC estimation could be validated easily from the discharge curve given by the manufacturer for ease of use [12, 14, 15]. Only the MSE (Mean-Square Error) indicator is used to verify the precision of the fitted data.

The remaining parts of the paper organized as follows: Sect. 2 describes various SOC estimation methods with their advantages and disadvantages. The proposed ECM-based SOC estimation method is given in Sect. 3. Section 4 discusses the simulation results and concludes in Sect. 5.

2 SOC Estimation

According to the statement of Plett [16], the SOC of the battery is the ratio of remaining charge to the total charge capacity of battery.

$$\text{SOC} = \frac{\text{Remaining Charge}}{\text{Total Charge}}$$

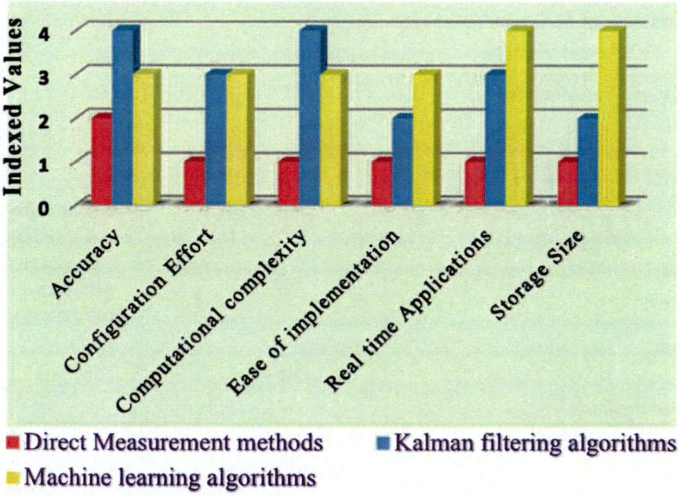

Fig. 1 Comparison of SOC estimation methods

More specifically, lower the accuracy of SOC reflects the lower the efficiency of the battery. Therefore, accuracy is crucial for SOC estimation to design a proper BMS for EVs.

The SOC can be calculated by direct measurement methods or estimated using Kalman filtering algorithms and Machine learning algorithms. Comparison between these three methods is shown in Fig. 1.

In the direct measurement methods, SOC is calculated by the measured voltage using open circuit voltage method (OCV) or measured current using coulomb counting method (CC). Although these methods are simple, accuracy of the methods depends on some constraints like CC method requires precise initial SOC and OCV method needs long rest time. It is not suitable for EV at running condition.

Due to self-correction nature of Kalman filter (KF), SOC estimation can be done accurately even with erroneous initial SOC. The various KF family algorithms are used in the literature. Depending upon the behavior of system, appropriate KF algorithm is chosen for applications.

As per the fast development in intelligence algorithms, some machine learning algorithms have been used to estimate SOC of the battery. These algorithms require lot of data for training and testing the data to estimate SOC without prior knowledge about battery chemistry.

Each method has its own merits and demerits. Table 1 summarizes the information about some algorithms used for SOC estimation. The main objective of the proposed work is to develop a method that must give good tradeoff between accuracy and simplicity.

Table 1 Summary of SOC estimation techniques

Method		Pros	Cons
Direct Measurement Methods			
CC method		• Simple • Low-cost method	• High precision • calculation of initial value of SOC • Initial value error, • Accumulated errors cause overcharging and deep-discharging of battery
OCV method		• Simple • Accurate	• More time consumption
Model-based methods			
Kalman filter algorithms	Kalman filter	• Optimal Estimator of linear dynamic systems • Self-corrective capability • Filter out the noises	• Computationally intensive filtering • Not suitable for nonlinear battery model
	Extended KF	• Suitable for nonlinear battery model • Recursive-based estimator	• Purely depends on linearization • Due to some uncertainties in linearization process, Substantial error occurs
	Particle filter	• Suitable for higher order model • No need to calculate Jacobian matrix	• Complex method
	Sigma point KF	• Can find mean and variance of state with minimum number of sample points • Eliminates the computational burden (No need to calculate analytical derivative) • Robust than EKF	• More computational cost
Machine learning algorithms	Fuzzy	• Easily expanded to any type and size of battery • Can identify the unknown parameters of Li-ion battery	• High storage • More computational time

(continued)

Table 1 (continued)

Method		Pros	Cons
	Artificial neural network	• Easy to implement • Does not required mathematical model	• High storage size
	Artificial neural network & Fuzzy Inference System	• High accuracy	• Need high storage size
	Support vector machine	• Powerful tool to solve regression algorithms • Perform well in higher order model	• High complex quadratic programming • Time consuming

3 ECM-Based SOC Estimation

Battery's SOC is calculated according to usable capacity value and battery current.

$$SOC = \mathrm{SOC}_0 - \int \frac{I_{bat}}{Q_u} dt \tag{1}$$

Here, SOC_0 represents the initial SOC, battery current is represented by I_{bat} and usable capacity is represented by Q_u.

For ECM-based SOC estimation approach, state space equations are derived from the model and an appropriate algorithm has been developed to assess the internal state of a system. The schematic of the ECM of battery, consisting of a controlled voltage source and a parallel RC (Resistor–Capacitor) network with serial resistance, is shown in Fig. 2.

All parameters are strongly depending on SOC. Therefore, estimated SOC is fed back into the battery model to correct the parameter values. So, only SOC is considered as a state of the system.

According to circuit theory, the equations of one RC-ECM for discharging scenario under constant current are as follows:

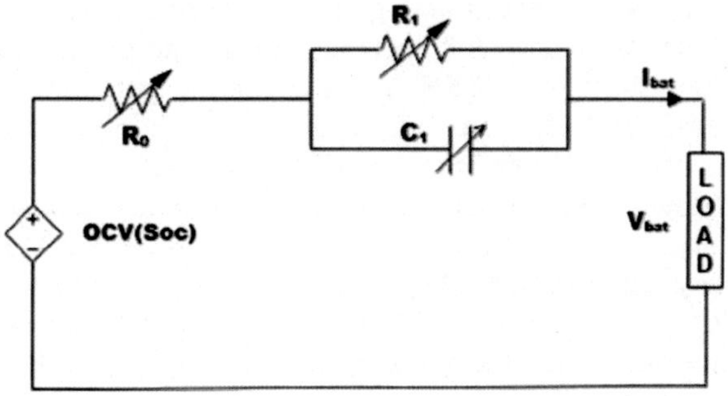

Fig. 2 One RC equivalent circuit model

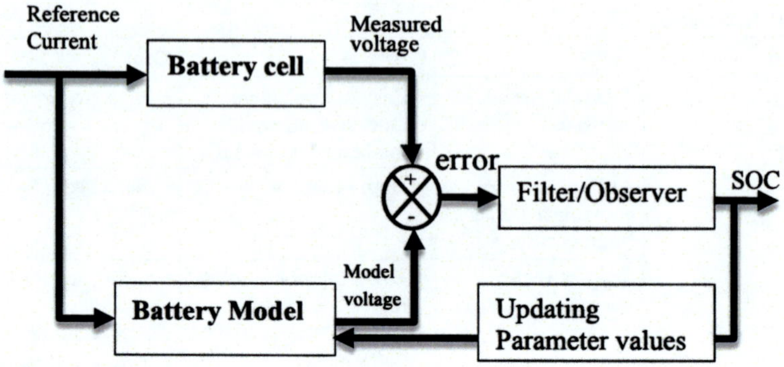

Fig. 3 Model-based SOC estimation

$$V_{bat} = OCV(SOC) - I_{bat} * R_0 - V_1 \tag{2}$$

$$I_{bat} = \frac{V_1}{R_1} + C_1\frac{dV_1}{dt} \tag{3}$$

$$V_1 = \left(\frac{Q_u}{C_1} + I_{bat}{*}R_1\right){*}exp\left(\frac{-t}{R_1{*}C_1}\right) - I_{bat}{*}R_0 \tag{4}$$

where,

V_{bat}– Output voltage of battery,
I_{bat}– Input current of battery,
V_1– Voltage across RC network,
OCV– Open circuit voltage of battery,
Q_u–Capacity of battery.
Block diagram of model-based SOC estimation is shown in Fig. 3.

Reference current is given to real battery and battery model to get measured voltage and model voltage. Error between measured and model is corrected by Filter/Observer to estimate SOC. Since the battery is a non-linear system, EKF is the right choice for estimating the SOC of the battery. EKF is a filter that estimates state based on statistical knowledge of the state and noises (process noise and measurement noise) [10]. EKF uses basic matrix operations for state estimation, so it can be easily embedded for onboard BMS, making it useful in EV applications.

The state estimation using EKF has two modes, i.e., prediction and correction as shown in Fig. 4. After initializing the state variables (X_{K-1}) and error covariance matrix value (P_{K-1}), state matrix (X_K) and error covariance matrix (P_K) is predicted in prediction mode. Then, Kalman gain (G_K) is calculated to correct the state matrix (X_{K+1}) and error covariance matrix (P_{K+1}) in correction mode. V_{model} represents predicted model output and V_{meas} defined as measured output.

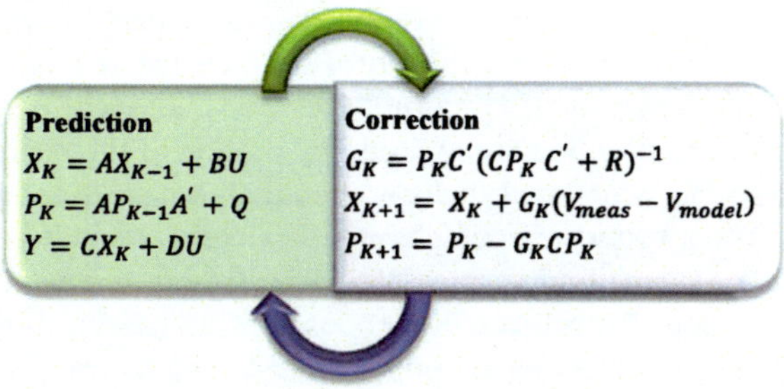

Fig. 4 EKF calculation process

At every sample, estimated SOC is given into battery model to update the model parameters for next state estimation.

4 Simulation Results and Discussion

The battery model is developed in Simulink and EKF algorithm is coded in M-file which is shown in Fig. 5. In the 'measurement' block, actual battery data such as voltage, current and discharge capacity obtained from data sheet are given through signal builder block which is available in Simulink. Additive white gaussian noises are added with measured values. In the 'state matrices' block, state transition matrices A, B, C are calculated using model parameters which are updated by SOC at every sample step. SOC and voltage are predicted in the 'state prediction' block. All state matrix values, noise covariance matrices, measured voltage and predicted SOC & voltage are given as an input to M-file, then SOC is updated using EKF algorithm. At every sample, updated SOC is feedback to state matrices block to modify model

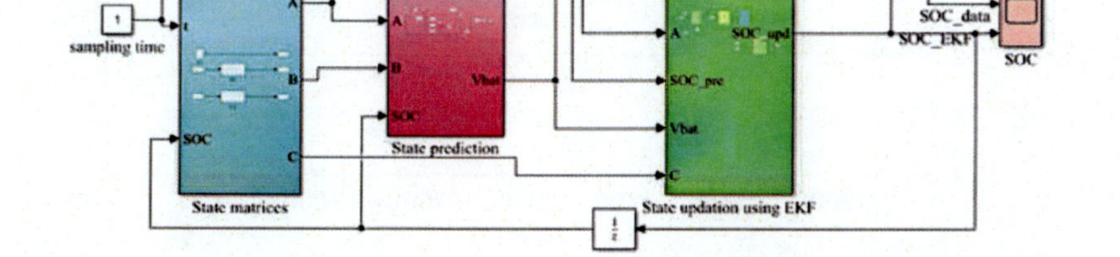

Fig. 5 SOC estimation using EKF

parameters like OCV, R0, R1, C1. Error between real SOC and estimated SOC is calculated to prove the ability of an algorithm. In this proposed method, all parameters are strongly depending on SOC; So, the separate algorithm for model parameter identification is not required. In addition to this, only SOC is considered as a state variable which makes the proposed algorithm is computationally efficient.

In a battery manufacturer's data sheet, standard charging and discharging rates are governed by C-rate. With 1 C, battery can completely charge or discharge in one hour. To validate the proposed ECM-based estimation algorithm, manufacturer (EEMB) data sheet of 2600 mAh Li-ion battery cell is used. In spec sheet the nominal capacity is 2600 mAh, so 0.2 C corresponds to 0.52 A. The simulation has been done for discharging state at 0.2 C which is the preferable discharging rate as mentioned in spec sheet and 1 C which is maximum allowable discharging rate.

4.1 Constant Current Discharge

The battery completely released its storage with 2.6 A constant current in one hour, it is shown in Fig. 6. The discharge voltage, SOC and SOC error with respect to time are shown in Fig. 6. It is noticed that battery reached its discharge cut off voltage 3 V in time.

Figure 7 shows the battery fully discharged at 0.52 A constant current rate in five hours. The discharge voltage, SOC and error with respect to time is shown in Fig. 7. This algorithm produces near optimal result for state of charge estimation with mean square error (MSE) of 0.0086.

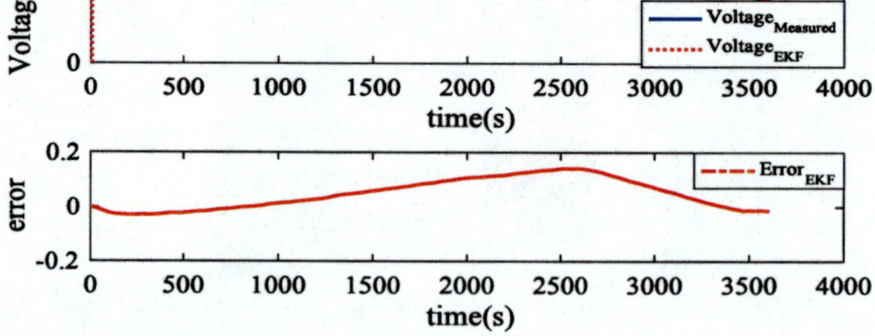

Fig. 6 Constant current discharging at 1C

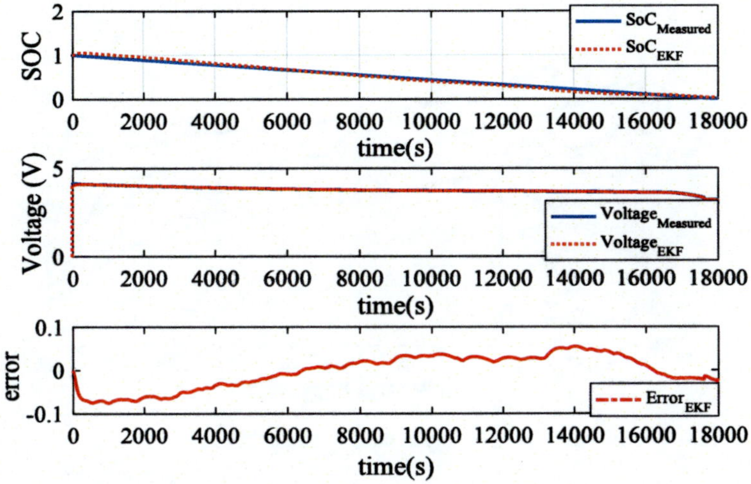

Fig. 7 Constant current discharging at 0.2C

Simulation is done for battery's discharging scenario at 2C-rate. Since C-rate increases, battery discharge time decreases, i.e., battery completely discharged with 2C-rate in half-hour as shown in Fig. 8.

Figure 9 shows the complete discharge characteristics (Discharge Capacity vs Voltage) of EEMB battery getting from proposed algorithm and spec sheet for validation. Based on the results, it is noticed that the discharge characteristic curve of

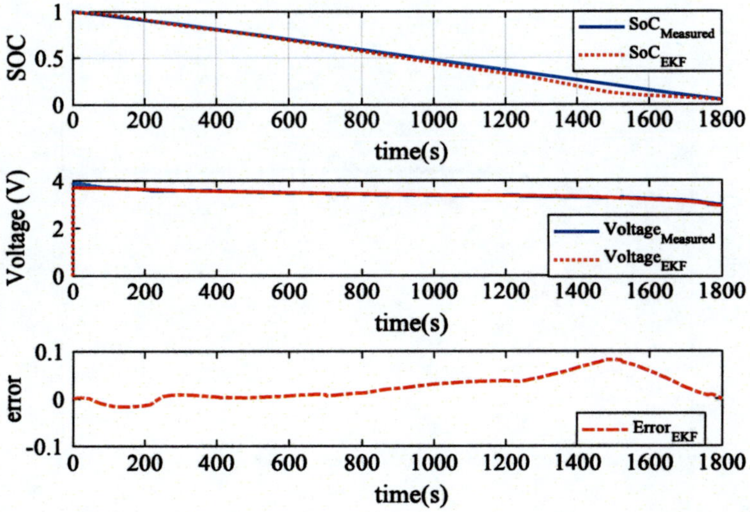

Fig. 8 Constant current discharging at 2C

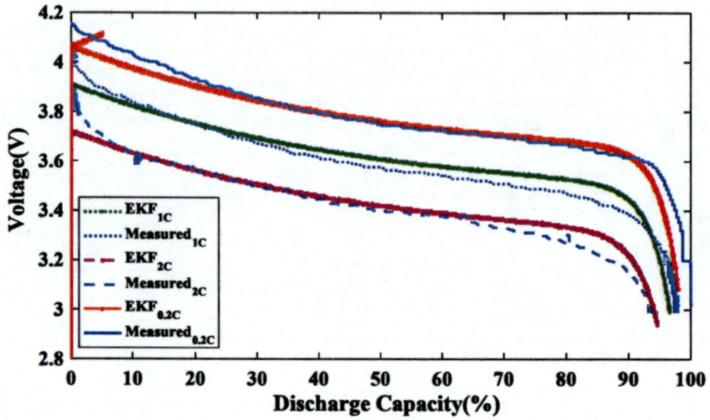

Fig. 9 Discharge characteristics of battery at three different C-rates

developed algorithm match very well with the curve from manufacturer's spec sheet. According to Fig. 9, increase in C-rate reduces the discharge capacity of a battery which degrades the battery.

4.2 Constant Current Constant Voltage Charge

As per data sheet, CCCV method of charging is used in this simulation. In Fig. 10, solid lines represent measured data and marker indicates result obtained from EKF

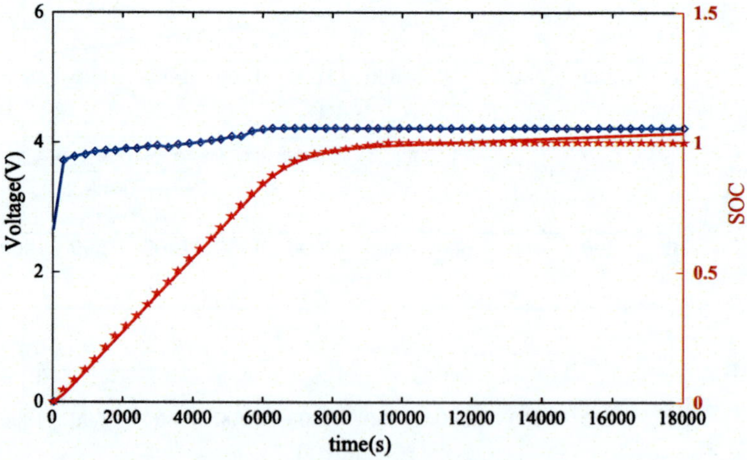

Fig. 10 Constant Current (1.3 A)-Constant Voltage (4.2 V) charging

algorithm. In a CC mode, 1.3 A (0.5 C) constant current is applied to a battery. During CC mode, voltage and SOC gradually increases to specific limit as shown in Fig. 10. When voltage reached 4.2 V, CV mode is ON. It is noticed that SOC reached to 84% at the end of CC mode (Time = 6000 s). Therefore, battery takes 4500 s to attain the remaining 16% of SOC during CV mode. Even though CCCV method is preferred for controlling the temperature rise, battery charging time is major concern for large scale adoption of electric vehicle.

Further research concentrates on SOC estimation of battery module using this algorithm with the consideration of cell balancing.

4.3 Drive Profile

The test drive profile [17] for Electric Two-Wheeler is taken for testing the proposed algorithm.

The result obtained from EKF algorithm for the drive profile is shown in Fig. 11.

Blue color solid line indicates measured data whereas red color dotted line represents data obtained from proposed SOC technique. For EV drive profile, this algorithm produces the result with MSE of 0.0880.

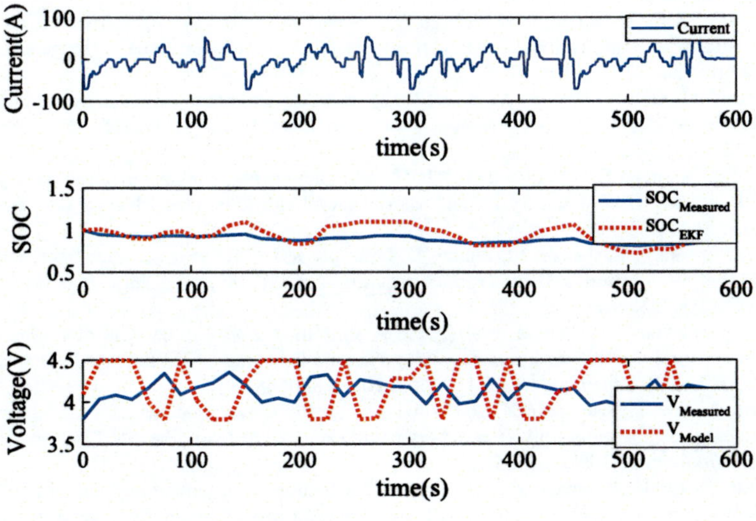

Fig. 11 Comparative result for EV drive profile

5 Conclusion

An accurate SOC is needed for driving range estimation of an electric vehicle. Accuracy of state depends on battery modelling and estimation algorithm. One RC equivalent circuit model of Li-ion battery cell is built with the consideration of modelling requirements. Measured data is collected from spec sheet. Then, SOC is estimated using extended Kalman filtering algorithm. Since model parameters are depending on SOC, separate identification algorithm is not required for this closed loop approach. At every sample, battery parameters are updated by SOC. The comparison has been done for the discharging scenario under constant current & test drive profile and CCCV charging of Li-ion battery. The comparative result shows the results from EKF closely matched with measured data from spec sheet. This approach is limited to operating EV in constant temperature condition only. But it can be extended to operate EV in different temperature conditions.

References

1. Al-Gabalawy M, Hosny NS, Dawson JA, Omar AI (2020) State of charge estimation of a Li-ion battery based on extended Kalman filtering and sensor bias. Int J Energy Res 1–19 https://doi.org/10.1002/er.6265
2. Ali MU, Zafar A, Nengroo SH et al (2019) Towards a smarter battery management system for electric vehicle applications: a critical review of lithium-ion battery state of charge estimation. Energies 12
3. Brandl M, Gall H, Wenger M et al (2012) Batteries and battery management systems for electric vehicles. Proceedings, design, automation, & test in Europe: dresden. IEEE, Germany, pp 12–16
4. Fotouhi A, Auger DJ, Member S et al (2018) Accuracy versus simplicity in online battery model identification. IEEE Trans Syst Man Cybern Syst 48:195–206. https://doi.org/10.17862/cranfield.rd.3545847
5. Kazhamiaka F, Keshav S, Rosenberg C, Pettinger KH (2018) Simple spec-based modeling of lithium-ion batteries. IEEE Trans Energy Convers 33(1757):1765. https://doi.org/10.1109/TEC.2018.2838441
6. Meng J, Stroe DI, Ricco M et al (2019) A simplified model-based state-of-charge estimation approach for lithium-ion battery with dynamic linear model IEEE Trans Industr Electron 66:7717–7727 https://doi.org/10.1109/TIE.2018.2880668
7. Pang H, Guo L, Wu L, Jin X (2020) An enhanced temperature-dependent model and state-of-charge estimation for a Li-Ion battery using extended Kalman filter Int J Energy Res 1–14 https://doi.org/10.1002/er.5435
8. Plett GL (2004) Extended Kalman filtering for battery management systems of LiPB-based HEV battery packs-Part 3. State and parameter estimation. J Power Sources. https://doi.org/10.1016/j.jpowsour.2004.02.033
9. Pop V, Bergveld HJ, Danilov D et al (2008) Battery management systems accurate state-of-charge indication for battery-powered applications. Springer
10. Sangwan V, Kumar R, Rathore AK (2017) Estimation of model parameters and state-of-charge for battery management system of Li-ion battery in EVs. In: 2017 IEEE transportation electrification conference (ITEC-India) estimation

11. Thirugnanam K, Ezhil Reena Joy TP, Singh M, Kumar P (2014) Mathematical modeling of li-ion battery using genetic algorithm approach for V2G applications IEEE Trans Energy Convers 29:332–343. https://doi.org/10.1109/TEC.2014.2298460
12. Ting TO, Man KL, Zhang N et al (2014) State-space battery modeling for smart battery management system. Lecture Notes Eng Comput Sci 2210:12–15
13. Tremblay O, Dessaint L-A (2009) Experimental validation of a battery dynamic model for EV applications World Electr Veh J 3(289):298
14. Yang J, Xia B, Shang Y et al (2017) Adaptive state-of-charge estimation based on a split battery model for electric vehicle applications. IEEE Trans Veh Technol 66(10889):10898. https://doi.org/10.1109/TVT.2017.2728806
15. Zhang C, Li K, McLoone S, Yang Z (2014) Battery modelling methods for electric vehicles -A review. In: 2014 European control conference, ECC 2014. Institute of Electrical and Electronics Engineers Inc., pp 2673–2678
16. Zhang R, Xia B, Li B et al (2018) State of the art of lithium-ion battery SOC estimation for electrical vehicles. Energies
17. Loganayaki S, Nageswari S, Renganathan S (2019) State of charge estimation of battery for electric two-wheeler. In: 2019 International conference on computer communication and power systems (ICCCPS-2020)

Interval Modeling of Riverol-Pilipovik Water Treatment System

Nitin Mathur, V. P. Meena, and V. P. Singh

Abstract Riverol-Pilipovik (RP) water treatment system is a reverse osmosis-based desalination plant which is represented by a set of transfer functions. At times, there may be variations in the parameters of the system which may change its behavior. In this paper, interval modeling of RP system is done to accommodate the effect of parametric variations. A comparative study of original system with its interval model is shown with the help of step response and impulse response. Also, the characteristics of the response are compared to demonstrate the resemblance of the interval model to the original system.

Keywords Interval modeling · Reverse osmosis · Riverol-Pilipovik water treatment system · System modeling

1 Introduction

With the alarming rise in demand of water, amount of freshwater in water bodies like lakes, rivers, ponds, etc., is decreasing day by day. With two-third of the earth's surface being covered by sea and ocean, desalination of sea water is a good alternative in order to meet this alarming demand. Reverse osmosis (RO) is one of the techniques of desalination. The existing literature shows that RO plant is generally a multi–input–multi–output (MIMO) system. The variables on input and output side are called manipulated and controlled variables, respectively. The most widely used models for RO process are Doha plant [1], Chaabene model [2] and Riverol-Pilipovik (RP) plant [3].

Mathematical modeling of a complex high-order dynamic system is the prime objective of control system. A mathematical model of any system is obtained such

N. Mathur · V. P. Meena (✉) · V. P. Singh
MNIT, Jaipur, India
e-mail: vmeena1@ee.iitr.ac.in

V. P. Singh
e-mail: vinay.ee@mnit.ac.in

A. R. Gupta et al. (eds.), *Power Electronics and High Voltage in Smart Grid*, Lecture Notes in Electrical Engineering 817,
https://doi.org/10.1007/978-981-16-7393-1_13

that it completely describes the dynamics of that system. In most of the cases for any physical system, a nominal model is deduced. Nominal model is simpler to analyze. Also, designing the controller for a nominal model is easy. However, there are several types of uncertainties that are present in the system which may lead to some perturbations in the system dynamics. The behavior of the system may change due to these uncertainties. The nominal model generally fails to perform satisfactorily when variations in parameters occur. For any system to be worked satisfactorily under parameter variations, interval modeling of that system may help. The coefficients of interval polynomial of system are represented in intervals which are bound by the upper limit and lower limit. Systems like oblique wing aircraft, DC shunt motor and cold rolling mill are represented in the form of interval systems [4].

The model order reduction of interval system can also be done if the order of system is high. Techniques like Routh approximation [5], γ-δ Routh approximation [6], direct truncation and factor division method [7], Mihailov criterion and factor division method [8], method based on alpha table and factor division method [9] are introduced for model order reduction of interval systems.

In this paper, interval model of Riverol-Pilipovik (RP) water treatment system is derived by taking 15\% deviation in the transfer function of the system. comparison of the original system with its interval model is shown using step and impulse responses. The remainder of the paper is arranged as follows. Section 2 describes the RP water treatment system and its equations. Interval model for this RP water system is derived in Sect. 3. Finally, the results are shown in Sect. 4 by plotting the step and impulse responses of both the original system and its interval model. Also, the comparison of the step response characteristics like peak value, maximum overshoot, settling time, rise time and peak time is shown in this section. At the end, article is concluded in Sect. 5.

2 System Description

The general layout of Riverol-Pilipovik (RP) water treatment system is shown in Fig. 1. This structure is divided into four segments: pre-treatment, high-pressure pump, membrane module and post-treatment. The manipulated variables for this system are pressure and pH value. However, the controlled variables are flux and conductivity at the permeate stream [10].

The mathematical representation of RP system is given as

$$\begin{bmatrix} F \\ C \end{bmatrix} = \begin{bmatrix} g_{11} & g_{12} \\ g_{21} & g_{22} \end{bmatrix} \begin{bmatrix} P \\ pH \end{bmatrix} \tag{1}$$

where P is pressure (kPa), F is flow rate $(m^2 d)$ and C is conductivity at the permeate stream $(\mu s/cm)$. The transfer function values of g_{11}, g_{12}, g_{21} and g_{22} are

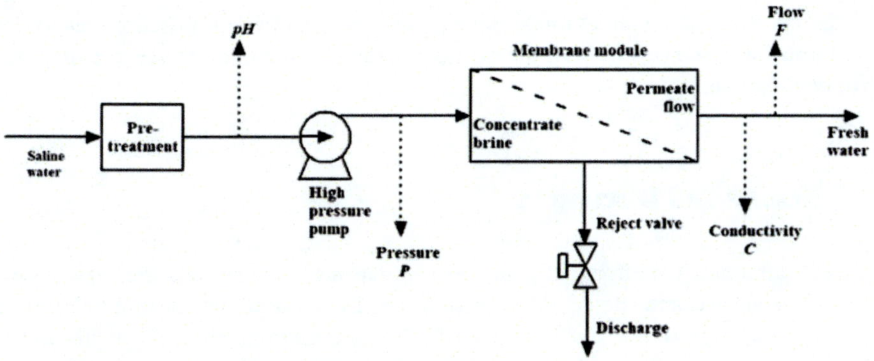

Fig. 1 Block diagram of RP water treatment plant

$$g_{11} = \frac{0.0045(0.104s + 1)}{0.012s^2 + s + 1} \tag{2}$$

$$g_{12} = 0 \tag{3}$$

$$g_{21} = \frac{-0.12s + 0.22}{0.1s^2 + 0.3s + 1} \tag{4}$$

$$g_{22} = \frac{10(-3s + 1)}{s^2 + 5s + 1} \tag{5}$$

3 Interval Modeling of RP Water Treatment System

The systems g_{11}, g_{12}, g_{21} and g_{22} are presented in terms of intervals which are bound by upper limits and lower limits. The limits are determined by taking 15\% deviation from the nominal values. Thus, transfer functions in (1)-(4) equation become (5)-(8), respectively.

$$g_{11} = \frac{[0.0003978, 0.0005382]s + [0.003825, 0.005175]}{[0.0102, 0.0138]s^2 + [0.85, 1.15]s + [0.85, 1.15]} \tag{6}$$

$$g_{12} = 0 \tag{7}$$

$$g_{21} = \frac{[-0.138, -0.102]s + [0.187, 0.253]}{[0.085, 0.115]s^2 + [0.255, 0.345]s + [0.85, 1.15]} \tag{8}$$

$$g_{22} = \frac{[-34.5, -25.5]s + [8.5, 11.5]}{[0.85, 1.15]s^2 + [4.25, 5.75]s + [0.85, 1.15]} \tag{9}$$

In this paper, the interval model for g_{22} is considered for analysis. Step responses and impulse responses of both the original system and interval model are obtained for comparison.

4 Results and Discussion

To accommodate the effects of parametric variations in the system, the system with 15\% deviation is considered. After considering 15\% variations, the interval models are derived in (6) to (9). For further analysis, g_{22} system given in (9) is taken. The original system, system with lower limits of interval model and system with upper limits of interval model are presented in (10), (11) and (12), respectively.

$$g_{22} = \frac{10(-3s + 1)}{s^2 + 5s + 1} \tag{10}$$

$$g_{22}(L) = \frac{-34.5s + 8.5}{0.85s^2 + 4.25s + 0.85} \tag{11}$$

$$g_{22}(U) = \frac{-25.5s + 11.5}{1.15s^2 + 5.75s + 1.15} \tag{12}$$

The step responses of systems (10), (11) and (12) are plotted for analysis. Figure 2 shows the step response of (10), whereas Figs. 3 and 4 show the step responses of (11) and (12), respectively.

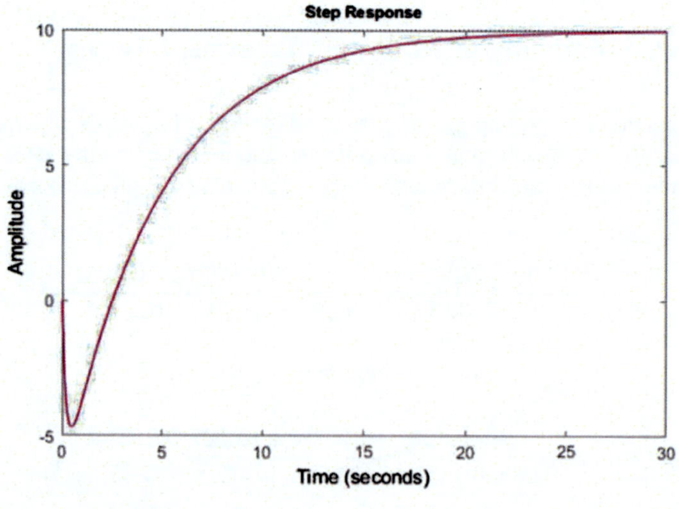

Fig. 2 Step response of (10)

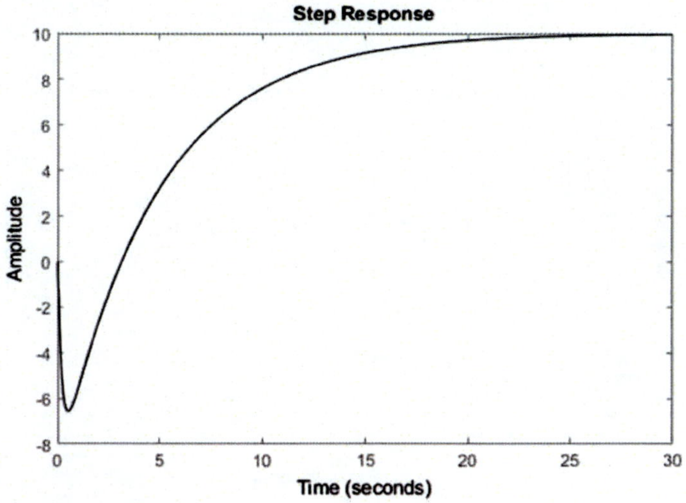

Fig. 3 Step response of (11)

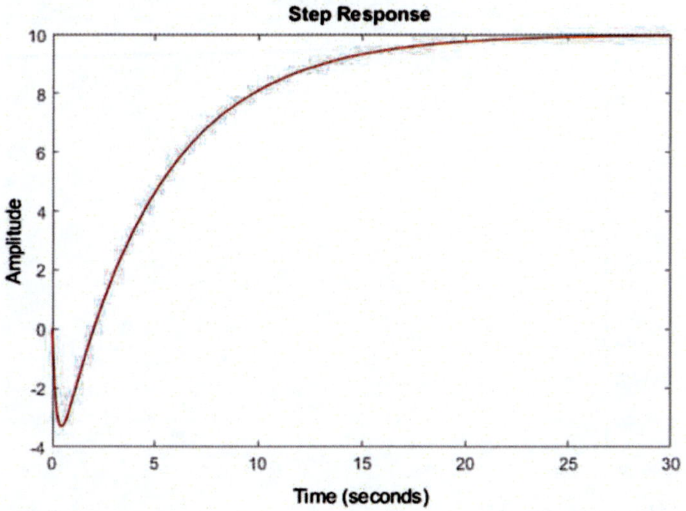

Fig. 4 Step response of (12)

Similarly the impulse response of (10) is shown in Fig. 5 and that of interval models, (11) and (12) are shown in Figs. 6 and 7, respectively. Table 1 gives the comparison of the characteristics of step response like peak value, maximum overshoot, rise time and peak time. The results reveal that the performance of interval model is similar to original system. In Table 1 the characteristics of (11) and (12) are very close to that of (10).

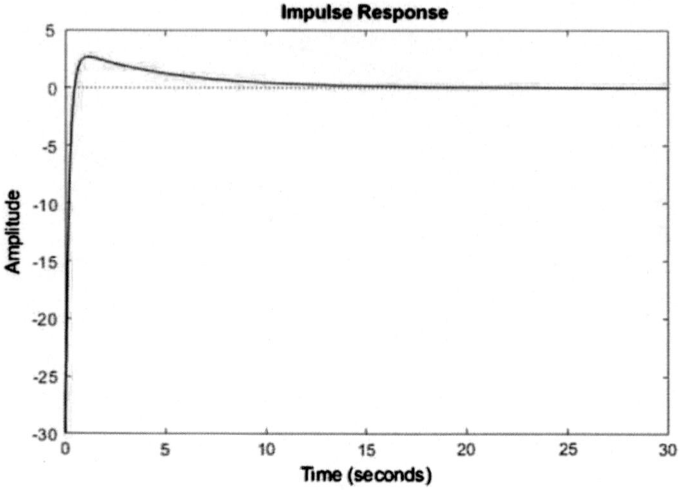

Fig. 5 Impulse response of (10)

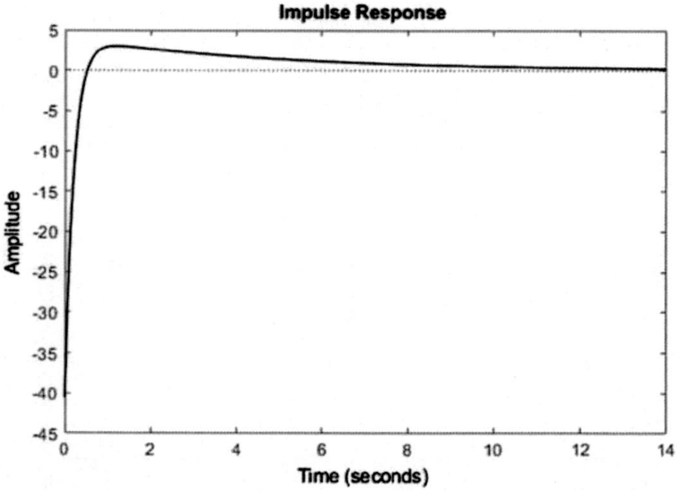

Fig. 6 Impulse response of (11)

5 Conclusion

RP water treatment system is a two-input-two-output system having pressure and \$pH\$ as manipulated variables and flow rate and conductivity as controlled variables. The mathematical representation of this system has transfer functions g_{11}, g_{12}, g_{21} and g_{22}. In this paper, the interval models for all transfer functions are derived by taking

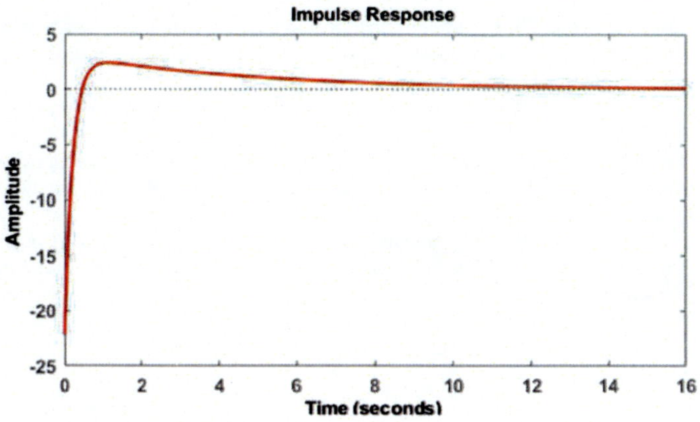

Fig. 7 Impulse response of (12)

Table 1 Step response characteristics

S. No.	System	Peak	Settling time (sec)	Rise time (sec)	Peak time (sec)
1	(10)	9.9828	19.4475	10.5275	33.0445
2	(11)	9.9805	19.4817	10.5275	33.0445
3	(12)	9.9857	19.4095	10.5275	33.4290

15\% deviation from the nominal values of system. The step and impulse responses of the original system as well as interval models are shown. The results clearly depict the resemblance of the interval model with the original system. In future, the derived interval model can be used for model order reduction and controller design for the system.

References

1. Rathore NS, Singh VP, Kumar B (2018) Controller design for doha water treatment plant using grey wolf optimization. J Intell Fuzzy Syst 35(5):5329–5336
2. Naresh P, Pattnaik S, Singh VP (2020) Salp swarm optimization based controller design for photovoltaic reverse osmosis plant. J Inf Optimiz Sci 41(2): 651–659
3. Chodavarapu MM, Singh VP, Devarapalli R (2020) Interval modeling of Riverol-Pilipovik water treatment plant and its model order reduction. In: Computing algorithms with applications in engineering, pp 361–367. Springer, Singapore
4. Choudhary AM, Nagar SK (2019) Order reduction in z-domain for interval system using an arithmetic operator. Circuits Syst Signal Process 38(3): 1023–1038
5. Bandyopadhyay B, Ismail O, Gorez R (1994) Routh-Pade approximation for interval systems. IEEE Trans Automatic Control 39:12:2454–2456
6. Bandyopadhyay B, Upadhye A, Ismail O /spl gamma/-/spl delta/Routh approximation for interval systems. IEEE Trans Automatic Control 42(8):1127–1130

7. Kumar DK, Nagar SK (2013) Direct truncation and factor division method for order reduction of interval systems. In: National conference on recent developments in control, automation and power engineering, pp 37–40
8. Kumar DK, Nagar SK, Tiwari JP (2011) Model order reduction of interval systems using mihailov criterion and factor division method. Int J Comput Appl 28(11):4–8
9. Kumar DK, Nagar SK, Tiwari JP (2013) Order reduction of interval systems using alpha and factor division method. In: Recent advancements in system modelling applications. Springer, India, pp 249–259
10. Riverol C, Pilipovik V (2005) Mathematical modeling of perfect decoupled control system and its application: a reverse osmosis desalination industrial-scale unit. J Autom Methods Manage Chem 2005(2):50–54

Interval Modeling of Cuk Converter

Surjeet Choudhary, V. P. Meena, and V. P. Singh

Abstract This paper proposes a method of interval modeling for the Cuk converter. State space averaging (SSA) technique is used to find the output to control transfer function for Cuk converter. The combined state space description is obtained by SSA technique. The deviation in physical parameters is present in system due to uncertainties and imperfect modeling. To consider the effect of parametric variations in system, the interval modeling for Cuk converter is done.

Keywords Cuk converter · Interval modeling · Parametric uncertainty · State space averaging · State space model

1 Introduction

A DC to DC converter changes DC voltage from one level to another level. In literature, there are many DC to DC converters available. However, Cuk converter is popular because of its ability to step up or step down the output voltage in comparison to input voltage. The main advantage of Cuk converter is that it forms LC-filters at output and input. The key benefit of LC-filters is that these provide smooth current waveforms. The LC-filters also reduce ripples in output voltage.

As far as modeling of DC to DC converters is concerned, the SSA method is utilized [1]. Averaging of state space descriptions is obtained over single switching time in SSA technique. The state space equations (SSEs) during ON and OFF switching conditions of switch can be obtained separately. A single matrix equation is obtained by mixing these SSEs in an average manner. The output to control transfer function for Cuk converter can be obtained easily without circuit analysis by using SSA technique.

S. Choudhary · V. P. Meena (✉) · V. P. Singh
MNIT, Jaipur, India
e-mail: vmeena1@ee.iitr.ac.in

V. P. Singh
e-mail: vinay.ee@mnit.ac.in

A. R. Gupta et al. (eds.), *Power Electronics and High Voltage in Smart Grid*, Lecture Notes in Electrical Engineering 817,
https://doi.org/10.1007/978-981-16-7393-1_14

The modeling of DC to DC converters can be performed by considering uncertainties. These uncertainties occur in modeling due to variations in system parameters. The uncertainties affect the system coefficients either due to aleatoric nature or epistemic nature. The random cause of uncertainties is called aleatoric and cognitive cause is known as epistemic [2]. The model which describes the system dynamics in absence of uncertainties is known as nominal model generally. However, uncertainties are always present in physical system. To include the effect of uncertainties in the design and analysis of system, the interval modeling of system can be performed. Some percentage of parametric variations can be considered in coefficients of system transfer function to define system in presence of uncertainties.

In literature, various examples are available which are designed and modeled as interval systems. The system like Doha water treatment plant, Riverol-Pilipovik water treatment plant [3], overlapping bar aircraft, d.c. motors and two tank system are modeled as interval systems. Some examples of physical systems, which are modeled as interval systems are:

(a) Cold rolling mill [4]

$$\frac{[4.2, 21]s^2 + [3, 16]s + [0.5, 2.6]}{s^4 + [3, 8]s^3 + [1, 2.5]s^2 + [0.05, 0.15]s} \tag{1}$$

(b) Oblique wing aircraft [5]

$$\frac{[54,74]s + [90, 166]}{s^4 + [2.8, 4.6]s^3 + [50.4, 80.8]s^2 + [30.1,33.9]s + [0.1,0.1]} \tag{2}$$

(c) Electric motors

$$\frac{50 \times 10^3}{[0.0000096,0.0000336]s^3 + [0.0012,0.0028]s^2 + [0.002025,0.002475]s} \tag{3}$$

In this paper, a method is proposed for interval modeling of Cuk converter. To obtain output to control transfer function for Cuk converter, the SSEs during both ON and OFF switching periods are required. SSA technique is applied to get combined state space description. The proposed interval model of Cuk converter is obtained from constant coefficient transfer function by considering $\pm 1\%$ parametric variations. These variations occur in parameters due to uncertainties present in physical system. The step and impulse responses for both constant coefficient model and interval model are plotted. To plot responses, upper and lower bounds of coefficients for interval model are considered. The system characteristics like settling time, rise time, peak time and overshoot are tabulated for fair analysis.

This paper is organized as follows: SSA technique is discussed in Sect. 2, system analysis of Cuk converter is explained in Sect. 3, interval model for Cuk converter is obtained in Sect. 4, simulation results are included in Sect. 5 and paper is concluded in Sect. 6.

2 SSA Technique

The modeling of switching converters is mainly based on two approaches. One is state space modeling and another is SSA technique. In this paper, SSA technique is used. The main benefit of SSA technique is that it provides unified description of all stages of converter. There are two circuit states of switch for Cuk converter operation in continuous conduction mode (CCM): one during ON switch for DT interval of time and another during OFF switch for $(1-D)T$ interval of time. The modeling of Cuk converter using SSA technique is obtained using following three steps:

2.1 State Variable Description

For Cuk converter SSEs during switch ON and switch OFF conditions are obtained as

$$\dot{X} = \begin{cases} a_1 X + b_1 V_s \ 0 < t < DT, \text{ during on switch} \\ a_2 X + b_2 V_s \ 0 < t < (1-D)T, \text{ during off switch} \end{cases} \tag{4}$$

$$V_0 = \begin{cases} c_1 X + e_1 V_s \text{ during } DT \\ c_2 X + e_2 V_s \text{ during } (1-DT) \end{cases} \tag{5}$$

where D is duty ratio, V_s is input voltage, V_o is output voltage, X is state vector, a_1 and a_2 are system matrices during switch ON and switch OFF, respectively, b_1 and b_2 are input matrices during switch ON and switch OFF, respectively, c_1 and c_2 are output matrices during switch ON and switch OFF, respectively and e_1 and e_2 are feed forward matrices during switch ON and switch OFF, respectively.

2.2 SSEs Averaging Using Duty Ratio

SSEs during ON switch (4), (5) are multiplied with duty ratio D and SSEs during OFF switch are multiplied with (1−D) to obtain combined state space description (6), (7).

$$\begin{aligned} \dot{X} &= [a_1\mathrm{D} + a_2(1-\mathrm{D})]\mathrm{X} + [b_1\mathrm{D} + \mathrm{b}_2(1-\mathrm{D})]\mathrm{V}_s \\ &= aX + bV_s \end{aligned} \tag{6}$$

$$
\begin{aligned}
V_0 &= [c_1\mathrm{D} + c_2(1 - \mathrm{D})]\mathrm{X} + [e_1\mathrm{D} + e_2(1 - \mathrm{D})]\mathrm{V}_s \\
&= cX + eV_s
\end{aligned} \tag{7}
$$

where a, b, c, d are system matrix, input matrix, output matrix and feed forward matrix of combined state space description, respectively.

2.3 *Transfer Function*

The output to input transfer function can be defined as

$$
H_{V_0V_s} = \frac{V_o}{V_s} = c(sI - a)^{-1}b + e \tag{8}
$$

Also the output to control transfer function is defined as

$$
H_{V_0D} = \frac{V_o}{D} = c(sI - a)^{-1}b_q + e_q \tag{9}
$$

where

$$
b_q = (a_1 - a_2)\mathrm{X} + (b_1 - b_2)\mathrm{V}_s \tag{10}
$$

$$
e_q = (c_1 - c_2)\mathrm{X} + (e_1 - e_2)\mathrm{V}_s \tag{11}
$$

3 System Analysis

The circuit of Cuk converter during ON and OFF switching periods is analyzed. The behavior of each circuit element like inductor, capacitor and diode during switch ON and switch OFF periods is analyzed. The SSEs of Cuk converter during both switching operations are obtained. The SSA method is used to find combined state space description of converter [6, 7].

3.1 *Modeling of Cuk Converter by State Space Technique*

The power circuit of Cuk converter contains one inductor L_1 with resistance r_{l1} in series with supply source and another inductor L_2 with resistance r_{l2} in series with

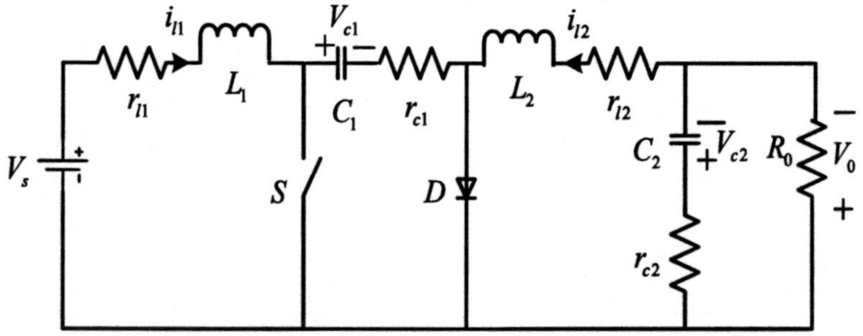

Fig. 1 Circuit diagram of Cuk converter

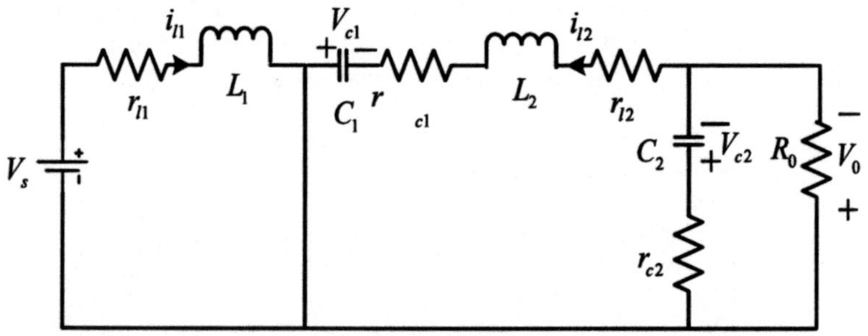

Fig. 2 Circuit diagram of Cuk converter during switch ON

load. The storage capacitor C_1 with series resistance r_{C1} and output capacitor C_2 with series resistance r_{C2}, respectively, switch S, diode D and load resistance R_0 are shown in Fig. 1. Initially, switch is OFF during $(1 - D)T$ period, diode conducts and inductor L_1 provides energy to output via capacitor C_1. The capacitor C_1 is fully charged at the end of $(1 - D)T$ period. When switch is ON during DT period, diode is reverse biased due to voltage polarity of capacitor C_1. The input voltage V_s feeds energy to inductor L_1 and inductor L_2 stores energy from capacitor C_1. Energy of capacitor C_1 gets dissipated through capacitor C_2 and load R_0 at output side (Figs. 2 and 3).

3.2 SSEs of Cuk Converter

SSEs are mathematical representation of physical system in form of differential equations. To write SSEs of Cuk converter, four state space variables are selected for four energy storage elements. The inductor currents and capacitor voltages are selected as state space variables. The state space variables are as follows: the current

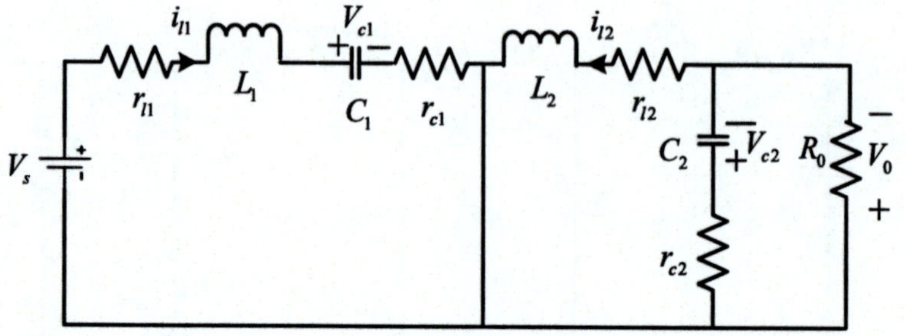

Fig. 3 Circuit diagram of Cuk converter during switch OFF

through the input inductor (i_{l1}), the voltage across the storage capacitor (V_{c1}), the current through output inductor (i_{l2}) and the output capacitor voltage (V_{c2}).

During ON switch, the SSEs of Cuk converter [8] are

$$\frac{di_{l1}}{dt} = \frac{V_s}{L_1} - \frac{r_{l1}i_{l1}}{L_1} \tag{12}$$

$$\frac{di_{l2}}{dt} = \frac{V_{c1}}{L_2} - \frac{R_0 V_{c2}}{(R_0 + r_{c2})L_2} - \frac{(r_{l2} + r_{c1} + r_{c2}\|R_0)i_{l2}}{L_2} \tag{13}$$

$$\frac{dV_{c1}}{dt} = -\frac{i_{l2}}{C_1} \tag{14}$$

$$\frac{dV_{c2}}{dt} = \frac{R_0 i_{l2}}{(R_0 + r_{c2})C_2} - \frac{V_{c2}}{(R_0 + r_{c2})C_2} \tag{15}$$

$$V_0 = \frac{r_{c2}R_0}{r_{c2} + R_0} i_{l2} + \frac{R_0}{r_{c2} + R_0} V_{c2} \tag{16}$$

During OFF switch, the SSEs of Cuk converter are

$$\frac{di_{l1}}{dt} = \frac{V_s}{L_1} - \frac{V_{c1}}{L_1} - \frac{(r_{l1} + r_{c1})i_{l1}}{L_1} \tag{17}$$

$$\frac{di_{l2}}{dt} = \frac{-R_0 V_{c2}}{(R_0 + r_{c2})L_2} \frac{(r_{l2} + r_{c2}\|R_0)i_{l2}}{L_2} \tag{18}$$

$$\frac{dV_{c1}}{dt} = \frac{i_{l1}}{C_1} \tag{19}$$

$$\frac{dV_{c2}}{dt} = \frac{R_0 i_{l2}}{(R_0 + r_{c2})C_2} - \frac{V_{c2}}{(R_0 + r_{c2})C_2} \tag{20}$$

The equations given in (12)–(21) can be represented in the matrices form as

$$a_1 = \begin{bmatrix} \frac{-r_{l1}}{L_1} & 0 & 0 & 0 \\ 0 & \frac{-(r_{l2}+r_{c1}+r_{c2}\|R_0)}{L_2} & \frac{1}{L_1} & \frac{-R_0}{(R_0+r_{c2})L_2} \\ 0 & \frac{-1}{C_1} & 0 & 0 \\ 0 & \frac{R_0}{(R_0+r_{c2})C_2} & 0 & \frac{-1}{(R_0+r_{c2})C_2} \end{bmatrix} \quad (21)$$

$$a_2 = \begin{bmatrix} \frac{r_{c1}-r_{l1}}{L_1} & 0 & \frac{-1}{L_1} & 0 \\ 0 & \frac{-(r_{l2}+r_{c2}\|R_0)}{L_2} & \frac{1}{L_1} & \frac{-R_0}{(R_0+r_{c2})L_2} \\ \frac{1}{C_1} & 0 & 0 & 0 \\ 0 & \frac{R_0}{(R_0+r_{c2})C_2} & 0 & \frac{-1}{(R_0+r_{c2})C_2} \end{bmatrix} \quad (22)$$

$$b_1 = b_2 = b = \begin{bmatrix} \frac{1}{L_1} \\ 0 \\ 0 \\ 0 \end{bmatrix} \quad (23)$$

$$c_1 = c_2 = c = \begin{bmatrix} 0 & \frac{r_{c2}R_0}{r_{c2}+R_0} & 0 & \frac{R_0}{r_{c2}+R_0} \end{bmatrix} \quad (24)$$

$$e_1 = e_2 = e_q = e = [0] \quad (25)$$

4 Derivation of Interval Model for Cuk Converter

The circuit parameters [9] of Cuk converter are as follows (Table 1).

For this converter output to control transfer function [10] is calculated by using Laplace transform as mentioned in (9)-(11).

Table 1 Parameters of cuk converter

S. no.	Parameter	
1	Supply voltage (V_s)	12 volts
2	Output voltage (V_0)	24 volts
3	Switching frequency (f_s)	100 khz
4	Load (R_0)	12 Ω
5	(L_1)	68.7 μH
6	(L_2)	2.2 mH
7	(C_1)	3.7 μF
8	(C_2)	984 μF
9	*Ripple*	±5%

$$H_{V_0 D} = \frac{V_o}{D} = \frac{-814.8s^3 + 2.456 \times 10^7 s^2 - 1.232 \times 10^{12} s + 2.145 \times 10^{16}}{s^4 + 149.4s^3 + 4.922 \times 10^8 s^2 + 6.25 \times 10^{10} s + 2.02 \times 10^{14}} \tag{26}$$

The physical parameters of systems are not known exactly because of uncertainties [11]. The causes of uncertainties in any physical system are un-modeled dynamics, perturbation, parametric variations, actuator constraints, sensor noises, etc. In this paper, ±1% parametric variation in each coefficient of numerator and denominator of transfer function is considered. The interval model for Cuk converter by considering uncertainty is defined as:

$$H_{cis} = \frac{\begin{array}{c}-[806.625, 822.948]s^3 + [2.432, 2.48] \times 10^7 s^2 \\ -\ [1.2197, 1.2443] \times 10^{12} s + [2.1236, 2.1665] \times 10^{16}\end{array}}{\begin{array}{c}[0.99, 1.01]s^4 + [147.91, 150.894]s^3 + [4.8727, 4.9712] \times 10^8 s^2 \\ +\ [6.1875, 6.3125] \times 10^{10} s + [1.9998, 2.0402] \times 10^{14}\end{array}} \tag{27}$$

5 Result and Discussion

Figure 4 shows the impulse and step response obtained for constant coefficient modeling of Cuk converter. Figures 5 and 6 show the impulse and step response obtained for interval modeling of Cuk converter for lower bound and upper bound of interval system, respectively. The step and impulse responses show the similar kind of responses when constant coefficient system is modeled as interval system.

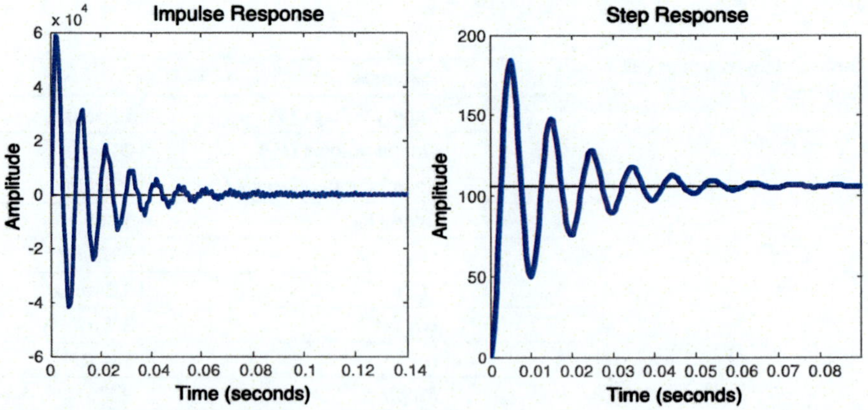

Fig. 4 Impulse and step response of original system

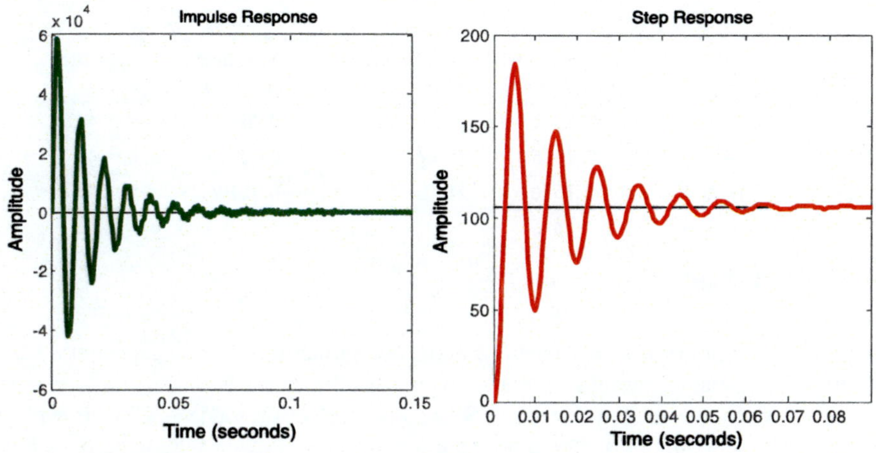

Fig. 5 Impulse and step response of interval system with lower bound

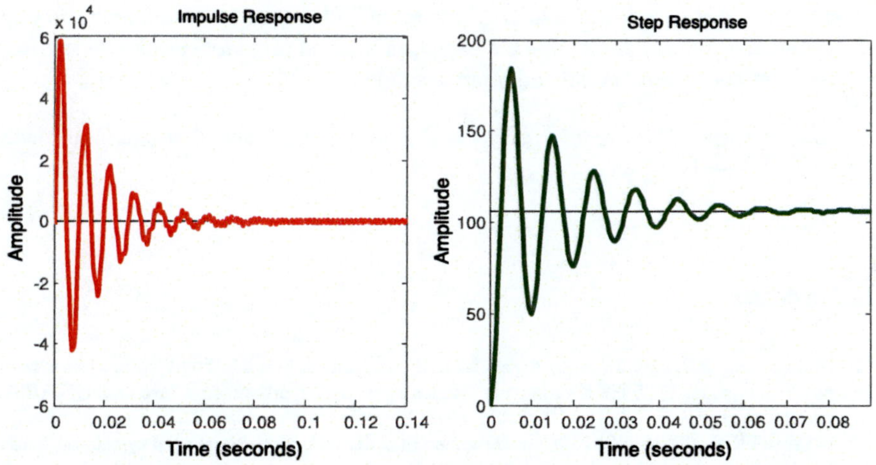

Fig. 6 Impulse and step response of interval system with lower bound

The comparison of characteristics for non-interval and interval models of Cuk converter is shown in Table 2. The characteristics considered for comparison are settling time, rise time, peak time, overshoot and peak value for interval model and constant coefficient model of Cuk converter. It is clear from Table 2 that interval modeling of Cuk converter provides similar kind of characteristics as of original constant coefficient model.

Table 2 Step response characteristics

System	Peak	Overshoot ($\%M_p$)	Settling time (t_s)(sec)	Rise time (t_r)(sec)	Peak time
H_{V_oD}	183.8712	73.1561	0.06	0.00176	0.0049
H_{cisLL}	183.8759	73.1565	0.06	0.00176	0.0049
H_{cisUL}	183.875	73.1556	0.06	0.00176	0.0049

6 Conclusion

The power circuit of Cuk converter contains two inductors and two capacitors. Due to these four energy storage elements, a fourth order model for Cuk converter is obtained. Cuk converter operates in continuous conduction mode (CCM), in which during one switching period of time the inductor current never falls to zero. In this paper, SSA technique is used to obtain combined state space descriptions. The deviations of physical parameters and imperfect modeling create parametric uncertainty in the system parameters. By considering $\pm 1\%$ parametric variation for each coefficient of transfer function, the interval model of Cuk converter is defined. The step and impulse responses depict that the proposed method provides matched responses for constant coefficient model and interval model.

Acknowledgements This work is supported by SERB, DST and Government of India (ECR/2017/000212).

References

1. Middlebrook RD, Cuk S (1976) A general unified approach to modelling switching-converter power stages. In: 1976 IEEE Power Electronics Specialists Conference. Cleveland, OH, USA, pp 18–34. https://doi.org/10.1109/PESC.1976.7072895
2. N Priya TK Sunilkumar 2013 A novel method of order reduction for interval system Int J Adv Res Electr Electron Instriment Eng 2 2278 8875
3. Chodavarapu MM, Shaw B, Singh VP (2020) GWO based controller design for Riverol-Pilipovik water treatment plant. J Stat Manag Syst 23(2):463 470
4. Choudhary AK, Nagar SK (2019) Order reduction in z-domain for interval system using an arithmetic operator. Circuits Syst Signal Process 38:1023–1038
5. Choudhary AM (2018) Shyam Krishna Nagar model order reduction of discrete-time interval systems by differentiation calculus. Aut Control Comp Sci 52(402):411. https://doi.org/10.3103/S0146411618050073
6. Marti JMV (2015) Analysis of duty cycle to output voltage transfer functions of cúk-like class dc-dc converters. Annual Seminar on Automation, Industrial Electronics and Instrumentation
7. Rim CT, Joung GB, Cho GH (1991) Practical switch based state-space modeling of DC-DC converters with all parasitics. IEEE Trans Power Electron 6(4):611–617. https://doi.org/10.1109/63.97759

8. Hsu S, Brown A, Rensink L, Middlebrook RD (1979) Modelling and analysis of switching DC-to-DC converters in constant-frequency current-programmed mode. In: 1979 IEEE power electronics specialists conference. San Diego, CA, USA, pp 284–301. https://doi.org/10.1109/PESC.1979.7081037
9. Kushwaha BK, Narain A (2012) Controller design for Cuk converter using model order reduction. In: 2012 2nd international conference on power, control and embedded systems, Allahabad, India, pp 1–5. https://doi.org/10.1109/ICPCES.2012.6508134
10. AK Prajapati R Prasad 2019 Order reduction of linear dynamic systems by improved routh approximation method IETE J Res 65 5 702 715
11. Matušů R, Prokop R, Korbel J (2009) Description and analysis of interval systems. Parameters 1:2

Investigation of High Voltage Circuit Breakers for Breaking Capacity Based on Simulation Models

V. L. Petrov, I. N. Morozov, N. M. Kuznetsov, and E. L. Soloviev

Abstract This article is devoted to the issue of using simulation modeling to study breaking capacity switches using the MatLab software environment. The basic diagrams of the breaking capacity of high voltage switches are described, a circuit with an oscillatory capacitance discharge, as well as a circuit with an increased breaking current, are proposed. For the proposed circuits, simulation models were compiled and transient graphs were obtained. The corresponding conclusions are drawn.

Keywords High voltage switches · Breaking capacity · Simulation · MatLab

1 Introduction

Modern industrial enterprises must be equipped with effective engineering systems that will ensure the implementation of technological processes in all modes of their operation. Among such engineering systems, electrical complexes associated with the provision of electrical energy to objects of technological processes are the most

V. L. Petrov (✉)
National Research Technological University "MISiS", Leninsky prospect 4, 119049 Moscow, Russian Federation
e-mail: petrovv@misis.ru

I. N. Morozov · N. M. Kuznetsov
Center for Physical and Technical Problems of Energy of the North of the Federal State Budgetary Institution of Science, Kola Scientific Center of the Russian Academy of Sciences, md. Akademgorodok 21A, 184209 Apatity, Murmansk region, Russian Federation
e-mail: moroz.84@mail.ru

N. M. Kuznetsov
e-mail: kuzn55@mail.ru

I. N. Morozov · E. L. Soloviev
Murmansk Arctic State University, st. Kapitan Egorova 15, 183038 Murmansk, Russian Federation
e-mail: moroz.84@mail.ru

A. R. Gupta et al. (eds.), *Power Electronics and High Voltage in Smart Grid*,
Lecture Notes in Electrical Engineering 817,
https://doi.org/10.1007/978-981-16-7393-1_15

important. The success of the enterprise depends on the reliability of the power supply systems and all its elements, which determines the relevance of the task of studying the efficiency of one of the key elements of the power supply systems of industrial enterprises—high voltage switches.

Obviously, manufacturers of high voltage circuit breakers conduct extensive testing to determine all characteristics and performance. At the same time, the actual operating conditions of these devices are quite diverse, the parameters of technological processes and equipment used in some types of industrial enterprises are prone to changes. Mining enterprises can be cited as an example of such industries [1–4]. All this determines the need to develop new approaches to study the characteristics of circuit breakers, the key of which is the ultimate breaking capacity. Using digital models to solve such a problem will allow it to be solved with minimal resource and time costs.

The development of high-tech electric grid mining complexes requires the introduction of modern energy-saving technologies and management of energy efficiency and the quality of electricity in electric grids [5–7]. When designing high voltage power lines, as well as to study the operating modes of individual elements of the network and the power system, it is necessary to perform the calculation of transients, analysis of the stability of electrical networks, analysis of short-circuit modes in networks, analysis of electromagnetic transients [8]. Increasing the efficiency of the power supply systems of mining and processing plants determines the need to improve the trouble-free and reliable operation of high voltage switches [9]. When choosing the parameters and characteristics of the switch, it is necessary to take into account the requirements of normal and emergency operation of the network, its ability to disconnect short-circuit currents [10]. The distortion of the shape of the supply voltage is affected by the operation of semiconductor converters as part of controlled electric drives due to the generation of higher harmonics in the electrical network [11].

The transient recovery voltage of the circuit breaker that occurs when the short circuit is turned off depends on many factors determined by the circuit and the network equipment (inductance, capacitance, active resistances, wave resistance, etc.), as well as the processes that occur when the circuit breaker is opened (arc voltage, residual follow-up conductivity, influence of shunt resistors and capacitors). The use of simulation modeling for the study of circuit breakers for breaking capacity using the MatLab software environment can improve the accuracy of calculations when choosing high voltage circuit breakers.

When designing high voltage power lines, as well as to study the operating modes of individual elements of the network and the power system as a whole, it is necessary to calculate transients, analyze the stability of electrical networks, analyze short-circuit modes in networks, and analyze electromagnetic transients. The construction of high voltage power lines, an increase in the load of the power system determines the need to increase the trouble-free and reliable operation of high voltage switches. When choosing the parameters and characteristics of the circuit breaker, it is necessary to take into account the requirements of normal and emergency operation of the network, its ability to disconnect short-circuit currents. The transient recovery voltage of the circuit breaker that occurs when the short circuit is turned off depends on many factors determined by the circuit and the network equipment (inductance, capacitance, active resistances, wave resistance, etc.), as well as the processes that occur when the circuit breaker is opened (arc voltage, residual follow-up conductivity, influence of shunt resistors and capacitors). The use of simulation modeling for the study of circuit breakers for breaking capacity using the MatLab software environment can improve the accuracy of calculations when choosing high voltage circuit breakers.

2 Analysis of Test Methods for High Voltage Circuit Breakers

For the development of high voltage circuit breakers, it is necessary to carry out a lot of experiments on prototypes, because one of the most important parts of the circuit breaker—the arc chutes cannot be reliably calculated. It also becomes necessary to take into account the requirements that contradict each other in terms of throughput, mechanical characteristics or insulation level.

One of the main tests of circuit breakers is the study of the switching capacity. In this study, circuit breakers are tested to make or break short-circuit currents.

The best performance is shown by network tests of high voltage circuit breakers. The advantage of these tests is that expensive installations are not required in real networks. But there are also serious drawbacks.

In the ultimate power trip test, the circuit breaker shall trip to trip a short-circuit current that can reach several hundred kiloamperes at rated voltage. It is necessary to draw up a circuit in which the test power is maintained. In practice, such tests are difficult because some industrial plants require uninterrupted power supply. In addition, connected large loads during testing may disrupt the normal operation of the power system.

Since the number of tests in the study of the structures of arc suppression devices is large, and the operation of the power system very rarely allows experiments to be carried out, the study of switches is not always possible. It is important to note that when developing circuit breakers included in future grids, the test results in existing grids may turn out to be unreliable, because this power system may lack the power or voltage required for conducting experiments.

To study the breaking capacity of the circuit breakers being developed, which are designed for higher currents than the rated short-circuit currents. in the existing network, laboratory installations are required, which must provide the required shape and magnitude of the curves of the breaking current and recovery voltage.

The simplest schematic diagram of the breaking capacity of high voltage circuit breakers, where a shock generator is used, is shown in Fig. 1. The MatLab software environment, widely used for modeling in electrical engineering, was chosen as an environment for creating simulation models [8, 12–15].

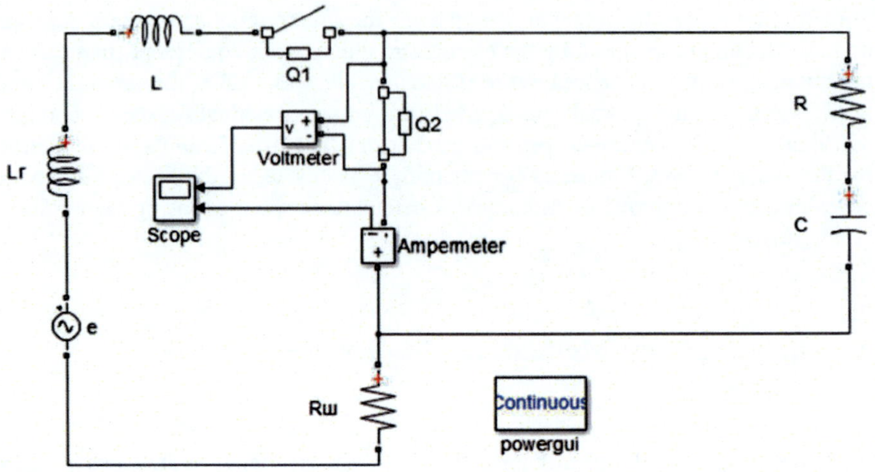

Fig. 1 Schematic diagram of the breaking capacity of high voltage switches in the Matlab environment (*e*—electromotive force of the generator, L_g—generator inductance, R_b—shunt resistance, *Q*—switch contacts, *L*—external inductance, *C*—capacitance, *R*—resistor)

Transient processes obtained as a result of the simulation are shown in Fig. 2.

The rotor of the generator is spun in advance by a low-power asynchronous motor to working speed. Due to this, there is a reserve of kinetic energy. At time t_0, when the contacts of the Q_1 switch are closed, a short-circuit current flows through the contacts of the Q_2 switch. The magnitude of this current, in accordance with the electromotive force of the generator, is determined by the inductance L_g, as well as

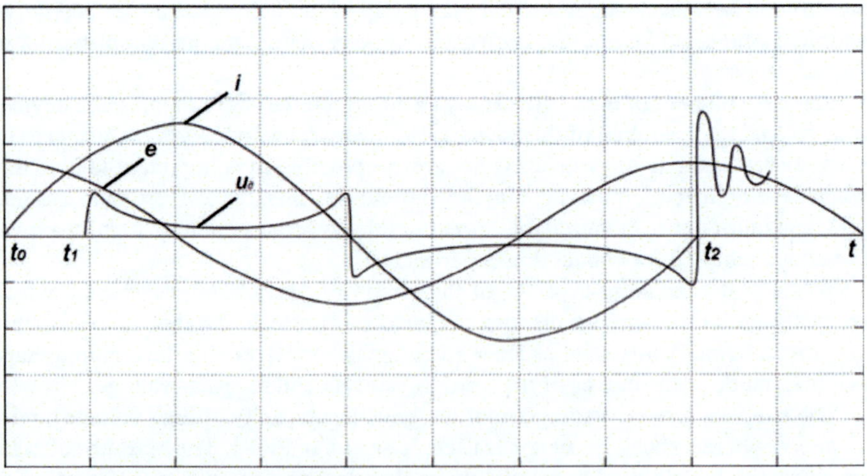

Fig. 2 Transient processes obtained as a result of modeling

by the external inductance L. At the moment t_1, the contacts of Q_2 open, and an arc appears at its ends, the voltage of which is equal to u_d. When the current passes the value of the time t_2 (zero value), the arc is extinguished due to the fact that the contacts of the tested switch are separated by a sufficient distance. As a result, the voltage is restored. How quickly the voltage will recover depends on the above inductances L_g and L, as well as the resistor and capacitance R, C included. The current readings are taken on the oscilloscope using the R_d bridge, and the voltage readings are taken using a voltage divider.

In order to increase the capacity of the installation, parallel connection of additional generators is used, but then the cost increases significantly, as does the complexity of installation and maintenance.

3 Simulation Models for Testing High Voltage Circuit Breakers

To test circuit breakers for breaking capacity, you can use a circuit with an oscillatory discharge of capacitance (a circuit called a Gorev oscillatory circuit (Fig. 3)). When a cluster of capacitors is discharged to a tested circuit breaker, it is possible to obtain the frequency of the oscillatory circuit, as well as the required current amplitude. The voltage waveform is regulated by the $R_0 - C_0$ circuit, and the $R_1 - R_4$ resistors are needed to protect the capacitor bank from destruction if a breakdown occurs in one of them.

Using this scheme, you can create a low-cost, powerful test rig. Operational costs are also reduced. But to obtain the limiting breaking currents, the capacitance of the capacitors must be very large, and since the arc voltage of the tested switch leads to

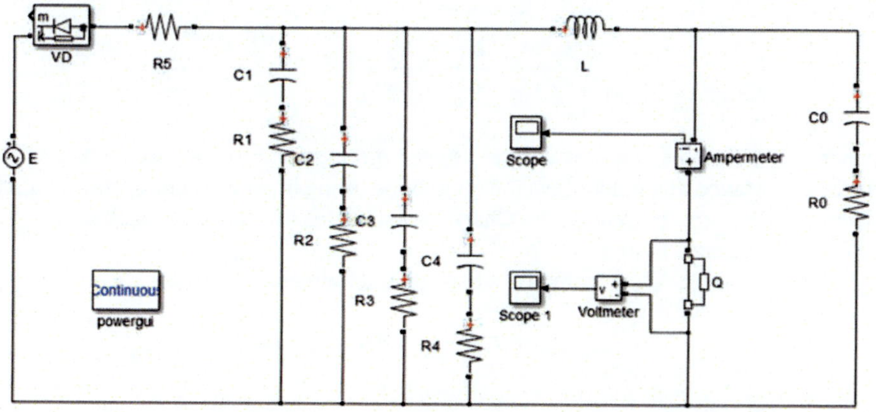

Fig. 3 Circuit with oscillatory capacitance discharge (*E*—electromotive force of the generator, *Q*—switch contacts, *L*—external inductance, *C*—capacitance, *R*—resistor, *VD*—diode)

an increase in the damping of the current passing in the oscillatory circuit, testing the switches for switching capacity becomes not an easy task. Based on the foregoing, a method has recently been developed for synthetic testing of high voltage switches for switching capacity.

As a rule, during the existence of the arc, the voltage between the contacts of the switch can be only a few percent of the nominal. And after the arc is extinguished, high voltage is restored, as a result of which the current in the circuit tends to zero. Then it becomes possible to use two synchronized low-power sources for testing. For example, a source with a rated voltage of 20 kV and a source with a current that is 3–6 times less than the breaking current.

In such a test, the source is a shock generator or a power system network, in which the tripping current is increased by connecting additional step-down transformers (Fig. 4). Inductance L_1 includes the equivalent inductance of the generator, power system, and power transformers. The source is the Gorev oscillatory circuit, in which the frequency is raised to values of 200–300 Hz.

At time t_0, the device Q_1 is turned on, as a result of which current i_1 passes through the contacts of switches Q_2 and Q_3. At time t_1, contacts Q_2 and Q_3 open. At time t_2, the arrester F is triggered, due to which a second current i_2 appears in the circuit $C_2 - L_2 - F - Q_2$. The capacitor cluster C_0 is charged in such a way that the current i_{12} will flow through the switch Q_2, which is equal to the difference between the currents i_1 and i_2. At time t_3, the current in Q_2 passes through zero and the arc is extinguished. But for some time, the switch Q_3, until the current in it goes to zero, remains connected to the serial circuit $L_1 - C_2 - L_2$, which consists of two circuits, the parameters of which are determined by the following expressions:

$$U = U_1 + U_2$$

$$L = L_1 + L_2$$

$$\frac{U_1}{L_1} = \frac{U_2}{L_2}$$

where: U – The highest operating phase voltage specified by the test conditions. L is the inductance that determines the magnitude of the breaker opening current, that is, the short-circuit current. U_2 – Charging voltage of the capacitor bank C_0. U_1 – voltage in the current loop i_1.

Derivatives $\frac{di_1}{dt}, \frac{di_2}{dt}, \frac{di}{dt}$ coincide when approaching zero, because:

$$\left.\frac{di_1}{dt}\right|_{i_1=0} = \frac{U_1}{L_1}$$

$$\left.\frac{di_2}{dt}\right|_{i_2=0} = \frac{U_2}{L_2}$$

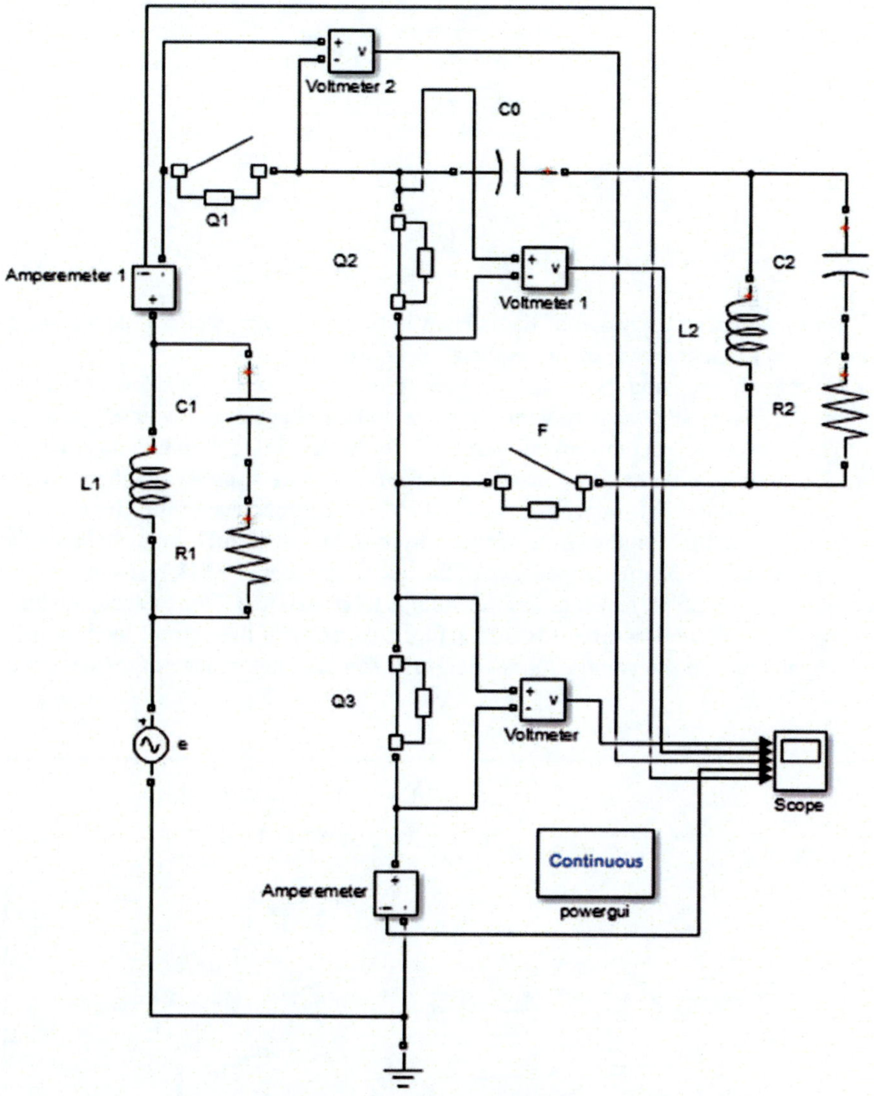

Fig. 4 Circuit with increased shutdown current (*e*—electromotive force of the generator, *Q*—switch contacts, *L*—inductance, *C*—capacitance, *R*—resistor)

$$\left.\frac{di}{dt}\right|_{i=0} = \frac{U_1 + U_2}{L_1 + L_2} = \frac{U_1}{L_1} = \frac{U_2}{L_2}$$

The current I_2 in the high voltage circuit $L_2 - C_0$ will be much less than the trip current. This is due to the expressions:

$$\frac{\omega_2}{\omega_1} = n$$

$$I_2 = \frac{U_2}{\omega \cdot L_2} = \frac{U_2}{n \cdot \omega_1 \cdot L_2}$$

or

$$I_2 = \frac{U_1}{n \cdot \omega \cdot L_2} = \frac{I_1}{n}$$

It turns out that the value of the current in the high voltage circuit decreases in proportion to the increase in its frequency (Fig. 5).

Circuits $R_1 - C_1$ and $R_2 - C_2$, which are connected in parallel to inductors L_1 and L_2, respectively, allow adjusting the slew rate and the shape of the recovery voltage.

The ratings used in the simulation models for testing 110 kV high voltage circuit breakers with 40 kA breaking current were taken as follows: supply voltage frequency —50 Hz; amplitude of supply voltage—20 kV; total inductance of the circuit of the test circuit—5.8 mH; inductance of the current loop —0.48 mH; inductance of the voltage circuit—5.32 mH; impedance of the test circuit loop—13 Ω; resistance of the current loop—9.35 Ω; resistance of the voltage loop—103.6 Ω; the total capacitance of the circuit of the test circuit is 0.25 μF; the capacity of the current loop—3 μF; the capacitance of the voltage loop—0.27 μF; the charging voltage of the capacitor bank —97 V.

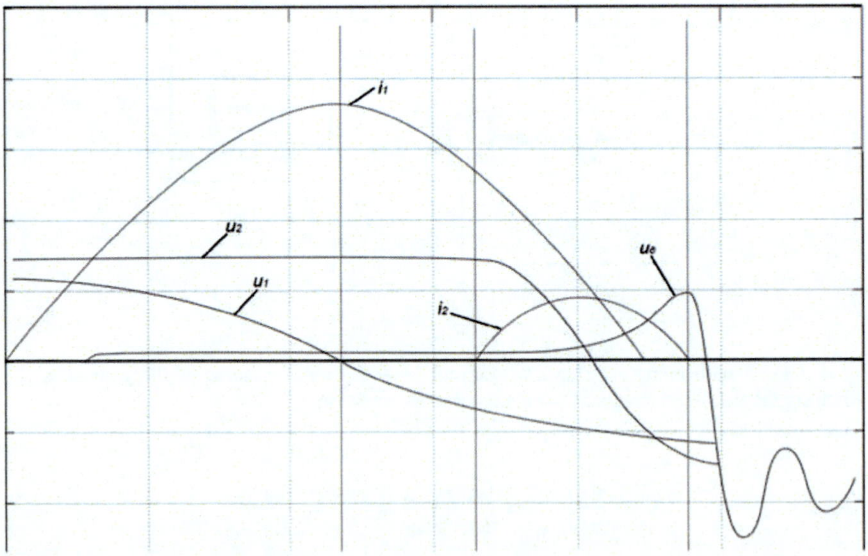

Fig. 5 Transient processes obtained as a result of simulation with increased trip current

Since the processes that determine the arc extinguishing conditions begin to develop several hundred microseconds before the current crosses zero, and the tested switch in the circuit is already in the same conditions as in the full power circuit, the synthetic circuit provides equivalent test conditions.

4 Conclusions

Thus, the work shows the possibility of using simulation modeling to study breakers for breaking capacity using the MatLab software environment. The results of the proposed models coincide with the data obtained during physical tests in the laboratory. In the course of the study, it was found that the use of simulation modeling significantly expands the capabilities of the operation and design of electrical devices. This factor will allow identifying problem areas of devices and, therefore, increasing the likelihood of equipment failure-free operation, as well as increasing other indicators of its reliability. The developed models have been tested on real mining facilities. The methodological support was successfully used in the educational programs for the training of mining engineers of electrical engineering specialization [16, 17].

References

1. Shpiganovich AN, Shpiganovich AA, Zatsepina VI, Zatsepin EP (2017) The problem of power supply system reliability. Gornye nauki i tekhnologii = Min Sci Technol (Russia) (3):47–79. https://doi.org/10.17073/2500-0632-2017-3-47-73
2. Yakonovskaya TB, Zhigulskaya AI (2021) Features of evaluating the economic security of peat industry enterprises in the Tver Region of Russia (the industry review). Min Sci Technol (Russia) 6(1):5–15. https://doi.org/10.17073/2500-0632-2021-1-5-15
3. Sadridinov AB (2020) Analysis of energy performance of heading sets of equipment at a coal mine. Gornye nauki i tekhnologii = Min Sci Technol (Russia) 5(4):367–375. https://doi.org/10.17073/2500-0632-2020-4-367-375
4. Potapova EV (2021) Typology of metro structures for the tasks of geotechnical risk classification. Min Sci Technol (Russia) 6(1):52–60. https://doi.org/10.17073/2500-0632-2021-1-52-60
5. Kuznetsov NM, Minin VA, Selivanov VN (2020) Razvitiye Kol'skoy energosistemy v interesakh gornopromyshlennogo kompleksa Murmanskoy oblasti. Gornyy zhurnal (9):96–100. https://doi.org/10.17580/gzh.2020.09.14
6. Kuznetsov NM (2020) Upravleniye energoeffektivnost'yu v regionakh arkticheskoy zony Rossiyskoy Federatsii. Izdatel'stvo FITS KNTS RAN, Apatity. https://doi.org/10.37614/978.5.91137.434.1
7. Tul'skiy VN, Kartashev II, Simutkin MG, Nazirov KhB, Kuznetsov NM (2012) Upravleniye kachestvom elektroenergii v elektricheskikh setyakh. Gornyy zhurnal 12:52–55
8. Morozov IN, Kuznetsov NM, Belova LA, YeD B (2020) Issledovaniye nestatsionarnykh rezhimov elektricheskoy podstantsii napryazheniyem 110 kV s ispol'zovaniyem imitatsionnogo modelirovaniya v srede MATLAB. Vestnik Chuvashskogo universiteta 1:113–122
9. Yefimov BV, Kuznetsov NM, Veselov AYe, Yaroshevich VV, Tokareva YeA (2008) Povysheniye effektivnosti raboty sistem elektrosnabzheniya gorno-obogatitel'nykh kombinatov. Gornyy informatsionno-analiticheskiy byulleten' (2):353–357

10. Yefimov BV, Kuznetsov NM, Nevretdinov YuM, Fastiy GP, Yaroshevich VV (2008) Dugovyye perenapryazheniya pri perekhode odnofaznykh zamykaniy na zemlyu v dvoynyye. Elektrika 5:8–11
11. Kartashev II, Tul'skiy VN, Kuznetsov NM, Simutkin MG, Nasyrov RR (2012) Otsenka raboty fil'trov v sisteme elektrosnabzheniya gornodobyvayushchego predpriyatiya po rezul'tatam kontrolya kachestva elektroenergii. Gornoye oborudovaniye i elektromekhanika 7:16–19
12. Zagolilo SA, Semenov AS (2020) Modelirovaniye tekhnicheskikh sistem v srede MATLAB: osobennosti razrabotki fizicheskoy modeli v SIMPOWERSYSTEMS. In: Informatsionnyye tekhnologii v elektrotekhnike i elektroenergetike. Materialy XII Vserossiyskoy nauchno-tekhnicheskoy konferentsii. Cheboksary, pp 484–489
13. Morozov IN, Kirillov IYe (2019) Modelirovaniye veroyatnosti perekrytiya girlyandy izolyatorov pri grozovykh udarakh v molniyeotvody. Promyshlennaya energetika (9):10–14
14. Lukutin B, Kadhim KH, Jbarah AN, Karrar O (2020) Energy systems modelling and simulation of behavior for agbwt in isolated network using simulink/MatLab. Test Eng Manage 83(5–6):15245–15249
15. Ike IS, Sigalas I, Iyuke SE (2017) Optimization of design parameters and operating conditions of electrochemical capacitors for high energy and power performance. J Electron Mater 46(3):1692–1713
16. Petrov VL (2016) Federal training and guideline Association on applied geology, mining, oil and gas production and geodesy-A new stage of government, academic community and industry cooperation. Gornyi Zhurnal (9):115–119. https://doi.org/10.17580/gzh.2016.09.23
17. Petrov VL (2017) Training of mineral dressing engineers at Russian Universities. Tsvetnye Metally (7):14–19. https://doi.org/10.17580/tsm.2017.07.02

Planning and Analysis of EV Charging Station with Solar-PV in Grid-Connected System of Durgapur

Aashish Kumar Bohre, Partha Sarathee Bhowmik, Baseem Khan, Tushar Kanti Bera, Irfan Ahmad, and Aniruddha Bhattacharya

Abstract The Electric Vehicles (EVs) are growing very rapidly worldwide as well as in Indian market, therefore the requirement of suitable charging stations is essential. Also, the adoption of EVs including solar system offers numerous benefits for example enhanced cost economy and decreased emission, etc. In this work the design and analysis based on minimum net present cost and cost of energy are presented for case study conducted on NIT Durgapur system. Hence, the proposed system of case study indicates the supplementary profits with respect to economic and technical parameters as compared to base system. Finally, the relative analysis shows that the proposed case study is economic in nature and efficient than current base system considering EV charging station.

Keywords Solar PV · Grid utility · EV fast charging station (EV-FCS) · Electric Vehicles (EVs)

1 Introduction

The renewable energy systems provide a number of advantages for reliable and efficient operation of system/micro-grid. The microgrid is generally found as either a grid-connected micro-grid system or isolated (off-grid) micro-grid system, which are self-sustainable to maintain reliable operation of the system. The off-grid system

A. K. Bohre (✉) · P. S. Bhowmik · T. K. Bera · I. Ahmad · A. Bhattacharya
National Institute of Technology (NIT), Durgapur, WB, India
e-mail: aashishkumar.bohre@ee.nitdgp.ac.in

P. S. Bhowmik
e-mail: psbhowmik@ee.nitdgp.ac.in

I. Ahmad
e-mail: ahmed.irfan@ee.nitdgp.ac.in

B. Khan
Hawassa University, Hawassa, Ethiopia

A. R. Gupta et al. (eds.), *Power Electronics and High Voltage in Smart Grid*,
Lecture Notes in Electrical Engineering 817,
https://doi.org/10.1007/978-981-16-7393-1_16

needs adequate backup power and the energy storage in the system to preserve reliability. The scrutiny of an off-grid hybrid micro-grid system including wind and solar-PV, to obtain optimal operation and configuration of micro-grid system concerning total system cost is discovered in [1]. The analysis and modeling of hybrid isolated system considering techno-economic measures for example LPSP (Loss-of-Power-Supply-Probability), energy cost, and pollutant emissions by particle swarm optimization (PSO) technique is presented in [2]. The valuation of many prospective advantages and incentives accessible through the hybrid renewable system like solar and wind systems is planned in [3]. The modern growth in the techno-socio-economic epoch, incentives, problems, system structure, and development approaches with distinct objectives for isolated and grid-connected hybrid systems are examined in [4], using distinct optimization approaches and also by HOMER-simulation-software in [5]. Optimum planning and analysis of distribution system including distinct kinds of renewable DGs considering multi-objective problems using GA and PSO are explored in [6]. The detail of electricity regulations and prices of Damodar Vally Corporation (DVC) tariffs for energy consumption are given in [7]. The EV charging structure expansion in the interpretation of lower emission to analyze future prospects in UK is presented in [8]. The inclusive reviews on EVs' current status, challenges, scopes, and trends are discovered in [9]. The planning of EVs discharging and charging strategies based on the multi-objective method employed in [10]. The EV station planning based on cost economy, including installation cost, equipment cost, and maintenance & operation cost constraints are examined in [11] using a grey-decision making approach. The optimal size and site planning of EVs charging infrastructure in various systems considering different objectives is suggested in [12, 13].

The review analysis presented above shows that the elated research works presented in the available literatures in this area, concentrate only the scrutiny of off-grid or grid-connected system consisting of solar, wind and battery systems. Also, the renewable energy generation has been utilized as the distributed generation process in the distribution grid and very few literatures considered the impacts of EV charging and EV load profiles with renewable energy systems. Therefore, this work presents a case study to target the research gap in this area, which includes the planning of EV charging station with solar system in the practical grid-connected system of NIT Durgapur (NITD) using HOMER-grid software for analysis and design of the proposed system.

2 The Data for Study Location

The availability of solar irradiation data for the projected location is evaluated through HOMER-grid software. In proposed work study, the NIT-Durgapur, WB of India has been chosen as the location of the case study. The longitude and latitude of the location are 87^0 17.4' E and 23^0 33.3' N. Figure 1 shows yearly average solar irradiance data with the yearly average radiation of 4.824 kW/m^2/day.

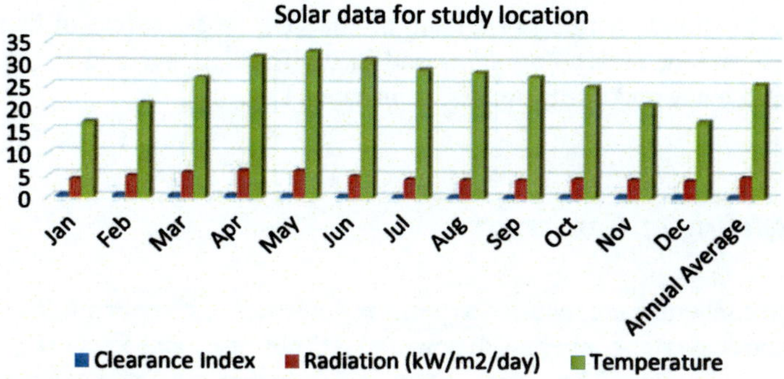

Fig. 1 Annual solar data

3 Modeling of System

The modeling of system components for proposed micro-grid systems is described in this section as:

3.1 Modeling of Solar

The solar-PV modeling based on output power with respect to available irradiation is described below [3–5].

$$P_{PV}^{t} = P_{PV}^{0} \times d_f \times \left(\frac{I_r^t}{I_r^{STC}}\right) \times \left\{1 + \tau_c\left(T^t - T^{STC}\right)\right\} \tag{1}$$

where, P_{PV}^{t} and P_{PV}^{0} are output power and rated capacity of PV. The d_f is derating factor. I_r^{STC} and I_r^t standard irradiation and solar irradiation at instance t. is T^{STC}, T^t and τ_c are standard (at STC 25 °C) temperature, the temperature at instance t and temperature coefficient, respectively.

3.2 Modeling of Battery Storage

The modeling of battery storage based on the SOC (state of charge) under discharging and charging conditions is described by [13]:

$$B_{SOC}^{t} = B_{SOC}^{t-1} \pm \left(\frac{V_{bat}^{t} \times \eta_{bat}}{C_{bat}}\right) \tag{2}$$

where sign ± shows charging (+ve) and discharging (-ve) process of battery. The battery SOCs at time *t* and *t-1* are B^{t}_{SOC} and B^{t-1}_{SOC}. The C_{bat}, η_{bat} and V^{t}_{bat} are battery capacity, efficiency and available SOC at instance t.

3.3 Modeling of Converters

The system converters are required to perform converter or inverter mode operations. The inverter/converter converts DC power to AC power or vice versa as per system operation requirement. The solar system mainly generates DC power; therefore, converter/inverter modeling is required to obtain the AC power at the PCC. Also, the system converters control the output power and frequency of variable frequency sources within the system operating frequency.

3.4 Modeling of Diesel Generator

To maintain the system's reliability and sustainability, the diesel generator is utilized to supply the backup power. The fuel consumption of diesel generator is assumed as [13]:

$$F_{cons} = k_0 W_0 + \mathrm{k}_1 W_{dg} \tag{3}$$

where, k_0 and k_1 are fuel consumption coefficients, W_{dg} and W_0 are output power and capacity of diesel generator.

3.5 Load Modeling

The NIT Durgapur's system load is considered as the commercial load. It has peakdemand of 1,488.52 kW, and energy consumption of 13,830.6kWh per day, 420.7MWh per month, and 5,048.2MWh per annum. The daily load profile of case study (NIT Durgapur) is shown in Fig. 2.

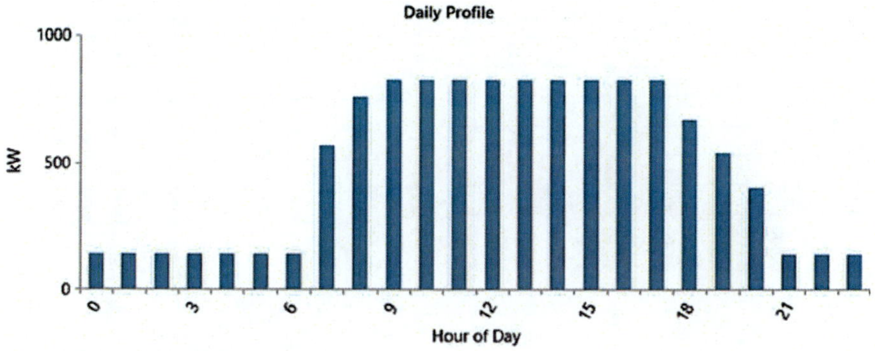

Fig. 2 Case study daily-load profile

3.6 *Electric Vehicle (EV) Fast Charging Station (EVFCS) Characteristics and Performances*

The presented EV fast-charging station's annual energy consumption is 1,60,068 kWh/yr and the peak load is 164 kW. The 22 charging sessions per day are supplied through 4 chargers, each capable of providing 100.0 kW maximum power output. The electric vehicles served by the EV fast-charging station are considered 30% Medium EVs with a maximum capacity of 100 kW and 70% Small EVs with a maximum capacity of 50 kW. The operational parameters for EV fast-charging stations are given in Table 1. The average daily and monthly EV load served by DC fast chargers in the charging station are demonstrated in Fig. 3.

4 Proposed Methodology

In this study, the planned grid-connected system comprises the solar system, battery storage, EV charger for charging station, and the diesel generator presented for the NITD system. The proposed system provides price (per-unit) reduction for energy consumptions. The utilization of fossil fuel and IC engine-based vehicles, huge conventional source-based power plants, and the presence of many industries in

Table 1 Operational parameters of EV fast-charging station

EV-FCS parameters	Values
Annual energy served	160 MWh
Peak load	164 kW
Energy per session	21.7 kWh
Charging sessions per day	20.2
Charging sessions per year	7,365

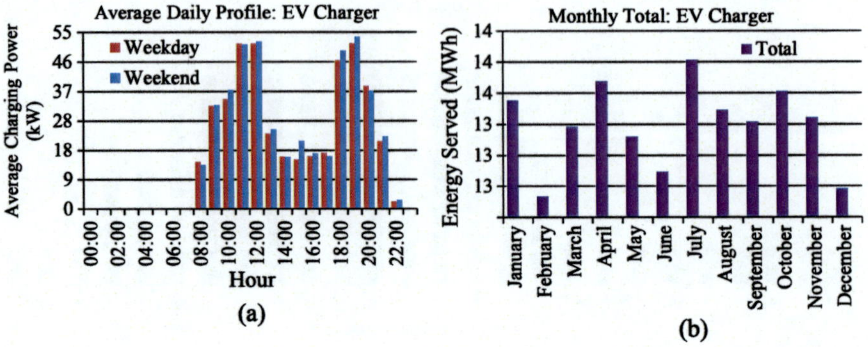

Fig. 3 **a** Average daily and **b** Monthly EV load served by DC fast chargers in charging station

Durgapur are increasing pollution day by day. Therefore the adoption of EVs is targeted by providing the suitable charging stations for EVs in Durgapur. Today, global demand is the green and clean campus infrastructure; therefore, the renewable sources such as solar-PV system is offered with the current NITD-grid-connected system. Thus, the proposed grid-connected hybrid system with EV charging stations is implemented through HOMER-grid software simulations. The minimization of net present cost and COE are achieved as the key objectives from this case study. Also, the different parameters of the projected grid-connected system with EV charging station are also observed which have been presented below.

4.1 The Cost of Energy (COE)

The determination of COE (cost of energy) is essential for cost-effective system analysis incorporating renewable sources. The average COE is given as [8–11]:

$$LCOE = \frac{C_{TNPC}}{\sum_{H=1}^{H=8760} P_{demand}} \times CRF \tag{4}$$

where P_{demand}, CRF and C_{TNPC} are Power consumption, capital recovery factor and net present cost. The CRF can be determined based on rate of interest (r), and project lifetime (n) as:

$$CRF = \frac{r(1+r)^n}{(1+r)^n - 1} \tag{5}$$

4.2 The Pollutant Emission Component

The components of pollutant emissions such as CO, CO_2, NO_x and SO_2 and PM (particular matter) are measured for the proposed system as compared to base system as social parameters. Generally, the electrical power comes to the grid through the thermal power plants, which emit a lot of pollutants. The pollutants containing two components $PM_{2.5}$ and PM_{10}badly affect the human health and the environment. In the present study, the emission factors of different pollutants are assumed as CO_2, SO_2, and NOx are assumed as 632.0 g/kWh, 2.74 g/kWh and 1.34 g/kWh respectively [11].

5 Results and Discussions

The results and their analysis have been conducted on the basis of two case studies. The first case includes present grid-connected system of NIT Durgapur (base case system) with DG (Diesel Generator). Second system includes Solar, Battery, DG (Diesel Generator) including EV charger for charging station with the grid-connected system which is the proposed system for case study.

Current System of NIT Durgapur (Base System): The electrical power requirement of the NIT, Durgapur, are met with a grid connection and 1,500 kW of generator capacity. Currently, the institute (NIT, Durgapur) is spending ₹24.8 M as an utility bill annually.

Proposed System of NIT Durgapur with EV Charging Station: This current study proposes integration of 500 kW solar-PV and 4,299 kWh of battery capacity in the current system with EV charging station. This will decrease yearly utility bills from ₹24.8–₹18.9 M and the investment has a payback of 0.59 years. The connection diagram of proposed system with EV-FCS is shown in Fig. 4a. The monthly utility bills for base/current system and proposed system with EV charging station are demonstrated in Fig. 4b, which clearly shows the monthly utility bills are reduced.

The total annual utility bills for electric energy supplied and the annual savings are tabulated in Table 2, which shows the savings on consumption charge and Demand Charge are ₹2.75 M and ₹3.24 M, respectively. The cumulative cash flow over project lifetime for current and proposed system is shown in Fig. 5. The output performances of system base/current system and proposed system with EV charging station are given in Fig. 6. The emissions for the projected system with EV charging station like CO_2 and PM are 23,92,516 kg/yr & 0.0998 kg/yr and for base/current system are 28,14,147 kg/yr & 0.861 kg/yr correspondingly as given in Table 3. Similarly, the analysis of other social and economic parameters for the current and proposed system is shown in Tables 4, 5, 6 and 7.

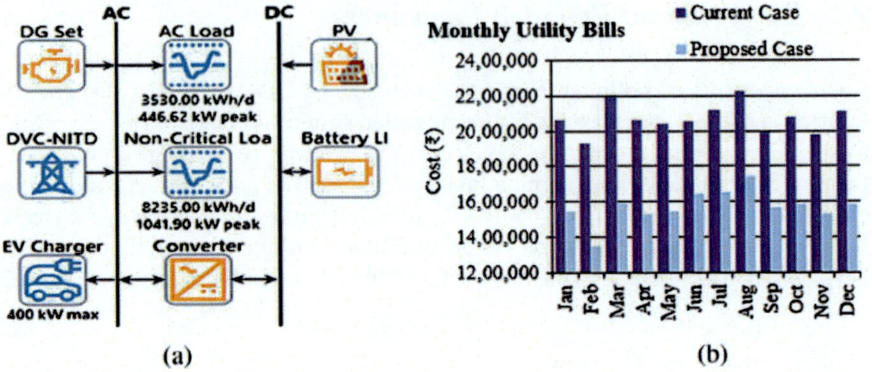

Fig. 4 **a** Connection diagram of proposed system with EV-FCS, **b** Monthly utility bills for base/current system and proposed system with EV charging station

Table 2 Annual utility bills and savings by category

System case/parameters	Consumption charge (M)	Demand charge (M)	Total (M)
Base Case	₹19.0	₹5.87	₹24.8
Proposed Case	₹16.2	₹2.63	₹18.9
Annual Savings	₹2.75	₹3.24	₹5.99

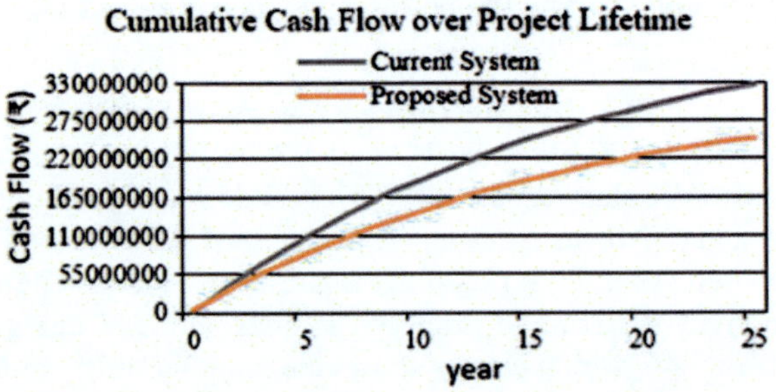

(Current System: - Grid-Connection and Generator. Proposed System:- Grid-Connection, Solar, Generator and Storage with EV charger for charging stations.)

Fig. 5 Cumulative cash flow over project lifetime for current and proposed system

6 Conclusions

This work presents the grid-connected system, including solar, battery, and DG, considering different performance parameters. The result analysis for the base system and proposed system with grid are given by HOMER-grid-software. The present expenditure of the NITD on the annual utility bill is ₹24.8 M, but it's significantly

Fig. 6 Output performance of systems **a** the current/base case, **b** the proposed case with EV charging station

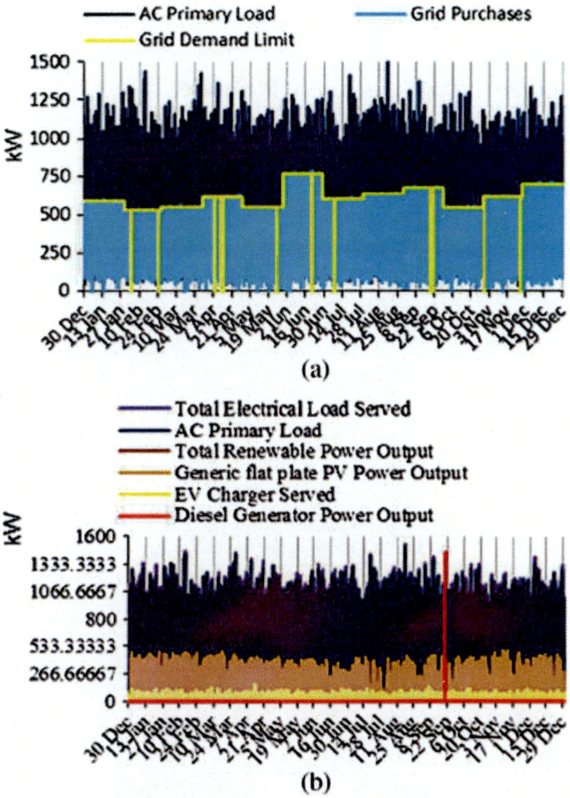

Table 3 Pollutants for the proposed case and base case system

Pollutant emission	Current system (kg/yr)	Proposed system with EV-FCS (kg/yr)
CO	28,14,147	**23,92,516**
CO_2	101	**11.7**
Unburn-hydrocarbon	5.35	**0.62**
PM	0.861	**0.0998**
SO_2	12,164	**10,368**
Nox	5,945	**5,070**

Table 4 Components installation detail for the proposed system

Component	Rating	Capital cost
Solar PV	500 kW	₹10,69,394
Storage battery	4,299 kWh	₹23,64,450
System converter	1172 kW	₹3,51,487
DG set	1500 kW	₹4,50,000

Table 5 Comparision of base case and proposed case system

Parameter	Current/base system	Proposed system with EV-FCS
COE	5.76₹/kWh	**4.36₹/kWh**
Net-present-cost(₹)	₹327,858,300.00	**₹250,104,000.00**
Operating cost(₹)	₹25,326,460.00	**₹19,019,020.00**

Table 6 Utility bill monthly summary—current system

Months	Energy purchase in kWh	Energy sold (kWh)	Net energy purchased (kWh)	Peak load (kW)	Energy charge (₹)	Demand charge (₹)	Total (₹)
January	3,72,002	0	3,72,002	1,339	1.59 M	4,76,026	2.06 M
February	3,29,963	0	3,29,963	1,462	1.41 M	5,19,737	1.93 M
March	3,89,876	0	3,89,876	1,445	1.68 M	5,13,804	2.19 M
April	3,65,853	0	3,65,853	1,376	1.58 M	4,89,298	2.07 M
May	3,70,649	0	3,70,649	1,246	1.60 M	4,42,900	2.04 M
June	3,67,826	0	3,67,826	1,328	1.59 M	4,72,048	2.06 M
July	3,74,661	0	3,74,661	1,427	1.61 M	5,07,348	2.11 M
August	3,92,561	0	3,92,561	1,515	1.68 M	5,38,732	2.22 M
September	3,47,180	0	3,47,180	1,407	1.49 M	5,00,246	1.99 M
October	3,74,139	0	3,74,139	1,332	1.61 M	4,73,639	2.08 M
November	3,60,293	0	3,60,293	1,240	1.54 M	4,41,066	1.98 M
December	3,76,955	0	3,76,955	1,397	1.61 M	4,96,703	2.11 M
Annual	**44,21,958**	**0**	**44,21,958**	**1,515**	**19.0 M**	**5.87 M**	**24.8 M**

reduced to ₹18.9 M for the proposed system, including EVs charging station and solar system. The base system's COE is ₹5.76/kWh, which is reduced to ₹4.36 ₹/kWh for the proposed system with the EV charging station. The base system's operating and net present cost are ₹327,858,300.00 and ₹25,326,460.00, which are reduced to ₹250,104,000.00 and ₹19,019,020.00, respectively for the proposed system including EVs charging station and solar systems. Also, the emission pollutant like CO_2 and SO_2 for base systems are 2814147 kg/yr and 12164 kg/yr, which are decreased to 2392516 kg/yr and 10368 kg/yr respectively. Similarly, the result analysis for net energy purchased, energy sold, demand charges, total annual utility bills, and other pollutants for proposed grid-connected-system by considering EV charring station are significantly reduced with respect to the current/base system.

Table 7 Utility bill monthly summary—proposed system

Months	Energy purchase in kWh	Energy sold (kWh)	Net Energy purchased (kWh)	Peak load (kW)	Energy charge (₹)	Demand charge (₹)	Total (₹)
January	3,12,607	0	3,12,607	1,339	1.33 M	2,09,997	1.54 M
February	2,72,336	0	2,72,336	1,462	1.16 M	1,88,523	1.35 M
March	3,23,677	0	3,23,677	1,445	1.40 M	1,95,754	1.59 M
April	3,03,331	0	3,03,331	1,376	1.31 M	2,18,830	1.53 M
May	3,13,067	1.16	3,13,066	1,246	1.35 M	1,95,383	1.54 M
June	3,18,782	0	3,18,782	1,328	1.37 M	2,73,149	1.65 M
July	3,35,375	0	3,35,375	1,427	1.44 M	2,15,094	1.65 M
August	3,52,614	0	3,52,614	1,515	1.51 M	2,27,194	1.74 M
September	3,09,519	0	3,09,519	1,407	1.33 M	2,40,519	1.57 M
October	3,22,414	0	3,22,414	1,332	1.38 M	1,94,952	1.58 M
November	3,05,702	0	3,05,702	1,240	1.31 M	2,19,765	1.53 M
December	3,12,631	0	3,12,631	1,397	1.33 M	2,48,998	1.58 M
Annual	**37,82,056**	**1.16**	**37,82,054**	**1,515**	**16.2 M**	**2.63 M**	**18.9 M**

References

1. Kumar, Prakash, and Dheeraj Kumar Palwalia (2018) Feasibility study of standalone hybrid wind-PV-battery microgrid operation. Technol Econ Smart Grids Sustain Energy 3(1):1–16
2. Sawle Y, Gupta SC, Bohre AK (2018) Review of hybrid renewable energy systems with comparative analysis of off-grid hybrid system. Renew Sustain Energy Rev 81:2217–2235
3. Babatunde OM, Munda JL, Hamam Y (2020) A comprehensive state-of-the-art survey on hybrid renewable energy system operations and planning. IEEE Access 8:75313–75346
4. HOMER-Energy, HOMER-grid v1.8 user manual, 2020. https://www.homerenergy.com/products/grid/docs/1.8/design.html
5. Go R, Kahrl F, Kolster C (2020) Planning for low-cost renewable energy. Electr J 33(2):106698(1–5)
6. Bohre AK, Agnihotri G, Dubey M (2016) Optimal sizing and sitting of DG with load models using soft computing techniques in practical distribution system. IET Gener Transm Distrib 10(11):2606–2621
7. DVC-Tariff-Order for the years 2015–16 and 2017–18. http://wberc.gov.in/sites/default/files/Tariff%20Order_DVC_2014-2017_pdf%20%281%29.pdf
8. Chen T et al (2020) A review on electric vehicle charging infrastructure development in the UK. J Mod Power Syst Clean Energy 8(2):193–205
9. Arias NB et al (2019) Distribution system services provided by electric vehicles: recent status, challenges, and future prospects. IEEE Trans Intell Transp Syst 20(12):4277–4296
10. Kasturi K, Nayak CK, Nayak MR (2019) Electric vehicles management enabling G2V and V2G in smart distribution system for maximizing profits using MOMVO. Int Trans Electr Energy Syst 29(6):e12013
11. Ren X, Zhang H, Hu R, Qiu Y (2019) Location of electric vehicle charging stations: a perspective using the grey decision-making model. Energy 173:548–553.https://doi.org/10.1016/j.energy.2019.02.015

12. AbuElrub A, Hamed F, Saadeh O (2020) Microgrid integrated electric vehicle charging algorithm with photovoltaic generation. Elsevier J Energy Storage 32(101858):1–11. https://doi.org/10.1016/j.est.2020.101858
13. Bouguerra S, Layeb SB (2019) Determining optimal deployment of electric vehicles charging stations: case of Tunis City, Tunisia. Elsevier Case Stud Transp Policy 7(3):628–642. https://doi.org/10.1016/j.cstp.2019.06.003

Study of Solar MPPT with Multi-level Cascaded Inverter Connected with Pumping System

Manoj Kumar Sahu, Chinmayee Sahoo, Jagan Mohana Rao Malla, and Siddhartha Behera

Abstract The paper presents photovoltaic system and MPPT topology used for centrifugal pump structure. Maximum power transfer from source requires the establishment maximum operating point in the characteristics of power-voltage. Photovoltaic cell parameters are fully nonlinearity characteristics in nature that are responsible for providing only one highest potential point on PV under constant insolation situation. With change in temperature and variation in insolation, the PV characteristics and maximum power point (MPP) are getting change to avail MPP from the array of PV. It can be able to take single input as that of PV curve and produce duty ratio of boost converter to mark the MPP by implementing the algorithm. Further, the water pumping system is employed using two different types of inverter design. The 2-level types along with 3-level type inverters are implemented and the validation is carried out in terms of motor speed and tank volume characteristics. The validation represents the better achievements of reduction of voltage harmonics, better motor speed, higher volume of outflow of water in case of 3-level type inverter as compared with former having same identical rating of voltage and current.

Keywords MPP · PV · Two levels · Three levels

1 Introduction

1.1 A Subsection Sample

Overall use of solar energy is very suitable for the production of electricity in rural areas located far from to the electrical network. Optimal use of solar energy is required

M. K. Sahu (✉) · C. Sahoo · J. M. R. Malla
Centre for Advanced Post Graduate Studies (CAPGS), Biju Patnaik University of Technology (BPUT), Rourkela, Odisha 769015, India
e-mail: capgs.mksahu@bput.ac.in

S. Behera
IGIT Saranga, BPUT, Talcher, Odisha, India

A. R. Gupta et al. (eds.), *Power Electronics and High Voltage in Smart Grid*, Lecture Notes in Electrical Engineering 817,
https://doi.org/10.1007/978-981-16-7393-1_17

to establish the operating point of the photovoltaic converter so as to ensure maximum power [1–5]. Growing power demand, increasing utilization of natural gas and fuels, giving realization to the human of global heating and environment contaminant have drawn observation toward alternating unconventional origin of power. Solar photovoltaic system can be considered available at negligible maintenance [6–8]. The primary major opposition in the solar PV structure is to achieve highest potential through MPPT technique under variation of temperature and solar irradiance [9–12]. Various methods are discussed for achieving MPPT from PV array, e.g. Perturbation and Observation (P&O) algorithm, $V_{oc,}$ I_{sc}, Incremental Conductance, etc. [13–15]. The inverter switches can be controlled and implemented through PWM topology, which controls the pump to work at higher range of torque even at lower speed by maintaining constant v/f ratio even when generator power decreases [16, 17]. The Perturbation and Observation technique is too easy to apply but drawbacks is that it exhibits vibration on every side of the last operating point and may break down below quickly in changeable environmental situations [14].

Application of solar panel for irrigation system is a vital requirement in the agricultural avenue. As, the availability of electricity is not always guaranteed for the remote operation of water pumps, the PV-based inverters may be helpful [9]. Hence, the present study is focused on solar MPPT-based inverter design for water pumping system. However, out of different configurations of inverters, presently, there are both 2-level and 3-level inverters are implemented [5]. Further, the rotational speed of an Induction motor was attached to that inverter and the water level of the tank is presented to compare their performances.

The primary purpose is to design the various models like: Designing a solar model, Designing different DC-DC converters and their analysis, Implementing the MPPT, Designing of inverters for the drive, Designing the water pumping system, Comparing water volume and motor speed between these inverters.

In the present context, as an initial attempt, solar modeling and their I ~ V and P ~ V characteristics are explored under both the symmetrical and asymmetrical irradiance situation. The implementation of DC to DC Converter, i.e. (Boost converter) is made for solar MPPT [11]. Also, the performance analyses about 2-level and 3-level inverters are designed and are integrated with water pumping system.

2 Inverters

There are different shortcomings of inverters, for example, less effectiveness, significant expense and high exchanging losses. AC power is very commonly used in all types of versatile applications, whereas DC power has some limitations of controlling in its higher range. So it is preferred toward research and development of implications of various recent developing stages of MLI topology. To overcome the demerits of different kinds of inverters, the inverters are generally worked in a pulse width modulation technique and they are switched between various circuit topologies, which imply that the inverter is nonlinear. Both the current control and voltage control are utilized in practical applications.

2.1 *Voltage Source Type Inverter*

The schematic chart of a PWM-VSI drive represents in Fig. 1. A PWM inverter regulates not only controls both the recurrence and the magnitude of the voltage output. In this case, a high carrier frequency brings about a basically sinusoidal current (in addition to a superimposed of smaller range of ripple having frequency of higher range) in the engine. Since the wave current through the dc transport capacitor is at the exchanging recurrence, the dc source impedance seen by the inverter would be at higher exchanging frequencies. The dc link capacitor must have the option to convey the wave current. A little capacitance over the diode rectifier additionally brings about a superior information current waveform drawn from the utility source. In any case, care ought to be taken in not letting the voltage swell in the dc voltage, which may become excessively enormous.

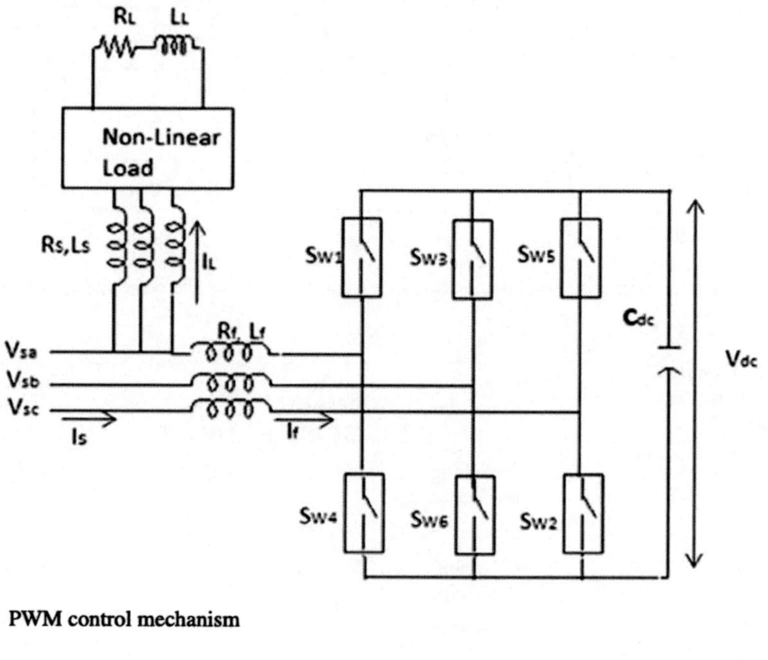

Fig. 1 PWM control mechanism

2.2 *Multilevel Inverter (MLI)*

The industrial demand for MLI's implementation is increasing rapidly due to its various advantages, which out ways the major principal demerits of classical two-level inverter. To get a feature about yield potential difference and electric current curve shape beyond the base measure about wave current, these are needed a large exchanging recurrence alongside different PWM methods. For the high and medium applications, these two-level inverters have a few drawbacks, similar to it working at high frequencies. Figure 2 represents the single leg of phase of an inverter with various levels.

Table 1 represents comparison of various converters of multilevel. Requirements of various items are demonstrated and from that various merits and demerits of preferring the particular MLI can easily be analyzed and selected for particular applications. Depending upon the need of applications and considering various factors, a particular MLI may be selected.

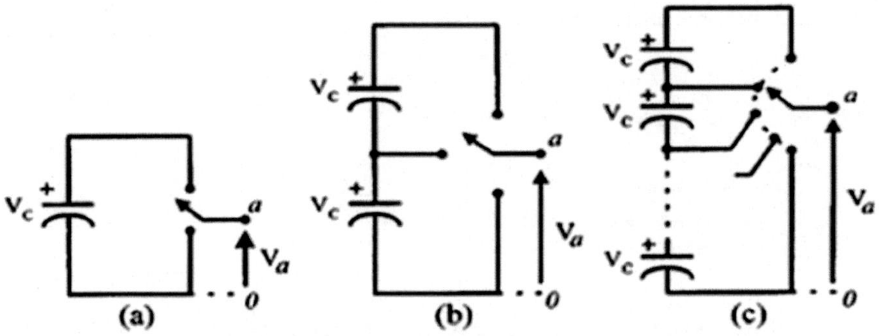

Fig. 2 Single leg of phase of an inverter having **a** 2-level type, **b** 3-level type, and **c** n-level type

Table 1 Comparison of various converters of multilevel

Configuration of converter	Diode clamped (DCMI)	Flying capacitors (FCMI)	Cascaded multilevel (CMI)
Switching devices (nos.)	2(b−1)	2(b−1)	2(b−1)
Diodes (nos.)	2(b−1)	2(b−1)	2(b−1)
Clamping diodes (nos.)	(b−1)(b−2)	0	0
Capacitors of DC_bus	(b−1)	(b−1)	(b−1)/2
Clamping capacitors	0	(b−1)(b−2)/2	0
Total	b^2+2b−3	(b^2+3b−8)/2	(9/2)(b−1)

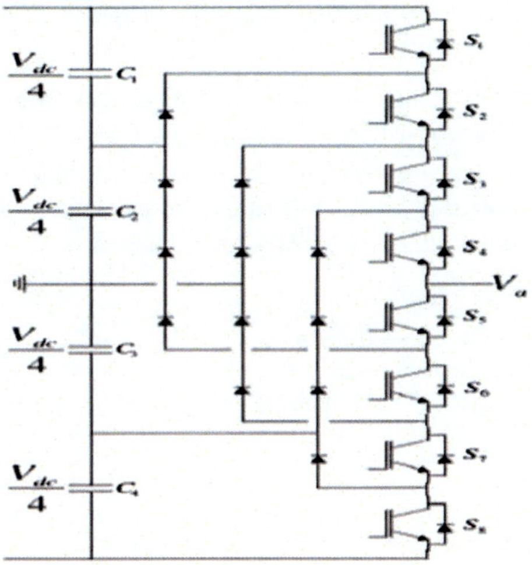

Fig. 3 A 5-level inverter having DCMLI types (of one leg of a phase)

2.3 Diode Clamped Type Multilevel Inverter

It was developed by Takahashi in the year of 1981 [16, 18]. Figure 3 demonstrates the 5-level inverter having DCMLI types (of one leg of a phase).

2.4 Operation of DCMLI

Table 2 represents the various levels of voltage and their relating states of the switching (represented as $A_1{}^1$–$A_8{}^1$). In Table 2, ON states of the switching are mentioned as ‘A’ and similarly the OFF states are mentioned as state ‘B’. For every leg of phase, a lot of four adjoining switches are turned ON at some random time.

Table 2 Levels of voltages of 5-level DCMLI

Voltage_level	$A_1{}^1$	$A_2{}^1$	$A_3{}^1$	$A_4{}^1$	$A_5{}^1$	$A_6{}^1$	$A_7{}^1$	$A_8{}^1$
$V_{dc}/2$	‘A’	‘A’	‘A’	‘A’	‘B’	‘B’	‘B’	‘B’
$V_{dc}/4$	‘A’	‘A’	‘A’	‘A’	‘A’	‘B’	‘B’	‘B’
0	‘B’	‘B’	‘A’	‘A’	‘A’	‘A’	‘B’	‘B’
$-V_{dc}/4$	‘B’	‘B’	‘B’	‘A’	‘A’	‘A’	‘A’	‘B’
$-V_{dc}/2$	‘B’	‘B’	‘B’	‘B’	‘A’	‘A’	‘A’	‘A’

2.5 *Capacitor Type Clamped_Multi_Level_inverter (CCMLI)*

The CCMLI has been developed during the year 1990s. Figure 4 represents the topology of Power circuit of CCMLI (5-level of one phase leg). The description of this figure is self-explanatory from the basic fundamental concept. Table 3 shows a potential switch mix of the voltage levels and their comparing switch states. Similarly, other states can easily be explained.

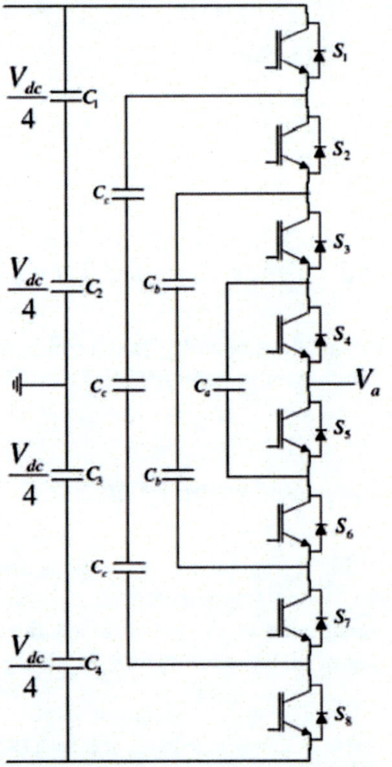

Fig. 4 Representation of topology of power circuit of CCMLI (5-level of one phase leg)

Table 3 Switching states and various levels of voltages of CCMLI (5-level)

Voltage_level	$A_1{}^1$	$A_2{}^1$	$A_3{}^1$	$A_4{}^1$	$A_5{}^1$	$A_6{}^1$	$A_7{}^1$	$A_8{}^1$
$V_{dc}/2$	'A'	'A'	'A'	'A'	'B'	'B'	'B'	'B'
$V_{dc}/4$	'A'	'A'	'A'	'B'	'A'	'B'	'B'	'B'
0	'A'	'A'	'B'	'B'	'A'	'A'	'B'	'B'
$-V_{dc}/4$	'A'	'B'	'B'	'B'	'A'	'A'	'A'	'B'
$-V_{dc}/2$	'B'	'B'	'B'	'B'	'A'	'A'	'A'	'A'

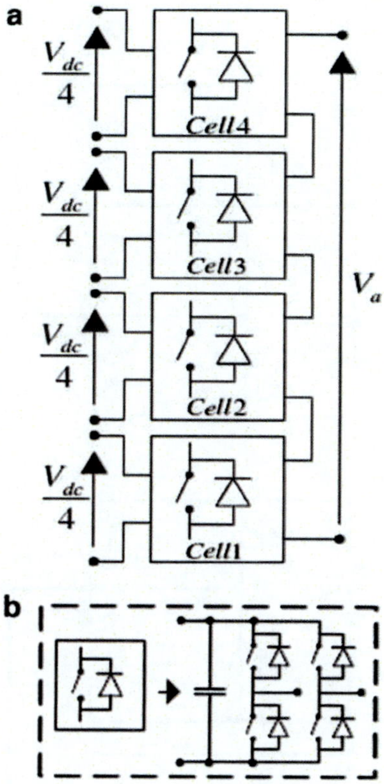

Fig. 5 **a** Single phase leg of structure of a CMI **b** individual H-bridge

The Table 3 represents the switching states and various levels of voltage of CCMLI (5-level). It represents that at output voltage of $V_{dc}/2$ and switching states from $A_1{}^1$ to $A_4{}^1$, the states are in ON states and from the switching states $A_5{}^1$–$A_8{}^1$, states are represented as OFF states. Similarly, other states can be explained (Fig. 5).

2.6 Cascaded Type H-Interface Multilevel_Inverter

Cascaded type Multilevel_Inverter (CMI) was first proposed in the year of 1975. The operation of CMLI is explained in terms of four modes. These are presented in Fig. 6 as (a)–(d), respectively.

Here the levels of output voltage can be represented as mentioned in Table 4. At the levels of voltage of $+V_{dc}$, at switching states of A_1, A_4, A_5, A_6 that states are considered as ON and in the states A_2, A_3, A_7 and A_8 the states are considered as OFF states. Similarly others states can be explained by referring Table 4 (Fig. 7).

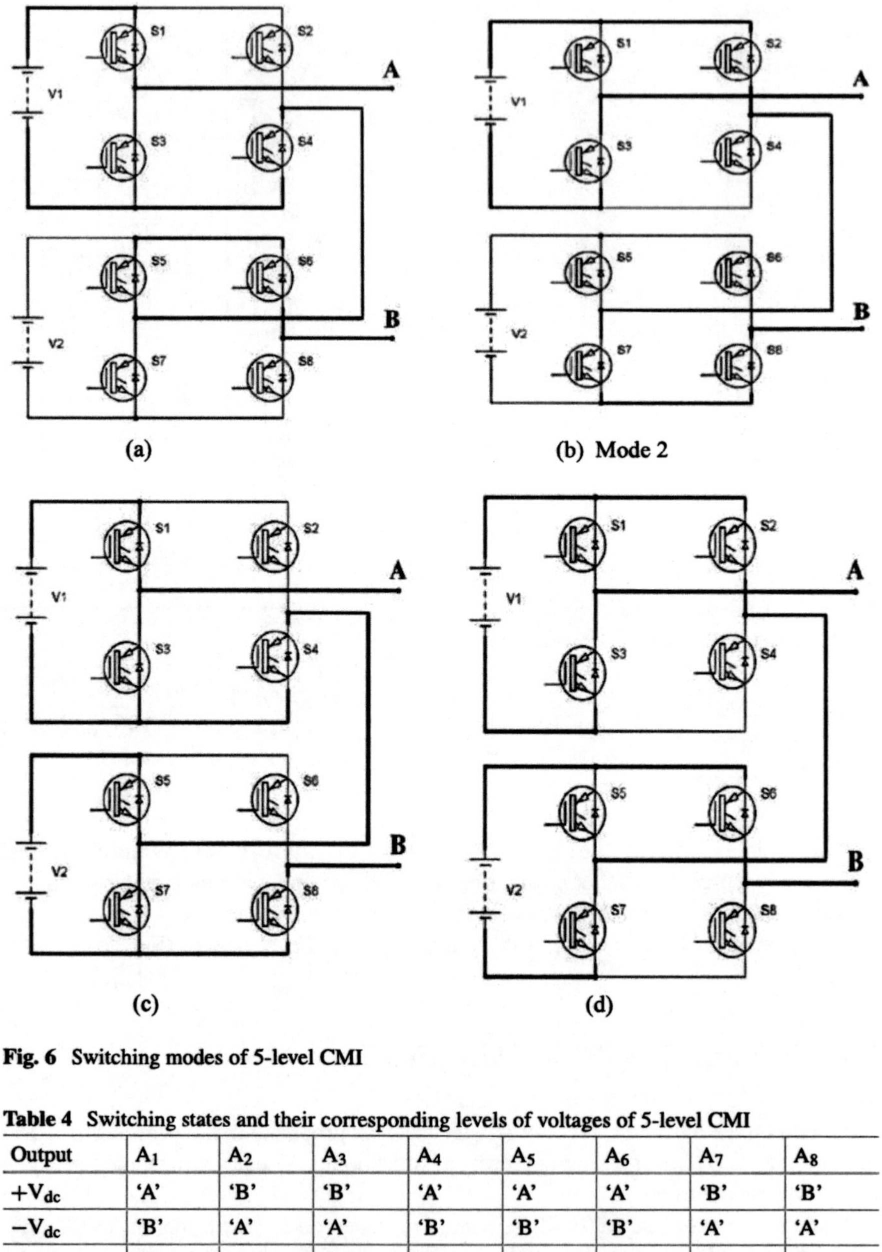

Fig. 6 Switching modes of 5-level CMI

Table 4 Switching states and their corresponding levels of voltages of 5-level CMI

Output	A_1	A_2	A_3	A_4	A_5	A_6	A_7	A_8
$+V_{dc}$	'A'	'B'	'B'	'A'	'A'	'A'	'B'	'B'
$-V_{dc}$	'B'	'A'	'A'	'B'	'B'	'B'	'A'	'A'
0	'B'	'B'	'B'	'B'	'B'	'B'	'B'	'B'
$+2V_{dc}$	'A'	'B'	'B'	'A'	'A'	'B'	'B'	'A'
$-2V_{dc}$	'B'	'A'	'A'	'B'	'B'	'A'	'A'	'B'

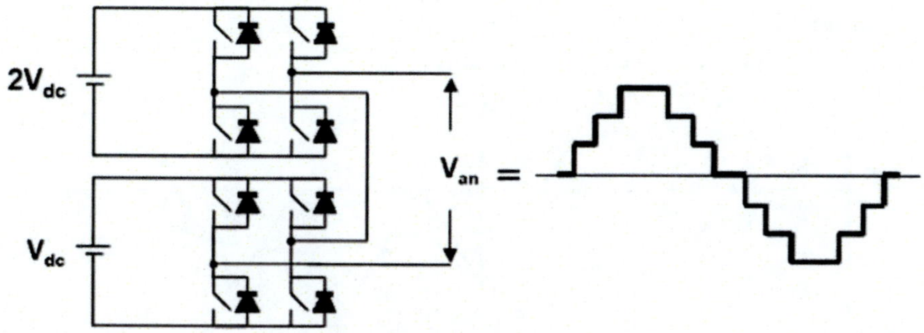

Fig. 7 Asymmetric cascaded H-interface multilevel inverter

3 Controlled Topology of Modulation

3.1 Pulse Width Modulation

PWM is a regulation method, which can change over a signal into a grouping of pulses having constant amplitude, yet the width is corresponding to amplitude of a sound signal. PWM is the method of controlling a computerized signal recreating a simple sign. The on–off conduct changes the normal intensity of sign. Yield signal shifts back and forth among on and off within indicated period. An optional utilization of PWM is to encode data for transmission (Figs. 8, 9 and 10).

Advantages and Disadvantages of SPWM are:

Merits

- Harmonics decreases
- Improved nature of voltage output
- THD reduces

Demerits

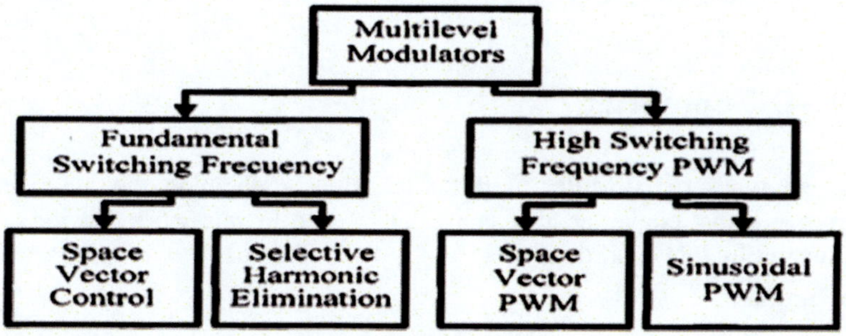

Fig. 8 Classification of multilevel modulation method

Sinusoidal PWM

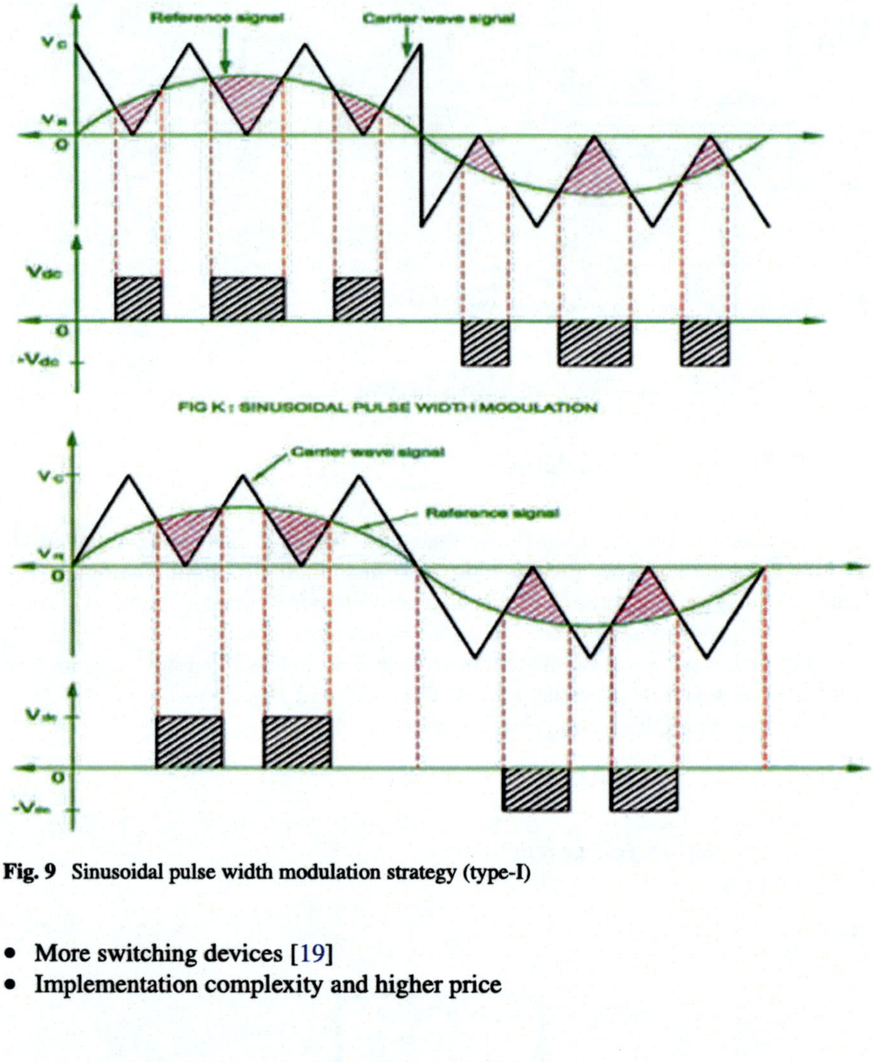

Fig. 9 Sinusoidal pulse width modulation strategy (type-I)

- More switching devices [19]
- Implementation complexity and higher price

4 DC–DC Converter

Among all types of converter used for real-time application the most suitable converter is Boost converter, that converter is used in this work. Power flowing in a converter by this condition:

$$P_{input} = P_{output} + P_{losses} \tag{1}$$

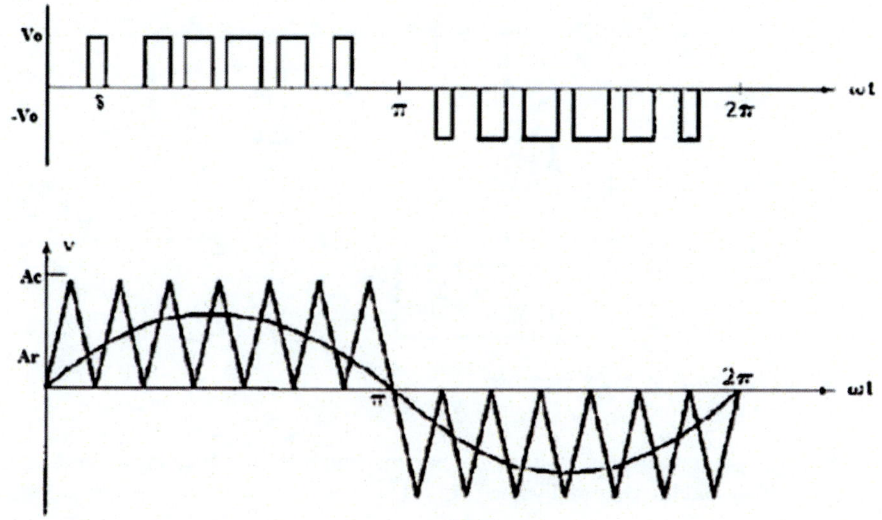

Fig. 10 Sinusoidal pulse width modulation strategy (type-II)

Here P_{input} is the power taken into the converter, P_{output} is the output power and P_{losses} is the power losses in this converter. It can be achieved as:

$$V_{in}I_{in} = V_{out}I_{out} \tag{2}$$

Again by rearranging,

$$\frac{V_{out}}{V_{in}} = \frac{I_{in}}{I_{out}} \tag{3}$$

Efficiency can be computed as:

$$\text{Efficiency} = \frac{P_{out}}{P_{in}} \tag{4}$$

4.1 Boost Type Converter

The Duty Cycle (D) of the boost type converter can be given as [20].

$$D = (1 - \frac{V_{in(\min)}}{V_{out}})efficiency \tag{5}$$

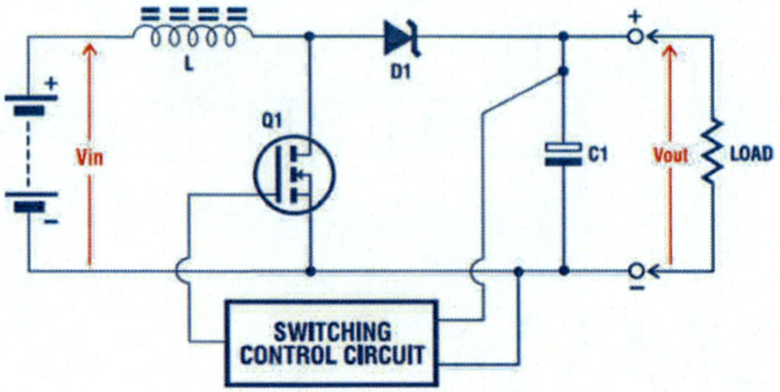

Fig. 11 Boost type of DC to DC converter

where, $V_{in(min)}$ is the input voltage having minimum value, V_{out} is the converter output voltage having required value (Fig. 11).

5 Simulation and Outcomes

5.1 Solar Panel and Its VI Characteristics

Figure 12 represents the single PV panel setup (Figs. 13, 14 and 15).

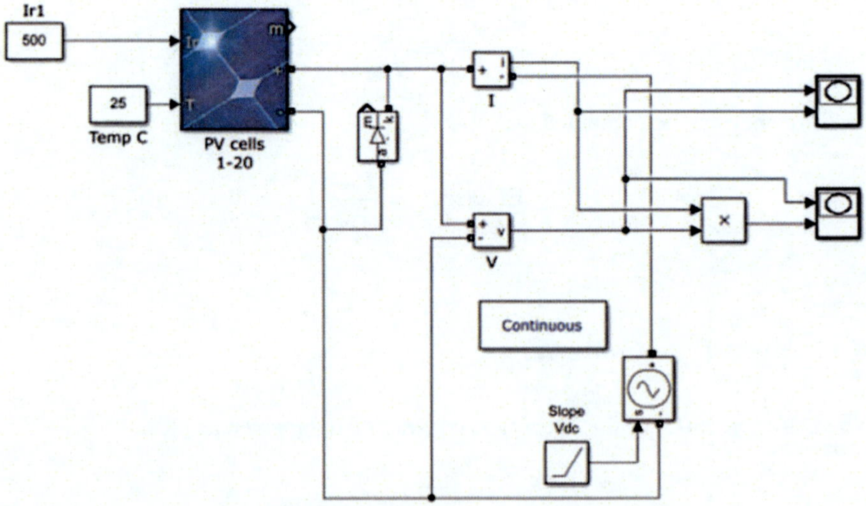

Fig. 12 Single PV panel setup

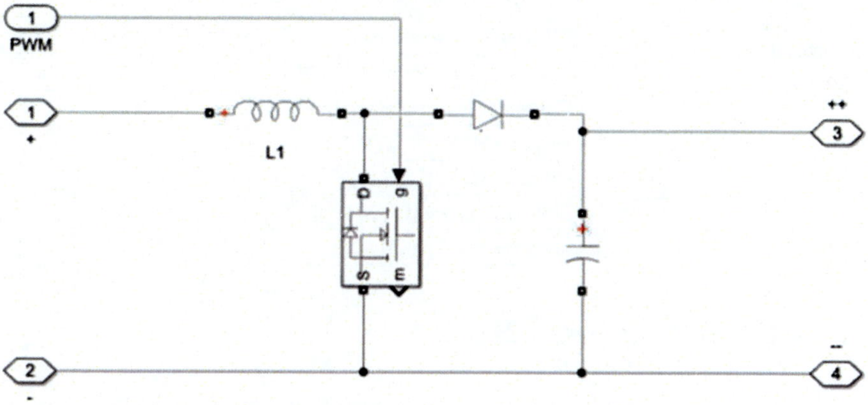

Fig. 13 DC to DC Boost type converter

Fig. 14 Single PV panel VI characteristics

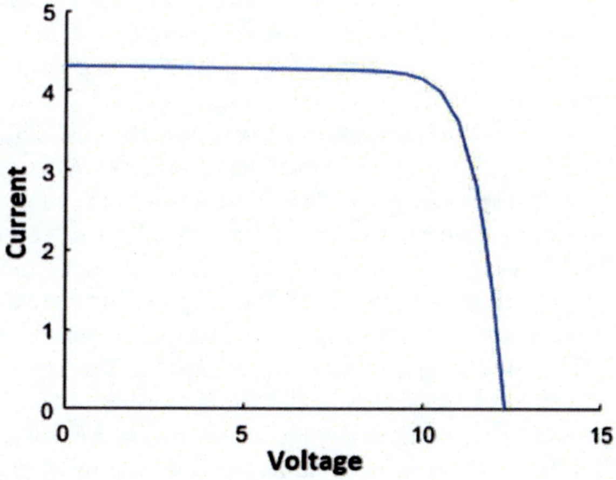

Fig. 15 Single photovoltaic PV characteristics

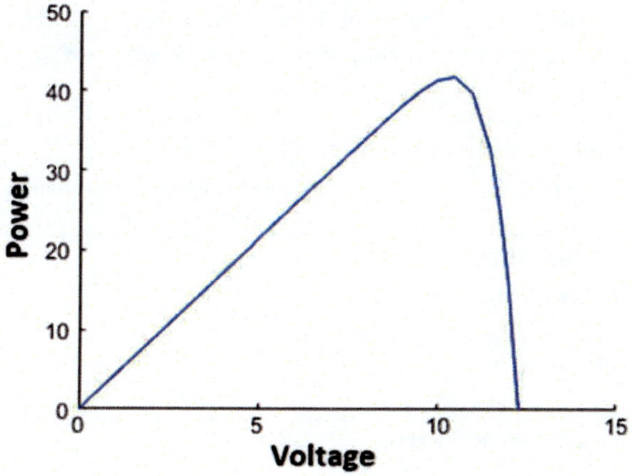

Table 5 PV panel characteristics

PV panel characteristics	Values
Voltage (open_circuit) [V_{oc} (V)]	36.8
Current (short_circuit) [I_{sc} (A)]	3.83
Voltage (MPP) [V_{mp} (V)]	150
Current (MPP) [I_{mp} (A)]	3.1
Solar_irradiance [G (W/m^2)]	1000
Temperature [T^0 (C)]	25

5.2 DC to DC Converter [21]

5.3 Perturb and Observe Algorithm Implementation

At present day the common and popular method is P&O method, which can be used for tracking optimum power. In this method, the sensing component is only voltage. Power output component of the system is verified by changing the supply voltage (Table 5).

The simulation work is carried out using MATLAB/Simulink environment at standard irradiance and temperature of 1000 W/m^2 and 25 °C, respectively, at an open circuit voltage of 36.8 V and short circuit current of 3.83A. The simulation results are mentioned in Table 6. The P&O algorithm implementation using Simulink is presented in Fig. 16. Cascaded-MLI simulation is presented in Fig. 17. Figure 18 represents the simulation of pumping system. Solar-based irrigation system is presented in Fig. 19. Similarly, the waveforms of voltage and current of DC to DC Boost converter are shown in Fig. 20. The signal of PWM Multilevel inverter of phase A is presented in Fig. 21. The other phases can also be implemented. The phase to phase output voltage waveforms of 3-level inverter is presented in Fig. 22. Discharge of water volume in terms of Gallon with respect to time in seconds is presented in Fig. 23. It is observed that within 1 s of duration, 8 gallons of volume of water discharged by using 2-level type inverter. But in case of 3-level type inverter, 27 gallons of water discharged by 3-level inverter within 1 s duration. The comparison of

Table 6 Simulation results [15, 22]

Parameter	Converter PV pumping system with 2-level inverter	Proposed PV pumping system with cascaded 3-level inverter
Voltage (THD in %)	13	7
Motor speed (rpm)	1170	1290
Water volume (Gal) over 1 s simulation	8	27
Output voltage (PV) (V)	145	145
Output current (PV) (A)	2.7	2.7

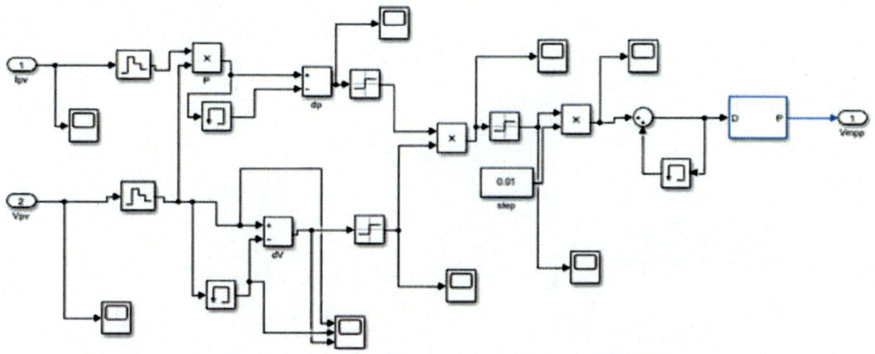

Fig. 16 P&O MPPT algorithm implementation using simulink

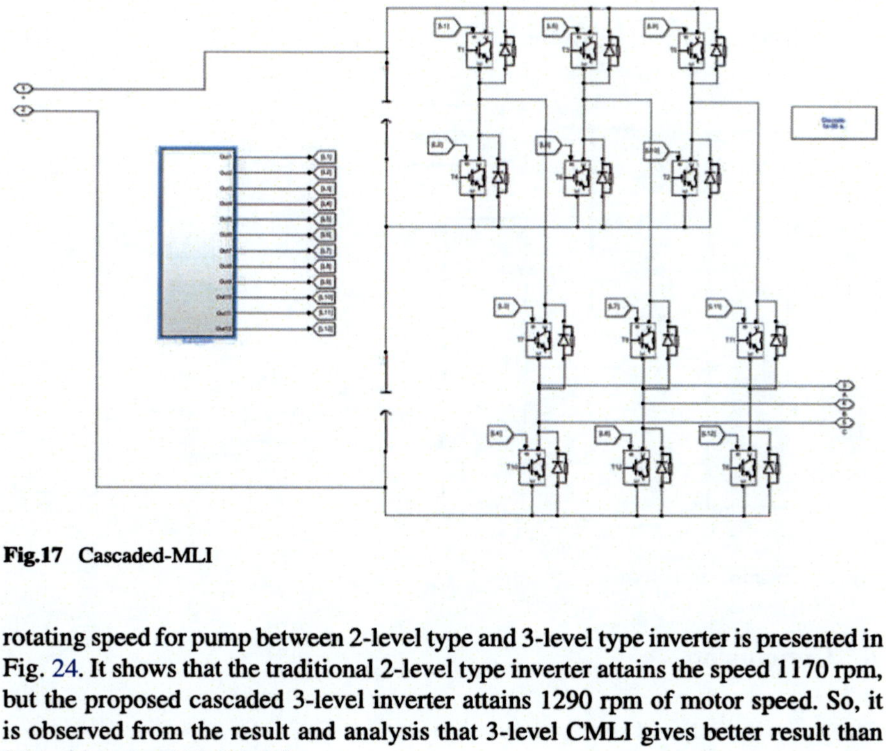

Fig.17 Cascaded-MLI

rotating speed for pump between 2-level type and 3-level type inverter is presented in Fig. 24. It shows that the traditional 2-level type inverter attains the speed 1170 rpm, but the proposed cascaded 3-level inverter attains 1290 rpm of motor speed. So, it is observed from the result and analysis that 3-level CMLI gives better result than 2-level inverter [15, 22, 23].

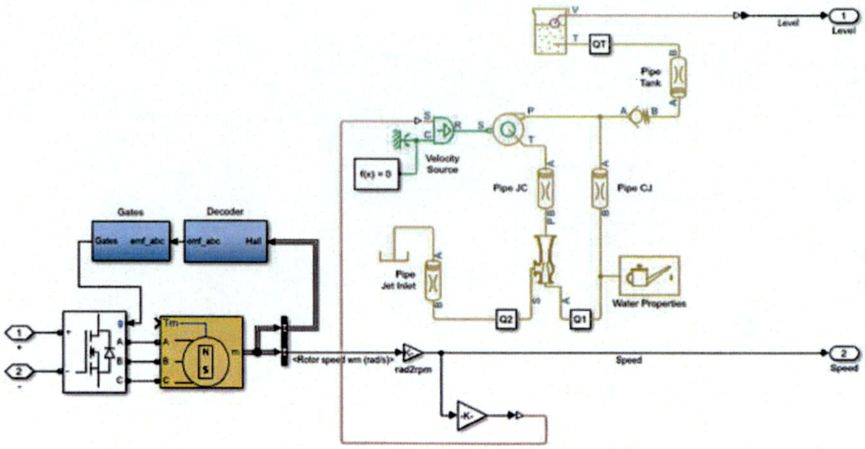

Fig. 18 Simulation of pumping system

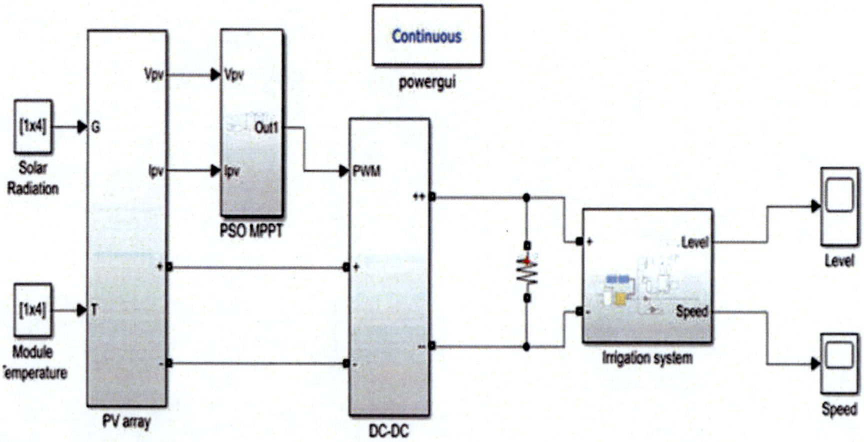

Fig. 19 Solar irrigation system

6 Conclusion

The analysis, design of solar panel is carried out to find their respective VI characteristics. DC-DC boost converter is implemented with P&O-based MPPT approach. 2-level and 3-level inverters are designed for driving the water pump for irrigation. The results are compared with the help of water volume and motor speed characteristics. So the three-level inverter is better suited in comparison to the other counterpart [15, 22]. However, further analysis on other multilevel topologies can be aimed as a future extension of the research work.

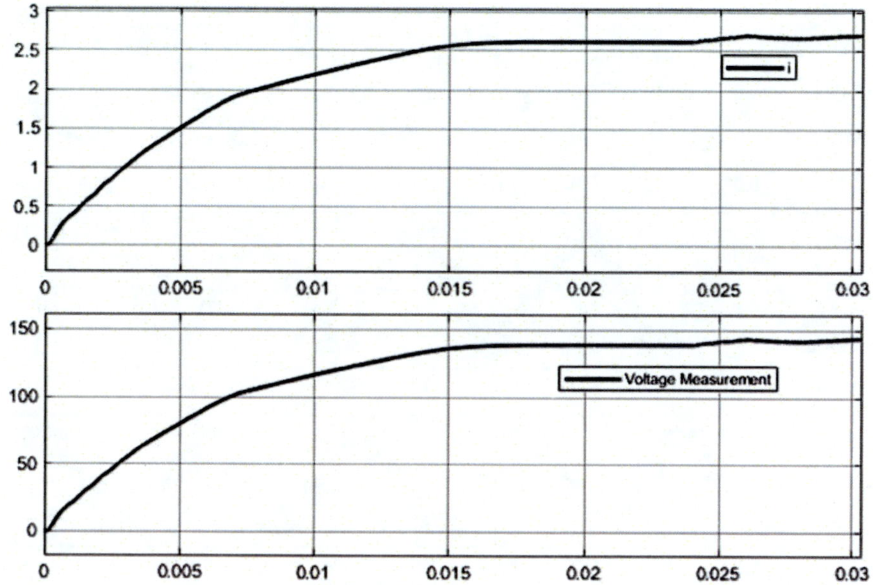

Fig. 20 DC-DC boost converter VI waveform

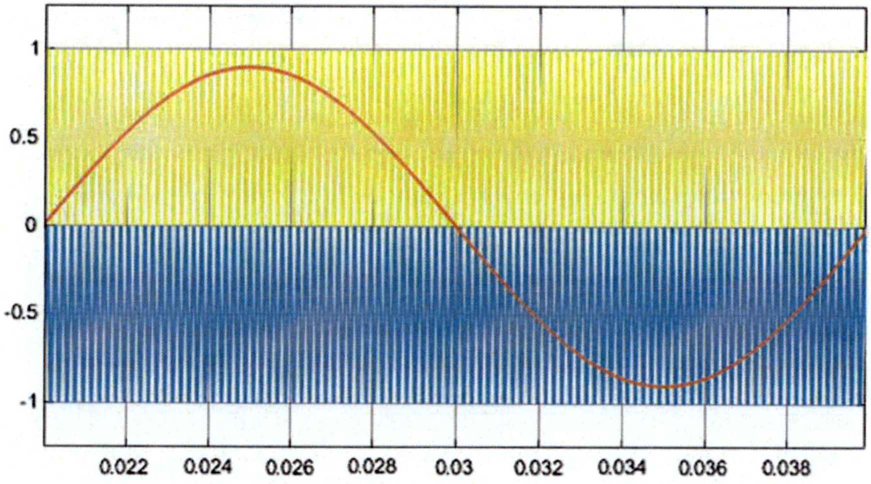

Fig. 21 Multilevel inverter PWM for phase A

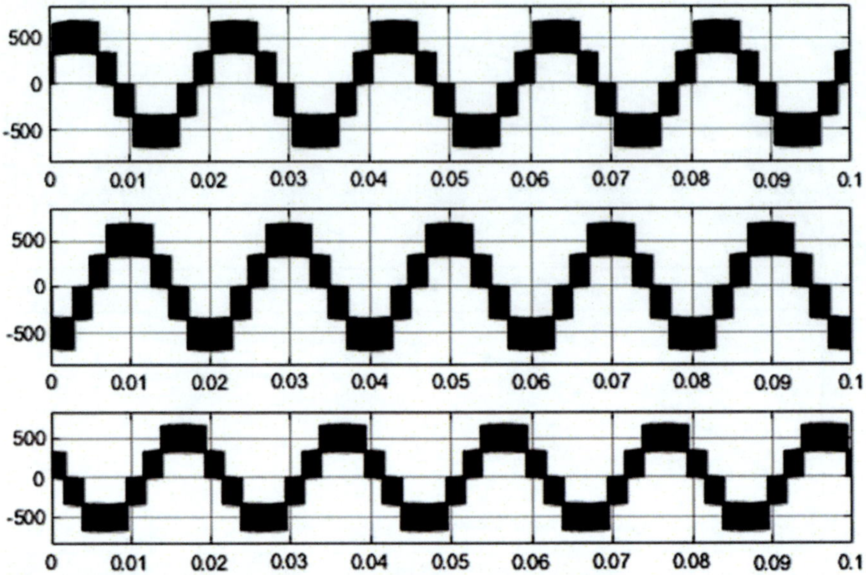

Fig. 22 Output waveform for the three-level inverter (phase to phase)

Fig. 23 Water volume comparison with 2-level type and 3-level type inverter [22]

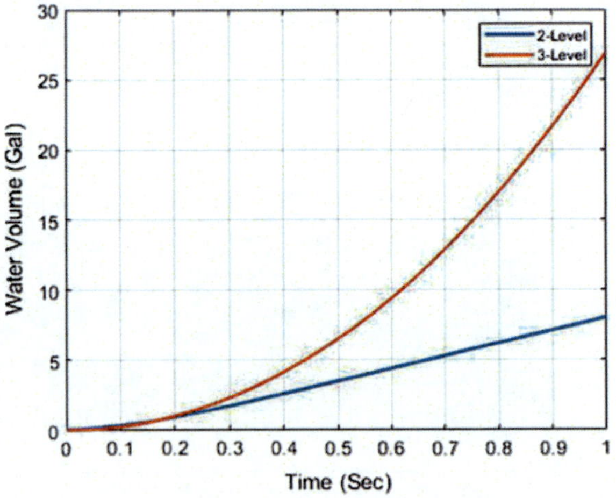

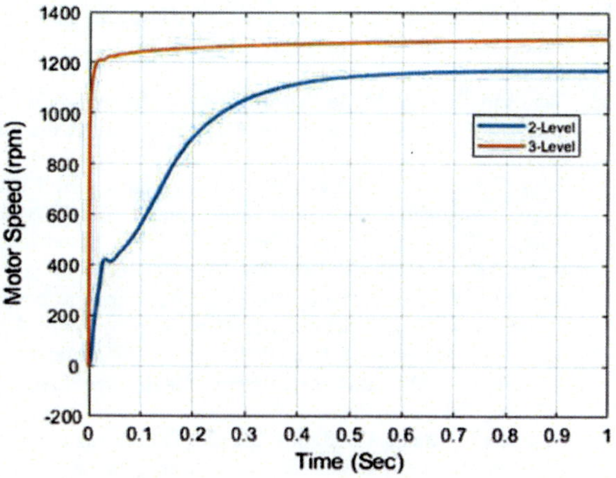

Fig. 24 Pump rotating speed comparison between 2-level type and 3-level type inverter [22]

References

1. Murshid S, Singh B (2019) Analysis and control of weak grid interfaced autonomous solar water pumping system for industrial and commercial applications. IEEE Trans Ind Appl 55(6):7207–7218. https://doi.org/10.1109/TIA.2019.2939705
2. Mishra AK, Singh B (2020) An efficient control scheme of self reliant solar-powered water pumping system using a three-level DC-DC converter. IEEE J Emerg Sel Top Power Electron 8(4):3669–3681. https://doi.org/10.1109/JESTPE.2019.2943203
3. Mudlapur A, Ramana VV, Damodaran RV, Balasubramanian V, Mishra S (2018) Effect of partial shading on PV fed induction motor water pumping systems. IEEE Trans Energy Convers 34(1):530–539. https://doi.org/10.1109/TEC.2018.2876132
4. Shukla S, Singh B (2019) Reduced current sensor based solar PV fed induction motor drive for water pumping. IEEE Trans Ind Inform 15(7):3973–3986. https://doi.org/10.1109/TII.2018.2885795
5. Mishra AK, Singh B (2019) High gain single ended primary inductor converter with ripple free input current for solar powered water pumping system utilizing cost-effective maximum power point tracking technique. IEEE Trans Ind Appl 55(6):6332–6343. https://doi.org/10.1109/TIA.2019.2929012
6. Rai R, Shukla S, Singh B (2020) Reactive power based MRAS for speed estimation of solar fed induction motor with improved feedback linearization for water pumping. IEEE Trans Ind Inform 16(7):4714–4725. https://doi.org/10.1109/TII.2019.2950094
7. Mishra AK, Singh B (2021) An improved control technique for grid interactive 4-phase SRM driven solar powered WPS using three-level boost converter. IEEE Trans Ind Inform 17(1):290–299. https://doi.org/10.1109/TII.2019.2957498
8. Shukla S, Singh B (2019) Reduced sensor based PV array-fed direct torque control induction motor drive for water pumping. IEEE Trans Power Electron 34(6):5400–5415. https://doi.org/10.1109/TPEL.2018.2868509
9. Rezk H, Al-Dhaifallah M, Hassan YB, Ziedan HA (2020) Optimization and energy management of hybrid photovoltaic-diesel-battery system to pump and desalinate water at isolated regions. IEEE Access 8:102512–102529. https://doi.org/10.1109/ACCESS.2020.2998720
10. Alghuwainem SM (1996) Speed control of a PV powered dc motor driving a self-excited 3-phase induction generator for maximum utilization efficiency. IEEE Trans Energy Convers 11(4):768–773. https://doi.org/10.1109/60.556377
11. Mishra AK, Singh B (2017) Solar photovoltaic array dependent dual output converter based water pumping using switched reluctance motor drive. IEEE Trans Ind Appl 53(6):5615–5623

12. Shukla S, Singh B (2018) Single-stage PV array fed speed sensorless vector control of induction motor drive for water pumping. IEEE Trans Ind Appl 54(4):3575–3585. https://doi.org/10.1109/TIA.2017.2732341
13. Sahu MK, Hota S, Malla JMR, Panda AK, Behera S (2019) Simulation of MPPT based standalone PV module design using GWO and PSO technique. Int J Innov Technol Explor Eng 8(11):2754–2759
14. Hota S, Sahu MK, Malla JMR (2017) A standalone PV system with a hybrid P&O MPPT optimization technique. Eng Technol Appl Sci Res ETASR 7(6):2109–2112
15. Malla JMR, Sahu MK, Subudhi PK (2018) PV based water pumping system for agricultural sector. J Elsevier Mater Today Proc 5:1008–1016
16. Sahu MK, Panda AK, Panigrahi BP (2012) Direct torque control for three-level neutral point clamped inverter-fed induction motor drive. Eng Technol Appl Sci Res 2:201–208
17. Sahu MK, Biswal M, Malla JMR (2020) THD analysis of a seven, nine, and eleven level cascaded H-bridge multilevel inverter for different loads. TEHNIČKI GLASNIK 14(4):513–522
18. Susheela N, Kumar PS, Reddy CH (2016) Performance analysis of four level NPC and NNPC inverters using capacitor voltage balancing method. In: 2016 IEEE Uttar Pradesh section international conference on electrical, computer and electronics engineering (UPCON), India
19. Carvalho AS (1990) Optimal PWM control strategy for an inverter drive in closed loop operation. In: IEEE fifth annual proceedings on applied power electronics conference and exposition, USA
20. Gomez JLD, Cervantes EG, Flores DRL, Enjeti PN, Palma L (2006) Analysis and evaluation of a series-combined connected boost and buck-boost DC-DC converter for photovoltaic application. In: Twenty first annual IEEE applied power electronics conference and exposition, 2006. APEC '06, USA
21. Sahbani A (2019) Chapter 42 single-phase grid connected photovoltaic system using fuzzy logic controller. Springer Science and Business Media LLC (2019)
22. Ramulu C, Sanjeevkumar P, Karampuri R, Jain S, Ertas AH, Fedak V (2016) A solar PV water pumping solution using a three-level cascaded inverter connected induction motor drive. Eng Sci Technol Int J Elsevier 19(4):1731–1741
23. Ronaki D, Perumal P (2016) A SVPWM for reduction in common mode and bearing currents applied to diode clamped three level inverter fed induction motor. In: 2016 IEEE international power electronics and motion control conference (PEMC), Bulgaria

Modelling and Analysis of Indirect Field-Oriented Vector Control of Induction Motor (IM)

Tanzeel Imtiyaz, Anupama Prakash, Farhad Ilahi Bakhsh, and Anjali Jain

Abstract High dynamic performance which is a characteristic feature of a dc motor is achievable in ac Induction Motor (IM) through advancements in different power electronic devices and development in various vector control strategies. The presented work is based on field-oriented vector-control of IM drive. For developing vector-controlled IM drive, first the IM is modelled in MATLAB/Simulink using space vector equivalent circuit in static reference frame. Then indirect field-oriented vector control (VC) is implemented in the developed IM model using rotating reference frame. In indirect field-oriented VC the current coming from the stator of the IM is decoupled into two orthogonal components of a vector where one component represents the magnetic flux of IM and the other component represents the torque. The analysis shows that by using indirect field-oriented VC the speed and torque of IM can be independently controlled.

Keywords Induction Motor (IM) · Vector control (VC) · Per-unit (pu) values · PI controllers · MATLAB/Simulink

Nomenclature

Us	Voltage of stator
Is	Current of stator
Ur	Voltage of rotor
Ir	Current of rotor
Rs	Stator winding resistance

T. Imtiyaz (✉) · A. Prakash · A. Jain
Department of Electrical Engineering, Amity University, Noida, UP 201313, India

A. Jain
e-mail: ajain5@amity.edu

F. I. Bakhsh
Department of Electrical Engineering, NIT Srinagar, Hazratbal, Srinagar, J&K 190006, India

A. R. Gupta et al. (eds.), *Power Electronics and High Voltage in Smart Grid*,
Lecture Notes in Electrical Engineering 817,
https://doi.org/10.1007/978-981-16-7393-1_18

Rr	Rotor winding resistance
Lm	Mutual inductance
$L_{s\sigma}$	Stator winding leakage inductance
$L_{r\sigma}$	Rotor winding leakage inductance
Ls	Inductance of stator winding
Lr	Inductance of rotor winding
Us (α, ß)	Stator voltage α,ß components
Ur (α, ß)	Rotor voltage α,ß components
Is (α, ß)	Stator current α,ß components
Ir (α, ß)	Rotor current α,ß components
ψ_s (α, ß)	Stator flux linkage α,ß components
ψ_r (α, ß)	Rotor flux linkage α,ß components
Uu (rated)	Motor rated voltage
I (rated)	Motor rated current
Drated	Frictional coefficient
Jrated	Rotating inertia coefficient
T	Electro-magnetic torque
TL	Load torque
ψ_s	Motor slip
ψ_r	Rotor speed
Is (d, q)	Stator phasor currents in d-q coordinates
ψ_{rd}	Rotor flux linkage in d co-ordinate
Ω	Mechanical angular velocity
τ	Time constant of rotor
Ki	Integral gain
Kp	Proportional gain

1 Introduction

The electrical dc drives are still extensively used in various industrial applications where variable speed operations are needed as they are simple, precise and have easy control techniques. But now-a-days ac drives find a place instead of dc drives as the former being more reliable. The ac drives are more expensive and have complex strategies. To decrease the cost of IM, the construction of IM is realized without the sliprings or brushes. This makes the motor relatively cheap and robust [1]. Also, IM can be used relatively for longer period of time without the need of maintenance. Thus, Induction Motor is used as an execution component for the development of ac electrical drives for various operations such as starting, braking, speed control, etc. The overall system performance is considerably dependent on the performance of the IM drive system.

The design strategies for ac drives are now-a-days very important topic for researchers. To design any control strategy for IM following steps are needed:

- To make a mathematical model representing the drive system for analysis purpose.
- To design an optimal regulator in order to get better response of the drive system in case of external perturbations.

There are two control strategies for IM.

- Analogue control: In this control, the machine parameters like the speed of rotor is directly measured and then compared with the reference.
- Digital control: In this control estimation of machine parameters is done using sensor less control schemes without direct measure of the rotor speed [2].

Another classification of control strategy for IM is:

- Scalar control: In this control, the air gap flux is kept constant by keeping frequency of supply fed to the machine and voltage magnitude constant. It includes:
- Voltage/frequency (V/f) control.
- Stator current and slip frequency control [3].
- Vector control: In this control, the machine's flux and electro-magnetic torque are decoupled in order to control them independently just like in dc drives. It includes
- Field-oriented control which further has direct method and indirect method.
- Stator flux and direct torque vector control [4].

Nema and Lodhi [5], have done modelling and analysis of both scalar control and vector control technique for IM and showed that vector control gives faster response and its speed regulation is good as compared to scalar control. Wang et al. [6], have presented a detailed analysis on different control strategies in IM and concluded that field-oriented vector control shows comparatively good performance as its total harmonic distortion (THD) of current is better and has less torque ripples.

Control of speed and torque in an IM is difficult because the motor currents and motor torque do not possess a linear relationship. Vector Control of IM makes it possible to decouple the flux and electro-magnetic torque of the motor and achieve separate control of both just as that of a dc motor [7].

In this paper, first the modelling of IM is done using space vector equivalent circuit in MATLAB/Simulink. For modelling all parameters are taken in per-unit (pu) for ease of calculations and then in the analysis stator currents, speed and torque of IM are obtained. Further indirect field-oriented vector control is realized in the developed model for independent control of speed and torque of IM. Own simulation model is build using fundamental equations of IM as self-created models have complete control and freedom and for control purpose the variables of PI controllers are self-tuned.

This paper is divided in six sections. Section 2 presents the modelling of IM using space vector equivalent circuit. Section 3 implements the developed model of IM in MATLAB/Simulink and presents its analysis. The vector control technique for the developed IM model is given in Sect. 4. Section 5 shows the results obtained using vector control strategy. Finally, Sect. 6 gives conclusion of the paper.

2 Modelling of Induction Motor (IM)

IM modelling is done in MATLAB/Simulink using the principal equations of an IM and then its control is done. For the dynamic study of IM, the first mathematical representation was formed on two real axis reference frame that was initially given for synchronous machines by Park [8]. Kovacs and Racz used the symmetric model of the IM and a model for steady state analysis was obtained by them [9]. Several assumptions were given when a complete set of equation is written to define continuous time linear model of Induction Motor:

- Electrical and geometric machine configuration is symmetrical.
- Harmonics in stator and rotor are neglected.
- There is no dependency of frequency and temperature on resistances and inductances
- Flux density distribution, currents and voltages are sinusoidal.
- No core loss and no iron loss.
- Constant magnetizing circuit saturation.

2.1 *Space Vector Equivalent Circuit of Induction Motor*

The IM space vector equivalent circuit is shown in Fig. 1. It should be noted that the squirrel cage rotor bars are short-circuited and thus voltage of rotor is zero. Also, all the bold quantities shown in Fig. 1 are complex vectors, i.e., $\mathbf{I_s} = \mathrm{I}_\alpha + \mathrm{jI}_\beta$.

The equations derived from Fig. 1 can be manipulated using a set of equations by Clarke and Inverse Clarke transformations. By this, the system can be used in discrete reference frame and state variables are selected as per the control plan to achieve the desired behaviour of the model.

The voltage, flux linkage and current equation from Fig. 1 are given as

$$Us\alpha = \frac{d\psi s\alpha}{dt} + RsIs\alpha \tag{1}$$

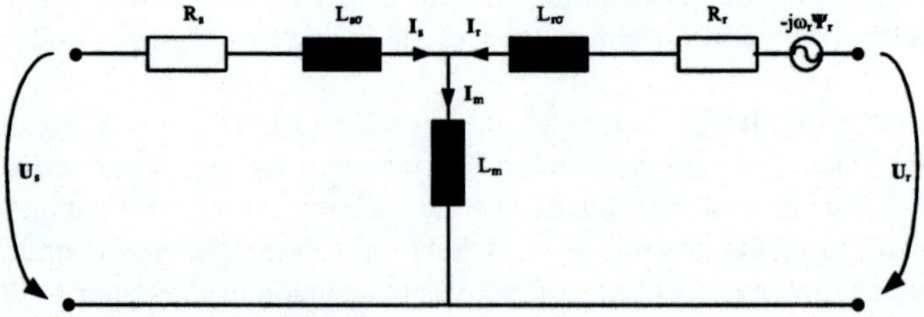

Fig. 1 IM space vector equivalent circuit

$$Us\beta = \frac{d\psi s\beta}{dt} + RsIs\beta \tag{2}$$

$$0 = \frac{d\psi r\alpha}{dt} + \omega r\psi r\alpha + RrIr\alpha \tag{3}$$

$$0 = \frac{d\psi r\beta}{dt} - \omega r\psi r\beta + RrIr\beta \tag{4}$$

$$\frac{d\psi s\alpha}{dt} = Us\alpha + \frac{RsLm}{LsLr - Lm^2}\psi r\alpha - \frac{RsLr}{LsLr - Lm^2}\psi s\alpha \tag{5}$$

$$\frac{d\psi s\beta}{dt} = Us\beta + \frac{RsLm}{LsLr - Lm^2}\psi r\beta - \frac{RsLr}{LsLr - Lm^2}\psi s\beta \tag{6}$$

$$\frac{d\psi r\alpha}{dt} = \frac{RrLm}{LsLr - Lm^2}\psi s\alpha - \frac{RrLs}{LsLr - Lm^2}\psi r\alpha - \omega r\psi r\beta \tag{7}$$

$$\frac{d\psi r\beta}{dt} = \frac{RrLm}{LsLr - Lm^2}\psi s\beta - \frac{RrLs}{LsLr - Lm^2}\psi r\beta + \omega r\psi r\alpha \tag{8}$$

$$Is\alpha = \frac{Lr}{LsLr - Lm^2}\psi s\alpha - \frac{Lm}{LsLr - Lm^2}\psi r\alpha \tag{9}$$

$$Is\beta = \frac{Lr}{LsLr - Lm^2}\psi s\beta - \frac{Lm}{LsLr - Lm^2}\psi r\beta \tag{10}$$

$$Ir\alpha = \frac{Ls}{LsLr - Lm^2}\psi r\alpha - \frac{Lm}{LsLr - Lm^2}\psi s\alpha \tag{11}$$

$$Ir\beta = \frac{Ls}{LsLr - Lm^2}\psi r\beta - \frac{Lm}{LsLr - Lm^2}\psi s\beta \tag{12}$$

The electromagnetic torque of the IM is specified as below

$$T = Tgain(\psi s\alpha Is\beta - \psi s\beta Is\alpha) \tag{13}$$

where, $Tgain$ is taken as 1 in pu system.

3 Implementation of Induction Motor in MATLAB/Simulink

For better simulation of IM in MATLAB/Simulink, the whole model is divided into three subsystems which are grid subsystem, IM model subsystem and the mechanics subsystem shown in Fig. 2.

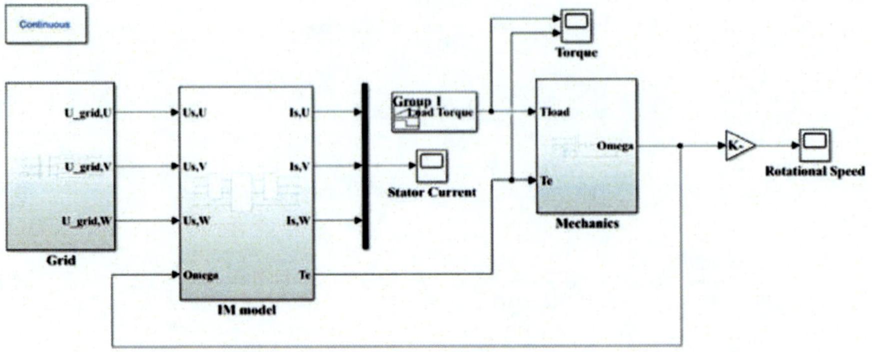

Fig. 2 Developed MATLAB simulation model of the system

The three-phase supply that is given to the IM is given from the grid which consists of initialized variables namely the frequency which is 50 Hz, voltage which is 400 $V_{rms\ (L\text{-}L)}$ and a digital clock. Figure 3 shows the subsystem of IM consisting of three blocks namely Clarke transformation block, Inverse Clarke transformation block and space vector equivalent model block. The inputs to the IM model are the voltages coming from the grid and the mechanical speed generated by the mechanics subsystem. The IM model outputs are three-phase currents and the electro-magnetic torque of the IM.

Figure 4 demonstrates the subsystem of space vector equivalent IM model in stationary reference frame. This model is modelled in Simulink with the help of Eqs. (5)–(12). Torque block is modelled with the help of Eq. (13).

Clarke transformation block is done by its respective matrix [10] and it converts the time-domain components of three-phase voltages into two orthogonal components (d-q) in stationary reference frame. Inverse Clarke transformation block converts

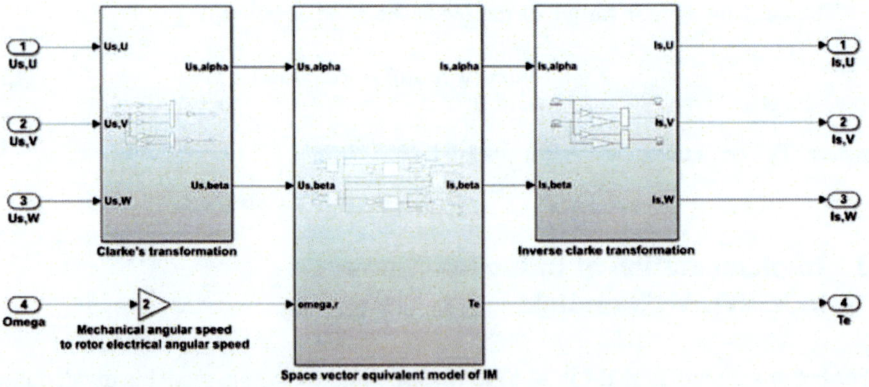

Fig. 3 Subsystem of IM model

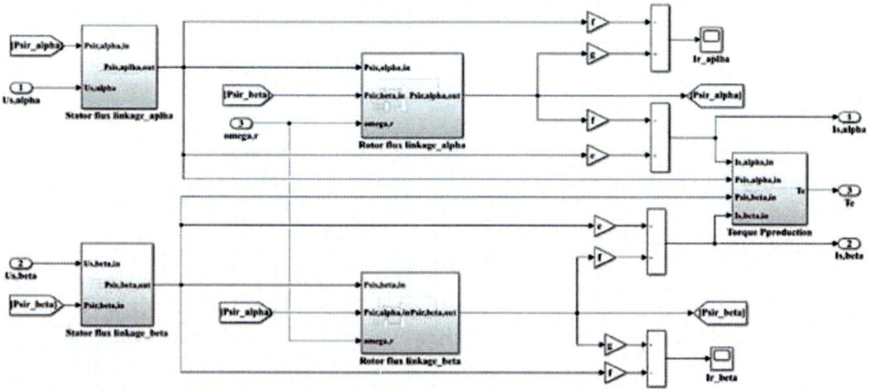

Fig. 4 Subsystem of space vector equivalent IM model

time-domain d-q components of voltages to three-phase voltages in stationary reference frame using Inverse Clarke matrix [10]. Simple *'abc to αβ0'* block and *'αβ0 to abc'* block can also be used.

For the simulation of mechanics Block the motion equation is given as

$$\frac{d\Omega}{dt} = \frac{T - TL - Drated\Omega}{Jrated} \tag{14}$$

3.1 Analysis

In this section, the analysis of the developed MATLAB model of the IM (Fig. 2) is done. From the analysis load torque, electromagnetic torque, stator currents and the speed of developed IM are achieved. Figure 5 shows IM load torque and electromagnetic torque. To analyse the developed model, at start zero load is applied to the motor at t = 0 s and then from t = 5 s to t = 6 s the load is linearly increased to achieve the rated torque. From t = 6 s to t = 10 s the torque is kept constant at rated value in pu. The developed model is run for 10 s.

Figures 6 and 7 show the speed and current waveforms of the IM, respectively. It is seen from Fig. 6 that the speed of motor initially increases and then it finally settles down to 1 pu at around t = 2 s. Under the transient condition, the stator currents of the motor are very high and it settles down to nominal value at t = 2 s.

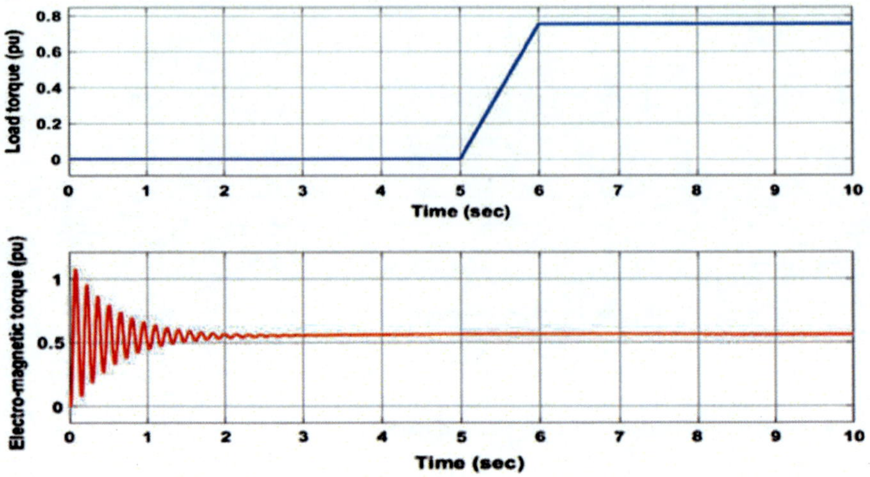

Fig. 5 IM load torque and electro-magnetic torque

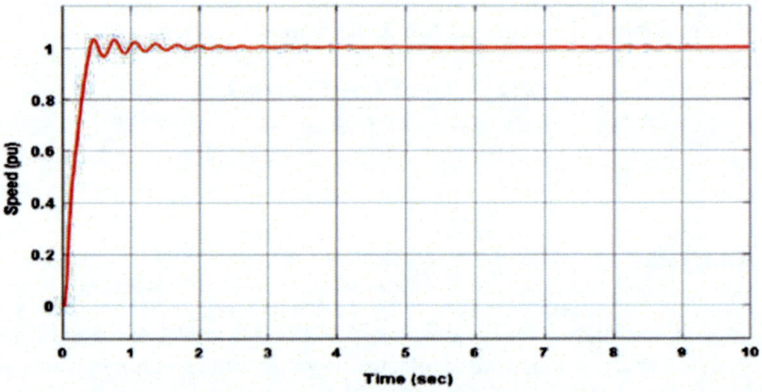

Fig. 6 IM rotational speed

4 Vector-Controlled Induction Motor

In this section, vector control technique of developed IM has been presented [11, 12]. The algorithm for indirect field-oriented vector control of IM is built on two elemental ideas [8]. The first one based on the torque and flux producing currents. The IM is modelled by use of two orthogonal currents instead of the known three phase currents that are supplied to the IM. The direct component (I_d) represents the flux of the motor and the quadrature component (I_q) represents the torque. The voltages applied and the resultant currents of the IM are in the three-phase stationary

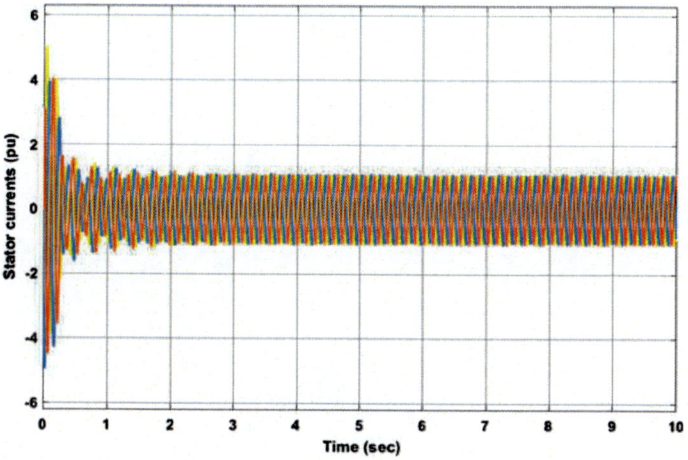

Fig. 7 Stator currents of the IM

system. Thus, going from stationary to rotating reference frame generates the second vector control elemental idea [8, 13].

The basic concept behind a rotating frame of reference is that a quantity which has sinusoidal nature in a particular frame can be converted to a fixed value in another frame provided both frames are at same frequency [13]. After this, the quantities are controlled independently with controllers. Conversion of three-phase quantities into two-phase quantities and moving from stationary to rotating reference frame is done by Clarke and Park transformation respectively [14].

In the control mechanism of IM, Proportional-Integrator (PI) controllers are used to get the desired output [15]. But these controllers have a major disadvantage of being sensitive to system parameters [16]. In the complete simulation model of the field-oriented VC of IM, inverter with an ideal modulator is used. This block helps in generating the desired reference voltages.

4.1 Calculating the Rotor Flux Angle

The calculation of rotor flux angle is required for transformation from stationary to rotating reference frame of rotor flux. In the reference frame of rotor flux, there are two components, i.e., the direct axis component and the quadrature axis component. The direct axis presents the rotor flux linkage component which is having certain magnitude and the quadrature axis component is zero. This is shown in Fig. 8.

The angle for rotor flux (θ) is calculated from the Eqs. (15)–(17)

$$\theta = \int (\omega s + \omega r)dt \tag{15}$$

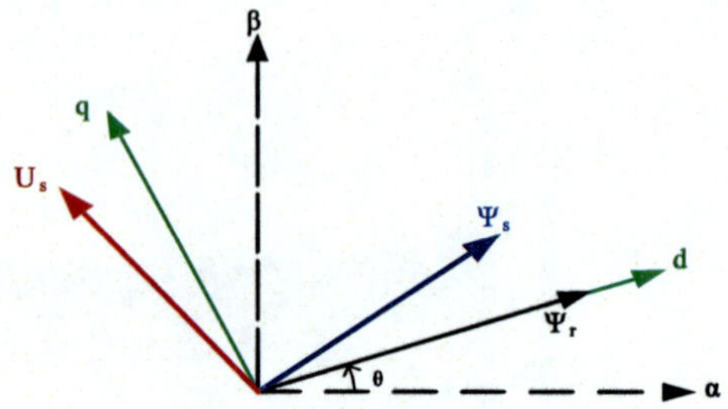

Fig. 8 Illustration of reference frame (d-q) of rotor flux linkage

where,

$$\omega s = \frac{LmIsq}{\tau \psi rd} \tag{16}$$

$$\omega r = p\Omega \tag{17}$$

4.2 Rotor Flux Linkage Estimation

The estimation of rotor flux linkage is obtained from Eq. (18)

$$\psi rd = \frac{IsdLm}{1 + s\tau} \tag{18}$$

The currents which produce the flux in direct axis and torque in quadrature axis are given in Eq. (19) and Eq. (20), respectively.

$$Isd = \frac{\psi rd}{Lm} \tag{19}$$

$$Isq = \frac{LrT}{TgainLm\psi rd} \tag{20}$$

Figure 9 shows the model of vector-controlled IM with PI controllers. Park transformation, Clarke transformation, Inverse Clarke transformation and Inverse Park transformation blocks are used. The Park transformation helps to move from stationary to rotating frame of reference and Inverse Park transformation helps to move from rotating to stationary frame of reference. These blocks are implemented with the help of their respective matrices [10].

Figure 10 shows the overall simulation model of vector-controlled IM. It consists of the space vector equivalent electric model of IM in stationary reference frame block, vector-controlled IM block, ideal inverter bridge block and finally the

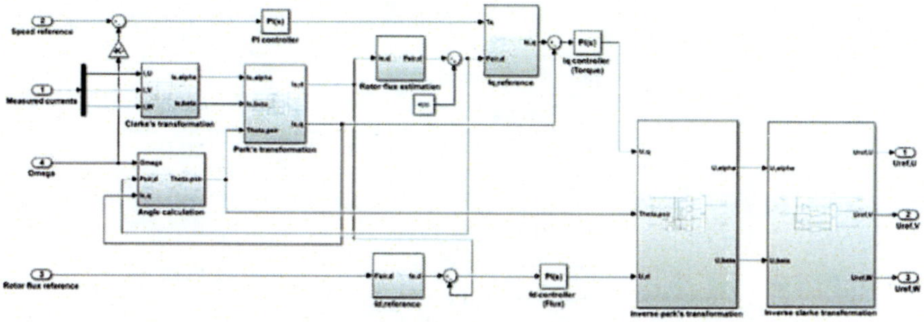

Fig. 9 Vector-controlled IM with PI controllers

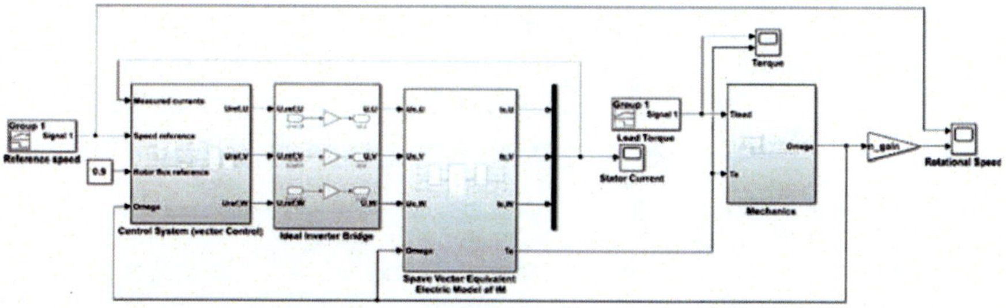

Fig. 10 Simulation model of vector-controlled IM

mechanics block. Three gain blocks, i.e., n_gain, I_gain and T_gain are used in the simulation so that it's easy to switch between real values and pu values. Also, a constant block is used for rotor flux reference and its value is taken as 0.9. Table 1 defines the simulation parameters of the vector-controlled IM.

5 Results

The result of the developed vector-controlled IM is obtained in Figs. 11 and 12. From Fig. 11, at t = 0 s the IM speed is zero and the speed is linearly increased to 0.8 pu in a time interval of 1 s. From t = 1 s to t = 3 s speed is kept constant at 0.8 pu and then from t = 3 s to t = 3.5 s the speed is linearly decreased from 0.8 pu to 0.5 pu and after t = 3.5 s speed is kept constant at 0.5 pu.

From Fig. 12, it is evident that load torque is zero from t = 0 s to t = 2 s and after t = 2 s rated torque (0.75 pu) is applied. The IM is made to run for a time duration of 5 s and the parameters of PI controllers are varied to get the desired output. The result of developed PI controllers shows that as the parameter K_p is increased, the system rise time decreases, overshot increases and there is minor effect on the settling

Table 1 Simulation parameters

Parameters	Values
Rated frequency	50 (Hz)
Rated current	72 (A)
Rated voltage	400 (V)
Rated power	37,000 (W)
Rated speed	737 (rpm)
Number of poles	8
Rotating inertia coefficient	0.01
Friction coefficient	1.85
Equivalent circuit parameters	
Resistance of rotor	0.089 Ω
Resistance of stator	0.063 Ω
Leakage inductance of rotor	0.00056 H
Leakage inductance of stator	0.00089 H
Magnetizing inductance	

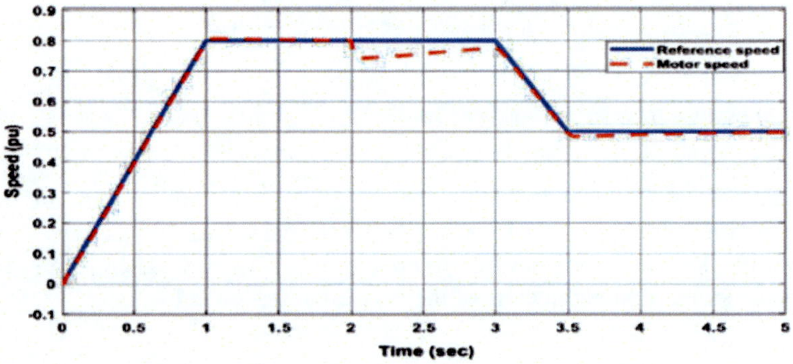

Fig. 11 Controlled motor speed

time. For increase in K_i, the system rise time decreases, overshoot will increase and settling time will increase drastically.

Therefore, by properly tuning the parameter of PI controllers, it is clear from Figs. 11 and 12 that the speed of motor is following the reference speed and the motor electro-magnetic torque is following the load torque, respectively in a quite good manner. Thus, speed and torque of IM are controlled independently.

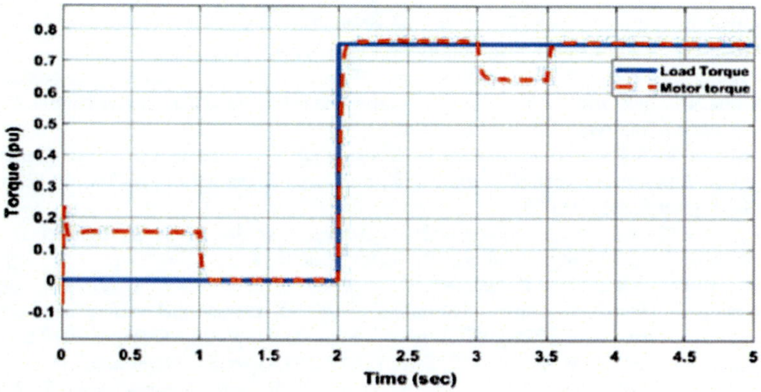

Fig. 12 Controlled motor torque

6 Conclusion

In this chapter, the vector control of three-phase IM is presented. For analysis of vector control of IM, first the IM model is developed in MATLAB/Simulink using space vector equivalent circuit and is analysed for electro-magnetic torque, speed and stator currents of IM. Then the indirect field-oriented VC technique is used for controlling the speed and torque of the developed IM model independently. The vector control is achieved using PI controllers. The analysis shows that from the developed vector-controlled scheme the IM speed and torque are following reference speed and load torque, respectively by proper tuning of PI controller variables. Hence, the IM speed and torque are separately controlled using the developed scheme.

The future scope of presented work is that the step response of IM speed can be plotted by considering various time specifications like rise time, peak time, delay time etc. Also, this IM model can be controlled with different artificial intelligence techniques.

References

1. Ouanjli NE, Aziz D, Ghzizal AE, Motahhir S, Chebabhi A, Mourabit YE, Taoussi M (2019) Modern improvement techniques of direct torque control for induction motor drives-a review. Prot Control Mod Power Syst 4(11). https://doi.org/10.1186/s41601-019-0125-5
2. Alwadie A (2018) A concise review of control techniques for reliable and efficient control of induction motor. Int J Power Electron Drive Syst 9(3). https://doi.org/10.11591/ijpeds.v9.i3.pp1124-1139
3. Liu C, Luo Y (2017) Overview of advanced control strategies for electric machines. Chin J Electr Eng 3(2):53–61. https://doi.org/10.23919/cjee.2017.8048412
4. Jisha LK, Thomas AAP (2013) A comparative study on scalar and vector control of Induction motor drives. In: 2013 International conference on circuits, controls and communications (CCUBE), Bengaluru. IEEE, pp 1–5. https://doi.org/10.1109/ccube.2013.6718554

5. Agarwal A, Lodhi RS, Nema P (2018) Comparison between scalar & vector control technique for induction motor drive. https://doi.org/10.11591/ijape.v8.i2.pp134-144
6. Wang F, Zhang Z, Mei X, Rodríguez J, Kennel R (2018) Advanced control strategies of induction machine: field oriented control, direct torque control and model predictive control. Energies 11(1). https://doi.org/10.3390/en11010120
7. Rafa S, Larabi A, Barazane L, Manceur M, Essounbouli N, Hamzaoui A (2014) Implementation of a new fuzzy vector control of induction motor. ISA Trans 53(3):744–754. https://doi.org/10.1016/j.isatra.2014.02.005
8. Popescu M (2000) Induction motor modelling for vector control purposes. Helsinki University of Technology, p 144. https://doi.org/10.1049/ic:20000384
9. Aamiovuori L, Kärkkäinen H, Niemelä M, Pyrhonen J, Alalibo B, Cao W (2018) Modelling a vector controlled induction motor in Simulink. In: 2018 X international conference on electrical power drive systems (ICEPDS). IEEE, pp 1–8 https://doi.org/10.1109/icepds.2018.8571659
10. O'Rourk CJ, Qasim MM, Overlin MR, Kirtley JL (2019) A geometric interpretation of reference frames and transformations: dq0, clarke, and park. IEEE Trans Energy Convers 34(4):2070–2083. https://doi.org/10.1109/tec.2019.2941175
11. Tao W, Liang Z (2011) Simulation of vector control frequency converter of induction motor based on Matlab/Simulink. In: 2011 Third international conference on measuring technology and mechatronics automation, vol 3. IEEE, pp 265–268. https://doi.org/10.1109/icmtma.2011.637
12. Khursheed M, Sarwar A, Islam S, Bakhsh FI Performance evaluation of an indirect vector controlled drive using synchronous current controller. Perform Eval 1(4):2062–2071. https://doi.org/10.1109/indcon.2011.6139558
13. Kan J, Zhang K, Wang Z (2015) Indirect vector control with simplified rotor resistance adaptation for induction machines. IET Power Electron 8(7):1284–1294. https://doi.org/10.1049/iet-pel.2014.0422
14. Al-Ghnimi S, Bakhsh FI, Tabrez M (2017) Position estimators for sensor-less vector control of a permanent magnet synchronous motor. In: 2017 International conference on intelligent sustainable systems (ICISS). IEEE, pp 371–374. https://doi.org/10.1109/iss1.2017.8389432
15. Habbi HMD, Ajeel HJ, Ali II (2016) Speed control of induction motor using PI and V/F scalar vector controllers. Int J Comput Appl 151(7):36–43. https://doi.org/10.5120/ijca2016911831
16. Yousfi L, Bouchemha A, Bechouat M, Boukrouche A (2014) Vector control of induction machine using PI controller optimized by genetic algorithms. In: 2014 16th international power electronics and motion control conference and exposition. IEEE, pp 1272–1277. https://doi.org/10.1109/epepemc.2014.6980687

A New Nine Level Multilevel Inverter Topology with 1:3 Source Configuration Using Unified Low Switching Frequency Control Scheme

Aratipamula Bhanuchandar and Bhagwan K. Murthy

Abstract This paper proposes a new nine-level Multilevel Inverter (MLI) topology with 1:3 source configuration. It possess 2 DC sources and nine switches in the form of level and polarity generator. In the nine-level output waveform, it has the capability to produce all plus and minus combinations of DC input levels. For generating switching/gate pulses, a new Unified Low Switching Frequency Control Scheme (ULSFCS) has been introduced and it is the modified version of the rounding method. This proposed control scheme works at a fundamental frequency of operation and it is applicable to any kind of MLI topology. The working operation of proposed topology has been validated with proposed control scheme through MATLAB/Simulink platform.

Keywords Multilevel Inverter · Rounding Method · Unified Low Switching Frequency Control Scheme

1 Introduction

In Renewable energy source integration, MLIs plays a very important/superior role in the application of DC to AC power conversion application [1]. In traditional MLI topologies, only Cascaded H-Bridge (CHB) MLI has high modularity as compared with Flying Capacitor (FC) and Neutral Point Clamped—A.Nabae (NPC)/Diode Clamped MLI. Nevertheless, it needs more number of isolated DC sources [2]. Generally, most of the hybrid configuration MLI topologies require large capacitors and complex control strategy for making capacitor balancing [3]. Based on the level and polarity generator concept, many MLI topologies are established in asymmetric source (unequal DC source magnitudes) configurations [4, 5]. In RSC MLIs, the reduction concept of independent DC sources has been reported in [6, 7]. With the same topology, both symmetric and asymmetric configurations are possible and it is reported in [8, 9]. A cascaded MLI topology with different configurations has been

A. Bhanuchandar · B. K. Murthy (✉)
National Institute of Technology Warangal, Hanamkonda, Telangana, India

A. R. Gupta et al. (eds.), *Power Electronics and High Voltage in Smart Grid*,
Lecture Notes in Electrical Engineering 817,
https://doi.org/10.1007/978-981-16-7393-1_19

reported in [10]. A switched ladder, square T-type MLI topologies have been developed in [11, 12] for producing higher levels. For generating desired output voltage level, reduced device count MLI topologies have been developed but these require more number of DC sources [13, 14]. The concept of reduction count in devices and DC sources has been well analyzed in [15, 16]. Based on switch count and the number of DC sources, a new asymmetric 1:3 source configuration MLI topology is proposed and it is compared with different types of source configurations. And also, it generates all plus and minus combinations of DC input sources in the load voltage waveform. For producing a particular level in the load voltage waveform, different modulation techniques are available in [17]. In this, a very few modulation schemes work at fundamental frequency of operation. A universal control scheme was developed in [18] and it is applicable to any MLI topology but it should require more number of high-frequency carriers and a separate lookup table is also required. For generating gate pulses under fundamental frequency of operation, rounding method or nearest level control scheme was developed in [19]. Based on this rounding method, a new Unified Low Switching Frequency Control Scheme (ULSFCS) has been introduced and it is also called as Unified Rounding Control Scheme (URCS). And it does not have any lookup table requirement, switching logic gate arrangement and not require any high-frequency carriers thereby switching losses greatly reduces.

The organization of the paper is as follows. In Sect. 2, the operation and control scheme of proposed topology has been explained. In Sect. 3, a comparative study has been explained with conventional topologies. In Sect. 4, the proposed control scheme has been validated through simulation results. And in Sect. 5, the overall conclusion points have been reported.

2 Proposed MLI Topology and Control/Switching Scheme

2.1 Proposed 9L-MLI Topology

The Proposed nine-level (9-Level) MLI topology is depicted in Fig. 1. It consists of five switches, 2 DC sources on level generator side, and four switches on polarity generator side. On the level generator side, S3 and S4 are bidirectional switches and these have bidirectional voltage blocking capability & bidirectional current-conducting capability. Here, DC source magnitudes are in asymmetric manner (Vdc1 = Vdc; Vdc2 = 3Vdc) for generating nine-level output voltage.

Table 1 shows switching arrangement for the proposed nine-level (9L) inverter topology and from switching table it is clearly observed that only four switches are conducting per output voltage level that means even switching action is present. And this table does not have any redundant states except 0Vdc. From Table 1, it is clearly observed that the proposed topology produces all plus and minus combinations of DC input sources in the load voltage waveform. Finally, the output levels are produced 0, ±1Vdc, ±2Vdc, ±3Vdc, and ±4Vdc. The load voltage is positive if S6 & S9

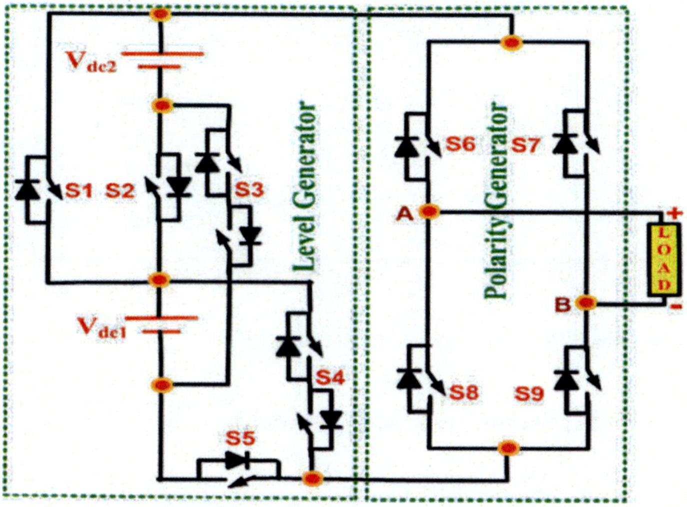

Fig. 1 Proposed 9L-inverter topology

Table 1 Switching table

Output voltage (1:3)	Switching table (S1–S9)
0Vdc	[1 0 0 1 0 1 0 0 1]/[1 0 0 1 0 0 0 1 1]
1Vdc = Vdc1	[1 0 0 0 1 1 0 0 1]
2Vdc = Vdc2–Vdc1	[0 0 1 1 0 1 0 0 1]
3Vdc = Vdc2	[0 1 0 1 0 1 0 0 1]
4Vdc = Vdc2 + Vdc1	[0 1 0 0 1 1 0 0 1]
−1Vdc = −Vdc1	[1 0 0 0 1 0 1 1 0]
−2Vdc = Vdc1–Vdc2	[0 0 1 1 0 0 1 1 0]
−3Vdc = −Vdc2	[0 1 0 1 0 0 1 1 0]
−4Vdc = −Vdc1–Vdc2	[0 1 0 0 1 0 1 1 0]

are switched on and similarly load voltage is negative if S7 & S8 are switched on in polarity generator. In order to avoid short circuits in the input DC sources, the switches S1 to S3 are not switched on and similarly, S4 & S5 are not switched on at the same time. Finally, Table 1 gives that there is no occurrence of short circuit in this topology. The PLECS software is the best one for finding switching losses of any topology.

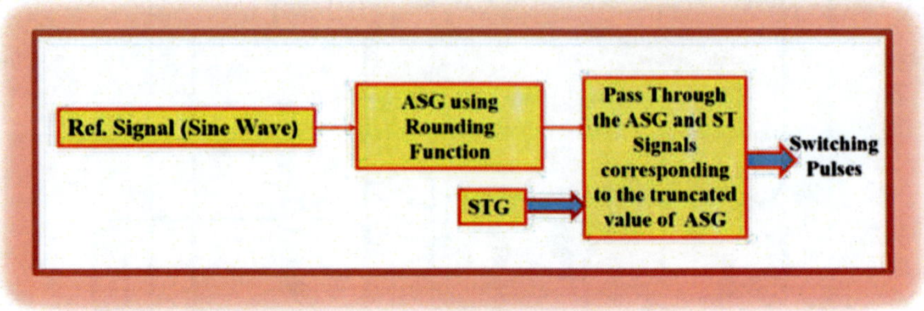

Fig. 2 Proposed control/switching scheme-ULSFCS/URCS

2.2 *Proposed Control/Switching Scheme*

Figure 2 shows Unified Low Switching Frequency Control Scheme (ULSFCS) and it is the modified version of the rounding method. Here, rounding function is used for generating Accumulated Signal (AS) with reference amplitude of four then the output will act as Control Port Signal (CPS). Switching Table Generation (STG) is taken from Table1, then output will act as Data Port Signal (DP). These two signals are truncated to each other and finally, gate pulses are generated. Here, the main aim of the rounding function is, it rounds each element to the nearest integer. The proposed control scheme is universally applicable to any MLI topology and the main advantage of this one is, only it operates with fundamental frequency thereby switching losses are greatly minimizes.

3 Comparative Study

Table 2 shows symmetric and asymmetric sequences of generalized formulae. It is observed that from type-1 source sequence configuration (1:1), the peak value of output voltage is 2Vdc, number of output levels = 5, and PIV = 13Vdc. Similarly, from type-2 to type-4 configurations (1:2), the peak value of output voltage is 3Vdc, number of output levels = 7, and PIV = 21Vdc. By taking 1:3 source configuration (type-5, proposed one), the peak value of output/load voltage is 4Vdc, the number of output levels = 9, and PIV = 65Vdc. Finally from 1:3 configuration, the maximum number (± 4Vdc) of load voltage level is obtained as compared with other types of source configurations with the same switching arrangement. Table 3 shows a comparison table of different conventional MLI topologies. From this table, it is concluded that, the proposed topology needs few switches, less number of DC sources as compared with conventional topologies then finally overall cost decreases (Table 4).

Table 2 Symmetric and asymmetric sequence configurations: generalized formulae

S. no	Source sequence configuration	Peak value of output voltage level (if n = 2)	Total number of levels	Peak inverse voltage
1	Type-1(1:1:1:....) $V_{dci} = V_{dc}$ Where i = 1,2...n	n V_{dc}=2Vdc	$2n + 1 = 5$	(8n-3)V_{dc}= 13Vdc
2	Type-2(1:2:3:...) $V_{dcj} = iV_{dc}$ Where i = 1,2...n	$\frac{n^2+n}{2} V_{dc} = 3Vdc$	$n^2 + n + 1 = 7$	$(4n^2 + 4n - 3)V_{dc} =$ 21Vdc
3	Type-3(1:2:4: ...) $V_{dcj} = 2^{i-1}V_{dc}$ Where i = 1,2...n	$(2^n - 1)V_{dc}$= 3Vdc	$(2^{n+1}) - 1 = 7$	(8X-3)V_{dc} = 21Vdc $X = (2^n - 1)$
4	Type-4 (1:2:6:....) $V_{dcj} =$ $\begin{cases} V_{dc} & i = 1 \\ 2 * 3^{i-2} & i \geq 2 \end{cases}$ Where i = 1,2...n	$(3^{n-1})V_{dc}$= 3Vdc	$1 + (2 * 3^{n-1}) =$ 7 Where n = 1, 2, 3...	(8X-3)V_{dc}= 21Vdc Where, $X = (3^{K-1})$
5	Type-5 (1:3:9:...) $V_{dci} = 3^{i-1}V_{dc}$ Where i = 1,2...n	$\frac{(3^n-1)}{2} V_{dc}$= 4Vdc	3^n= 9	$(X + 4X)V_{dc}$= 65Vdc Where, X = $(2(3^{n-1})) - 3$

Table 3 Comparison table for different conventional MLI topologies

Refs. no	Number of DC sources	Number of required switches	Number of drivers	Number of levels
[4]	$3S + 1$	$5S + 6$	$5S + 6$	$6S + 3$
[11]	$2S + 2$	$4S + 6$	$2S + 6$	$4S + 5$
[12]	4S	12S	9S	$16S + 1$
[13]	S	$2S + 2$	$2S + 2$	$2^{S+1} - 1$
[14]	S	$2S + 4$	$2S + 4$	$2^{S+1} - 1$
[15]	$4S + 2$	$4S + 6$	$3S + 6$	$8S + 5$
[16]	2S	$S + 4$	$S + 4$	$2S + 3$
[20]	S	$\begin{cases} 2S + 4, n = 1 \\ (n + 2) + 4, n \geq 2 \end{cases}$	$\begin{cases} 2S + 4, S = 1 \\ (S + 2) + 4, S \geq 2 \end{cases}$	$2S + 1$
[21]	S	$2S + 4$	$2S + 4$	$2S + 1$
[22]	S	$2S + 2$	$2S + 2$	4S-1
[23]	S	$2S + 2$	$2S + 2$	$S^2 + S + 1$

Table 4 Simulation parameters

S. no	Simulation parameters
1	$V_{0,\,peak}$ = 400 V (Assume)
2	Vdc1 = Vdc = 100 V; Vdc2 = 3Vdc = 300 V (1:3 configuration)
3	R = 100Ω, L = 30 mH
4	Switching frequency = 50 Hz
5	Modulation index = 1

4 Simulation Results

Figure 3 shows ASG, switching pulses, load voltage, and load current waveforms under 1:3 configuration. Here, the pulses are produced under fundamental frequency of operation. For getting the peak value of output voltage 400 V, the input Vdc = 100 V is selected from switching table peak value then the proposed control scheme is applied to proposed topology. Finally, load voltage is obtained in terms of 0, $\pm 100V$, $\pm 200V$, $\pm 300V$, and $\pm 400V$. The ULSFCS/URCS operates only with fundamental frequency thereby switching losses are drastically decreased and also if the devices are made up with SiC-MOSFETs then switching losses are greatly minimized. By using PLECS software, it is possible to get switching losses directly. The load current waveform completely depends on R, L values, and here, R = 100Ω, L = 30 mH is considered in simulation. Figure 4 shows output voltage waveform of type-1 configuration and it produces five-level output with peak value of 200 V. Similarly, Fig. 5 shows output voltage waveform of type-2 to type-4 configurations and it produces seven level output with peak value of 300 V. Figure 6 shows the harmonic spectrum of nine-level output voltage waveform with THD of 9.36% and because of proposed control scheme the DC bus utilization increases from 400 to 405.4 V.

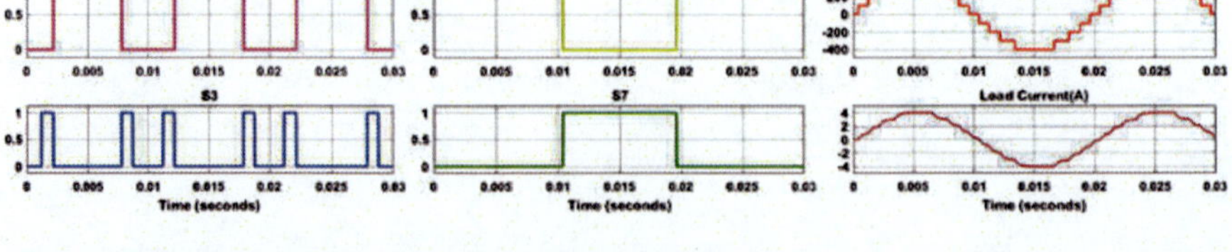

Fig. 3 Accumulated signal generation, switching pulses from S1 to S9, load voltage, and load current waveforms-1:3 source configuration (Type-5)

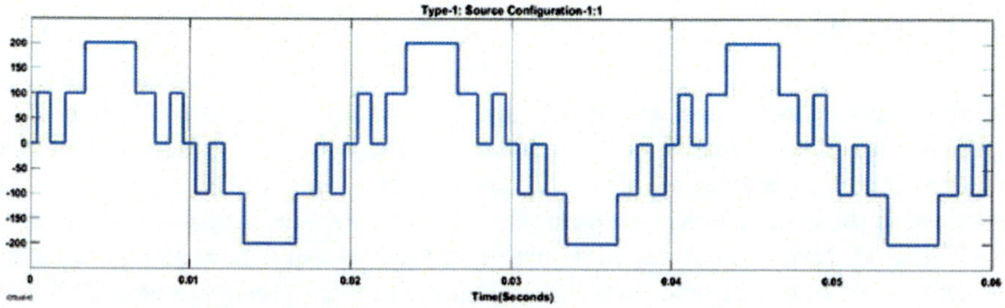

Fig. 4 Output voltage waveform of Type-1 configuration (1:1)

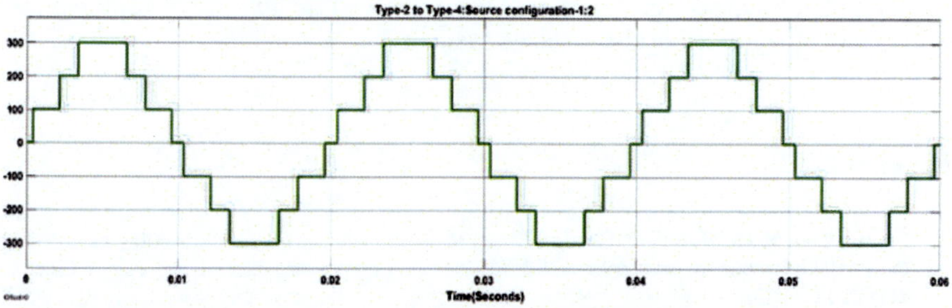

Fig. 5 Output voltage waveform of Type-2 to Type-4 configurations (1:2)

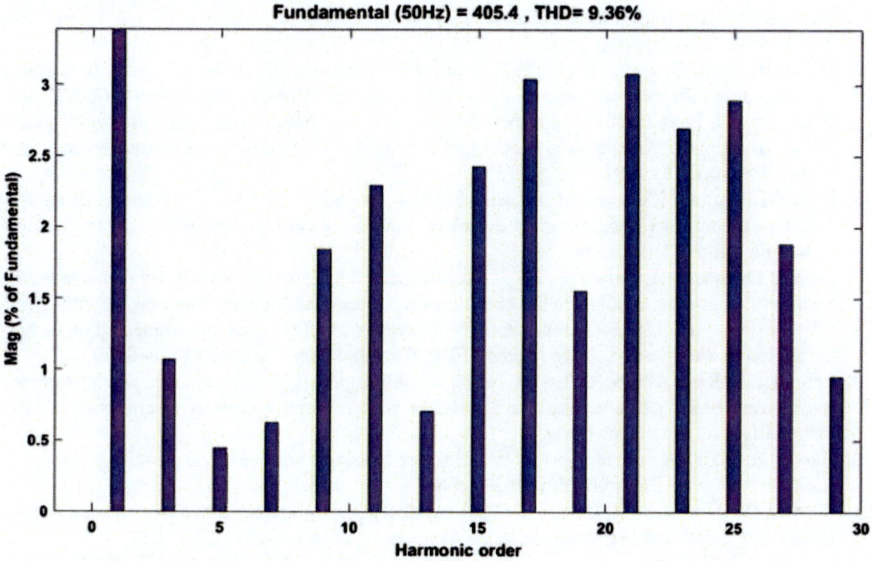

Fig. 6 Harmonic spectrum-9 level output voltage

5 Conclusion

In this paper, a new nine-level MLI topology with 1:3 source configuration is presented and it is compared with different types of source configurations. And also this configuration enables all plus and minus combinations of DC input levels are obtained in the load voltage waveform with only nine switches and two DC sources. Reduction of device count makes reduction of total cost of MLI topology in terms of gate driver, heat sink, and protection circuits. For this topology, a new ULSFCS has been proposed and of course, it is applicable to any kind of MLI topology under fundamental frequency of operation. A comparative study also done with conventional topologies. In application point of view, it is well suitable and easily integrated with photovoltaic-based grid-connected systems.

References

1. Zhao J, Han Y, He X, Tan C, Cheng J, Zhao R (2011) Multilevel circuit topologies based on the switched-capacitor converter and diode-clamped converter. IEEE Trans Power Electron 26(8):2127–2136
2. Huang J, Corzine KA (2006) Extended operation of flying capacitor multilevel inverters. IEEE Trans Power Electron 21(1):140–147
3. Prabaharan N, Palanisamy K (2017) A comprehensive review on reduced switch multilevel inverter topologies, modulation techniques and applications. Renew Sustain Energy Rev 76C:1248–1282
4. Babaei E, Laali S, Bayat Z (2015) A single-phase cascaded multilevel inverter based on a newbasic unitwith reduced number of power switches. IEEE Trans Ind Electron 62(2):922–929
5. Wu F, Duan J, Feng F (2015) Modified single-carrier multilevel sinusoidal pulse width modulation for asymmetrical insulated gate bipolar transistor clampedgrid-connected inverter. IET Power Electron 8(8):1531–1541
6. Babaei E, Farhadi Kangarlu M, Sabahi M (2014) Extended multilevel converters: an attempt to reduce the number of independent dc voltage sources in cascaded multilevel converters. IET Power Electron 7(1):157–166
7. Jannati Oskuee MR, Salary E, Najafi-Ravadanegh S (2015) Creative design of symmetric multilevel converter to enhance the circuit's performance. IET Power Electron 8(1):96–102
8. Choi JS, Kang FS (2015) Seven-level PWM inverter employing series connected capacitors paralleled to a single dc voltage source. IEEE Trans Ind Electron 62(6):3448–3459
9. Farakhor A, Reza Ahrabi R, Ardi H, Najafi Ravadanegh S (2015) Symmetric and asymmetric transformer based cascaded multilevel inverter with minimum number of components. IET Power Electron 8(6):1052–1060
10. Babaei E (2008) A cascade multilevel converter topology with reduced number of switches. IEEE Trans Power Electron 23(6):2657–2664
11. Alishah RS, Hosseini SH, Babaei E, Sabahi M (2017) Optimal design of new cascaded switch-ladder multilevel inverter structure. IEEE Trans Ind Electron 64(3):2072–2081
12. Samadaei E, Sheikholeslami A, Asghar Gholamian S, Adabi J (2018) A square T-type (ST-Type) module for asymmetrical multilevel inverters. IEEE Trans Power Electron 33(2):987–997
13. Babadi AN, Salari O, Jafar Mojibian M, Bina M (2018) Modified multilevel inverterswith reduced structures based on packed U-cell. IEEE J Emerg Sel Topics Power Electron 6(2):874–887

14. Vemuganti HP, Sreenivasarao D, Siva Kumar G, Spandana AS (2018) Reduced carrier PWM scheme with unified logical expressions for reduced switch count multilevel inverters. IET Power Electron 11(5):912–921, 2018
15. Avanaki HN, Barzegarkhoo R, Zamiri E, Yang Y, Blaabjerg F (2019) Reduced switch-count structure for symmetric multilevel inverters with a novel switched-dc-source submodule. IET Power Electron 12(2):311–321
16. Panda KP, Lee SS, Panda G (2019) Reduced switch cascaded multilevel inverter with new selective harmonic elimination control for standalone renewable energy system. IEEE Trans Ind Appl. https://doi.org/10.1109/TIA.2019.2904923
17. Rodriguez J et al (2009) Multilevel converters: an enabling technology for high-power applications. Proc IEEE 97(11):1786–1817. https://doi.org/10.1109/JPROC.2009.2030235
18. Gupta KK, Jain S (2012) A novel universal control scheme for multilevel inverters. In: 6th IET international conference on power electronics, machines and drives (PEMD 2012). Bristol, pp 1–6. https://doi.org/10.1049/cp.2012.0176
19. Kouro S, Bernal R, Miranda H, Silva CA, Rodriguez J (2007) High-performance torque and flux control for multilevel inverter fed induction motors. IEEE Trans Power Electron 22(6):2116–2123. https://doi.org/10.1109/TPEL.2007.909189
20. Kangarlu MF, Babaei E (2013) A generalized cascaded multilevel inverter using series connection of submultilevel inverters. IEEE Trans Power Electron 28(2):625–636
21. Lee SS, Chu B, Nik Idris NR, Hwang Goh H, Heng YE (2016) Switched-battery boost-multilevel inverter with GA optimized SHEPWMfor standalone application. IEEE Trans Ind Electron 63(6):3582–3594
22. Banaei MR, Fazel Bakhshayesh A (2014) Selection of DC voltage magnitude using Fibonacci series for new hybrid asymmetrical multilevel inverter with minimum PIV. Alexandria Eng J 53:529–535
23. Gupta KK, Jain S (2014) Comprehensive review of a recently proposed multilevel inverter. IET Power Electron 7(3):467–479

Practical Distribution System Analysis with Multiple PV and DSTATCOM Using Flower Pollination Algorithm

Nandola Maitrey Bharatbhai and Atma Ram Gupta

Abstract Solar Photovoltaic (SPV) is one of the most attractive types of renewable energy sources for the generation of electricity in order to accomplish targets such as the elimination of carbon dioxide emissions, energy sustainability, and increased efficiency of infrastructure. Therefore, by means of incentives and financial options, many countries are adopting policies to improve solar energy exploitation. There are various problems for distribution network operators (DNOs) with the growing number of SPVs connecting to Distribution Networks (DN), such as voltage variation, power losses, voltage consistency, and reliability. In this paper, we have done analysis of Portuguese 94 bus radial practical distribution system by means of power loss reduction and voltage profile improvement with installing PV and DSTATCOM at a suitable place with a suitable size. Here, Flower Pollination Algorithm (FPA) is used to find the optimum size and location of PV and DSTATCOM. FPA is meta-heuristic search algorithm. It is based on pollination techniques of flower which is completed by birds, insects, and other animals. The outcomes verify that installation of PV and DSTATCOM makes the system more reliable by means of voltage profile improvement and reduction in power loss. Seasonwise analysis is done for DN.

Keywords Distribution network operator · Distribution generation · Distribution system · Optimization technique · PV modelling · Load modelling

1 Introduction

Solar Photovoltaic (SPV) is one of the most attractive types of renewable energy sources for the generation of electricity in order to accomplish targets such as the elimination of carbon dioxide emissions, energy sustainability, and increased efficiency of

N. M. Bharatbhai (✉) · A. R. Gupta
Department of Electrical Engineering, National Institute of Technology Kurukshetra, Kurukshetra, Haryana, India
e-mail: nandola_31904219@nitkkr.ac.in

A. R. Gupta
e-mail: argupta@nitkkr.ac.in

A. R. Gupta et al. (eds.), *Power Electronics and High Voltage in Smart Grid*, Lecture Notes in Electrical Engineering 817,
https://doi.org/10.1007/978-981-16-7393-1_20

infrastructure. Therefore, by means of incentives and financial options, many countries are adopting policies to improve solar energy exploitation. There are various problems for distribution network operators (DNOs) with the growing number of SPVs connecting to Distribution Networks (DN), such as voltage variation, power losses, voltage consistency, and reliability [1, 2]. The cause is the incompatibility between the positioning of SPVs and the capacity of the local network to assign the new Distributed Generators (DGs) into the network. Distribution STATic COMpensator (DSTATCOM) is another solution to improve voltage profile of the system. It is a FACTS device which is fast and continuous in operation. The DSTATCOM provides sufficient support for reactive power and reduces radial distribution system losses. Both DG and DSTATCOM implementation gives better results to reduce active power losses and voltage fluctuations of the system. A DNO must establish a reasonable operating plan that takes into account the dispatch of DGs and DSTATCOM, the disruption of loads, and the buying power from the wholesale market while retaining system protection, as its key function is to supply loads at an appropriate voltage and loading level. In certain situations, DNOs play the position of retailers who purchase power at volatile rates on the wholesale market and sell it again to small customers at fixed tariffs. Separate business players with distinct purposes, networks, and sizes are DNOs and retailers [3, 4].

Generally, distribution systems are considered as a radial form. So, the integration of DGs or DSTATCOMs will change the power flow path in feeders. To obtain an appropriate objective, an appropriate management system is required like the best position and size of DGs and DTATCOMs in the system [5].

Many researchers are working on various optimization techniques which help DNOs to find the best location and best sizes of DG and DSTATCOM to obtain certain objectives. They proposed various techniques to find real-world problems using analytical, numerical, or hybrid approaches. The limitation of numerical approach is the initial guess. So, because of the incorrect initial guess, it could not hit the optimum solution. Heuristic and metaheuristic optimization methods have increasingly emerged in various fields in recent years to solve complex problems with diverse parameters, multiple variables, and objective functions. The population-dependent algorithms of metaheuristic optimization are divided into many categories based on the source of motivation [6]. These types are: (1) Based on swarm intellect, such as particle swarm optimization (PSO) [7, 8], ant colony optimization (ACO) [9], artificial bee colony optimization (ABC) [10], (2) Evolution-based algorithm, like genetic algorithm (GA) [11], differential evolution (DE) [12] and Least Error Iterative Method [13] (3) Biological algorithm such as, bacterial colony foraging optimization (BCFO) [14] and grey wolf optimizer (GWO) [15], Even if it is multidimensional, non-continuous, or non-differentiable, all artificial intelligence optimization techniques will address any engineering challenge.

In this paper, the analysis of DN has been carried out after installing multiple PV and DSTATCOM in the system. Location and the best size of PV and DSTATCOM found using Flower Pollination Algorithm to get minimum losses and voltage profile improvement. FPA is meta-heuristic search algorithm which is proposed by Yang and Deb [16]. A review of variants of FPA is presented in [17]. FPA is based on pollination

Table 1 Parameters for photovoltaic cell

Parameters	Value
Normal working temperature (Ta)	44 °C
Maximum power point current (Impp)	8.28 A
Maximum power point voltage (Vmpp)	30.20 V
Ki (Mean temperature ratio)	0.0045 A/°C
Kv (Temperature stroke coefficient)	0.1241 V/°C
Short circuit current (Isc)	8.67 A
Open-ended voltage (Vop)	

techniques of flowers which is done by birds, insects, bats, and other animals. For the analysis, Portuguese 94 bus Radial Distribution System (RDS) is used to analyze reduction in power loss and changes in voltage profile after implementing PV-DG + DSTATCOM in the system. The load data and bus data of Portuguese 94 bus RDS are taken from [18] and to analyze the effect of using Solar DG on loss minimization TMY (Typical Meteorological Year) data of solar irradiance is taken from National Renewable Energy Laboratory (given in Table 1) and applied on Portuguese 94 bus RDS. Solar power output is calculated from the TMY data as given in Table 1 and then this output is applied as a DG source in Portuguese 94 bus RDS.

The formulation of this paper is as follows, Sect. 2 covers Problem formulation, Sect. 3 includes FPA, results, and discussions are covered in Sect. 4, and finally, the conclusion is covered in Sect. 5.

2 Problem Formulation

2.1 Load Flow Study

In this paper, a direct approach-based load flow analysis method is used. This method is first proposed by Teng [19]. By this voltage profile at each node and power loss at each branch are calculated.

2.2 Solar PV Modelling

To analyze the effect of using Solar DG on loss minimization TMY (Typical Meteorological Year) data of solar irradiance is taken from National Renewable Energy Laboratory (NREL) and applied on Portuguese 94 bus RDS. Solar power output is calculated from the TMY data as given in Table 2 and then, this output is applied as a DG source in 94 bus RDS practical located in Portuguese.

Table 2 Seasonwise avg. solar irradiance (SI) and solar PV output (SPVo)

Time in hours	Winter		Spring		Summer		Autumn	
	SI (kW/m^2)	SPVo (kW)	SI (kW/m^2)	SPVo (kW)	SI (kW/m^2)	SPVo (kW)	SI (kW/m^2)	SPVo (kW)
1–6	–	–	–	–	–	–	–	–
7	0.014	14	0.0162	16.2	0.015	15	0.0232	23.2
8	0.0499	49.9	0.0557	55.7	0.0522	52.2	0.0613	61.3
9	0.0867	86.7	0.0938	93.8	0.09	90	0.0997	99.7
10	0.1226	122.6	0.1286	128.6	0.1225	122.5	0.1359	135.9
11	0.1507	150.7	0.1552	155.2	0.1492	149.2	0.1548	154.8
12	0.164	164	0.1611	161.1	0.1612	161.2	0.1616	161.6
13	0.1667	166.7	0.1651	165.1	0.1667	166.7	0.1624	162.4
14	0.1498	149.8	0.1553	155.3	0.1511	151.1	0.149	149
15	0.13	130	0.1299	129.9	0.1276	127.6	0.1241	124.1
16	0.0965	96.5	0.0951	95.1	0.0933	93.3	0.0929	92.9
17	0.057	57	0.0548	54.8	0.0544	54.4	0.0502	50.2
18	0.0198	19.8	0.0175	17.5	0.0191	19.1	0.0121	12.1
19–24	–	–	–	–	–	–	–	–

Table 2 shows hourly solar irradiance data of every month. From 1 a.m to 12 a.m, solar irradiance is zero in 1 a.m to 6 a.m and 7 p.m to 12 a.m. Solar irradiance data is taken from NREL (Solar irradiance and temperature data). Parameters specified for photovoltaic cell is given in Table 1. An SPV system is an independent energy system, which provides power for different types of devices [20]. The output power of a PV module is given as follows [21]. The flow chart for obtaining SPV output is shown in Fig. 1.

$$\mathrm{Ppv} = \mathrm{N} * \mathrm{FF} * \mathrm{Vy} * \mathrm{Iy} \tag{1}$$

$$\mathrm{FF} = \frac{\mathrm{Vmpp} * \mathrm{Impp}}{Voc * Isc} \tag{2}$$

$$\mathrm{Vy} = \mathrm{Voc} - \mathrm{Kv} * \mathrm{Tcy} \tag{3}$$

$$\mathrm{Iy} = (\mathrm{Isc} + \mathrm{Ki}(\mathrm{Tcy} - 25)) * \mathrm{s} \tag{4}$$

$$\mathrm{Tcy} = \mathrm{Ta} + \mathrm{s} * \frac{\mathrm{Not} - 20}{0.8} \tag{5}$$

where,

Fig. 1 Flow chart to find SPV output

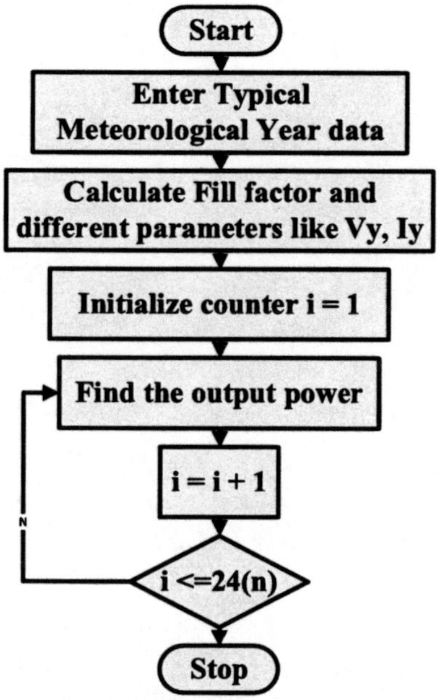

- Tcy is cell temperature in °C and Ta is ambient temperature in °C
- Not is the normal operating temperature in °C
- Ki and Kv are temperature coefficients of current in A/°C and voltage V/°C, respectively
- Isc is short circuit current in A and Voc are open-circuit voltage in V
- FF is Fill factor
- Vmpp and Impp are maximum power point voltage in V and current in A, respectively
- s is solar irradiance in kW/m^2.

Using Eqs. (1–5), SPV output is calculated and shown in Table 1.

2.3 *Load Modelling*

The load and weather data must be evaluated for the required integration of DG within the network. The hourly load pattern data for each season are taken from [18], since the time-varying constant load model has been considered for this analysis.

Time-varying load, affects system voltage. Minimum system voltage changes hourly as load changes hourly in different seasons. In the base case, the maximum load demand of the system is 521.1 KVA. In each season, load demand varies, Fig. 2 shows the maximum load demand of the system in each hour with different seasons.

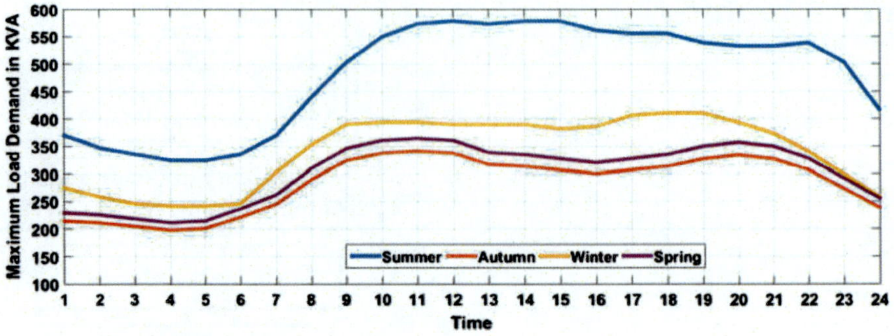

Fig. 2 Maximum load demand at different seasons

2.4 *Annual Energy Savings (AES) and Annual Energy Cost Saving (ACS)*

Annual Energy Saving (AES) is a difference between energy loss before installation of PV-DG and DSTATCOM and energy loss after installation of PV, DSTATCOM [22]. It can be formulated as (1).

$$\mathrm{AES} = \left(\mathrm{TPL} - \mathrm{TPL}'\right)\mathrm{TkWh} \tag{6}$$

$$\mathrm{ACS} = \mathrm{Ke}(\mathrm{AES}) \tag{7}$$

where, TPL is total active power loss in kW before installation and TPL' is total active power loss in kW after installation of DG. T is time duration in hours. Ke is the energy rate in per kWh. For the calculation, the values of Ke = 0.06 /KWh, T = 8760 (hours),

3 Flower Pollination Algorithm (FPA)

FPA is meta-heuristic search algorithm. This method is proposed by Yang and Deb [16]. FPA is based on pollination techniques of flower which is completed by birds, insects, and other animals. Pollination is often divided into two forms, organic and inorganic. Generally, 90% of pollination is finished by pollens so it's called organic and 10% of pollination is inorganic means no pollen is required. Wind and water diffusion in this inorganic pollination allows the pollination of certain plants. Organic pollination is done in two ways, pollination by self and cross-pollination.

Organic and cross-pollination may have counted in global pollination because in this, pollination can be done by long-distance travelling of bees, birds, and other animals. Study in [17], says that jump or distance covered by pollinators can be formulated as L'evy distribution. Hence birds, bees may behave as L'evy flight behavior. Steps in this distribution can be found using the similarity or difference of two flowers. Based on this, global pollination can be mathematically represented as [16],

$$\mathrm{Pi}^{t+1} = \mathrm{Pi}^{t} + \lambda L(\gamma)\left(\mathrm{G} - \mathrm{Pi}^{t}\right) \tag{8}$$

where P_i^{t+1} is the solution vector of P_i at iteration $t + 1$. G is the best solution at iteration $t + 1$. λ is a scaling factor, $L(\gamma)$ is the strength of the pollination, it is basically a step size since pollinators move for a long distance in steps. From [17], to imitate these functions, we can use L'evy travel (L), which is $L > 0$ from the L'evy distribution.

$$L(\gamma) = \frac{\lambda\Gamma(\gamma)\sin\left(pi * \frac{\lambda}{2}\right)}{pi} * \left(\frac{1}{s^{1+\gamma}}\right)s \gg s0 > 0 \tag{9}$$

$\Gamma(\gamma)$ is that standard gamma function and this distribution is used for larger steps $s > 0$. From [23], in this analysis, we have used $\lambda = 1.5$.

For the local pollination,

$$\mathrm{Pi}^{t+1} = \mathrm{Pi}^{t} + \varepsilon\left(\mathrm{Pj}^{t} - \mathrm{Pk}^{t}\right) \tag{10}$$

where Pjt and Pkt are pollen produced from separate flowers of the same genus of plants. ε is uniformly distributed in [0, 1]. A detailed review of variants of FPA is given in [17].

In real-life pollination action, the plant will contain several flowers and each flower patch usually emits millions or even billions of pollen gametes. In order to simplify the algorithm, it was believed that each plant would have a single flower and that each flower released only one pollen gamete. In this paper, optimum location and size of PV is found using FPA. FPA relies on flower pollination process. Number of populations N, and switch probability p, are the sole parameters that make FPA simple, and when N is fixed, the values of p control the pollinators. For the results, the values of $n = 20$ and $p = 0.8$ are taken. Figure 3, presents the flow chart of FPA [17].

4 Results and Discussion

Optimum location and size of PV is found using FPA. FPA relies on flower pollination process. Number of populations N, and switch probability p, are the sole parameters that make FPA simple, and when N is fixed, the values of p control the pollinators. The value of $n = 20$ and $p = 0.8$ is taken here, and PV range and position are found in Portuguese 94 bus RDS. Voltage profile and power loss of the system are found

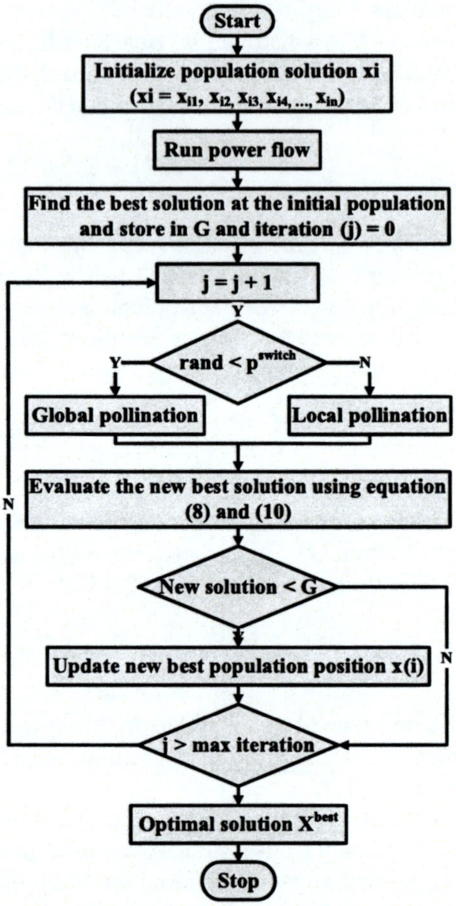

Fig. 3 Flow chart of FPA

using direct approach-based load flow method. Parameters associated with FPA are shown in Table 3. FPA is coded in MATLAB R2015a and simulation results for the Portuguese 94 bus practical RDS [18] are performed in Asus laptop with Intel core i5 @3.4 GHz. Results for different seasons are shown below.

In this section, the optimal location of PV-DG and DSTATCOM between bus number 2 and 94 are found. First, results of multiple PV-DG with various cases are calculated like, case 1 is base case (without PV-DG), case 2 is with single PV-DG, case 3 for two PV-DGs, and case 4 for three PV-DG. For the load flow parameters, we take

Table 3 Parameters of FPA

No. of populations	20
Probability	0.8
Max. iterations	1500

1 MVA as a base power and 15 kV as a base voltage. Total active and reactive power demand is 4.797 MW and 2.3239 MVAr, respectively, for the baseload condition. Results for case 2 and case 4 are given in Tables 4 and 5, respectively. Here, all the results are calculated considering maximum seasonal load.

After finding results for multiple PV, we have searched optimal location and size of DSTATCOM. The results for DSTATCOM are shown in Table 6 and results for PV + DSTATCOM are shown in Table 7. Here, all the results are calculated considering maximum seasonal load conditions. Here we took, case 1 for base case (without PV-DG and DSTATCOM), case 2 for with PV-DG only, case 3 for with DSTATCOM only, and case 4 is for with PV-DG and DSTATCOM both.

From Tables 6 and 7, we can see that only DSTATCOM or PV with DSTATCOM can improve voltage profile much better rather than multiple PV-DG. Here, we have done various analysis like % power loss reduction, minimum bus voltage, annual energy saving, and annual energy cost saving. Changes in system voltage profiles are shown in Figs. 4, 5, 6, and 7 for summer, autumn, winter, and spring, respectively.

From Fig. 4, we can say that, in the case of single PV + DSTATCOM, minimum bus voltage improves from 0.8508 to 0.9004 pu with PV size 166.7 kW and for DSTATCOM size 1000 KVAr. In autumn, the minimum bus voltage increases to 0.9534 pu from 0.9174 pu as shown in Fig. 5. As same, minimum bus voltage increases to 0.9422 pu from 0.8989 pu in winter and 0.9498 pu from 0.9113 pu in sping as shown in Fig. 6 and 7, respectively.

Table 4 Season wise optimal solution using FPA for single PV-DG allocation

Parameters	Summer	Autumn	Winter	Spring
Location of PV-DG @bus	94	28	28	28
Size of PV-DG (kW)	166.7	162.4	166.7	165.1
Maximum loading in %	100	59	71	63
TPL without PV kW	357.8774	111.8167	166.7586	128.7262
TPL with PV kW	321.0199	95.0073	144.5715	110.0833
% Reduction in TPL	10.29	15.03	13.31	14.48
Vmin (p.u) without PV-DG	0.8508	0.9174	0.8989	0.9113
Vmin (p.u) with PV-DG	0.8613	0.9263	0.9084	0.9205
Annual energy loss without PV (KWh)	3,135,006.024	979,514.292	1,460,805.336	1,127,898.964
Annual energy loss with PV-DG (KWh)	2,812,134.324	832,263.948	1,266,446.34	964,329.708
AES in KWh	322,871.7	147,250.344	194,358.996	163,569.256
ACS in $	19,372.302	8835.021	11,661.54	9814.155

Table 5 Season wise optimal solution using FPA for three PV-DG allocation

Parameters	Summer	Autumn	Winter	Spring
Locations of PV-DG @bus	94, 25, 24	28, 25, 85	28, 25, 85	28, 25, 85
Size of PV-DG (kW)	166.7, 166.7, 166.7	162.4, 162.4, 162.4	166.7, 166.7, 166.7	165.1, 165.1, 165.1
Maximum loading in %	100	59	71	63
TPL without PV kW	357.8774	111.8167	166.7586	128.7262
TPL with PV kW	262.535	70.4164	111.1215	82.5341
% Reduction in TPL	26.64	37.025	33.36	35.88
Vmin (p.u) without PV-DG	0.8508	0.9174	0.8989	0.9113
Vmin (p.u) with PV-DG	0.8805	0.94	0.9229	0.9346
Annual energy loss without PV (KWh)	3,135,006.024	979,514.292	1,460,805.336	1,127,898.964
Annual energy loss with PV-DG (KWh)	2,299,806.6	616,847.664	973,424.34	722,998.716
AES in KWh	835,199.424	362,666.628	487,380.996	404,642.796
ACS in $	50,111.96	21,759.99	29,242.86	24,278.56

5 Conclusions

In this paper, distribution system analysis on Portuguese 94 bus RDS is done. In this analysis, multiple PV-DG and single PV with DSTATCOM are installed in the system to get minimum active power loss and to improve the system voltage profile. For calculation of PV size, one-year hourly data of solar irradiance are taken from NREL and by PV modelling equations, DC power of PV-DG system has calculated. As PV system is dependent on weather conditions, seasonwise performance has been investigated. To find the best location and size of PV-DG and DSTATCOM to get minimum active power loss, a meta-heuristic search algorithm named Flower Pollination Algorithm (FPA) is used. Outcome concludes,

- Reduction in power loss with multiple PV is 26.64%, 37.025%, 33.36%, and 35.88% for summer, autumn, winter, and spring, respectively. For single PV and DSTATCOM, it is 27.79%, 31.33%, 30.25%, and 30.98% for summer, autumn, winter, and spring, respectively.
- Minimum system voltage improves drastically in the case of single PV and DSTATCOM. In case of multiple PV, minimum bus voltage improves to 0.8805 pu, 0.9400 pu, 0.9229 pu, and 0.9646 pu for summer, autumn, winter, and spring, respectively. For single PV and DSTATCOM, it improves to 0.9004 pu, 0.9534 pu,0.9422 pu, and 0.9498 pu for summer, autumn, winter, and spring, respectively.

Table 6 Season wise optimal solution using FPA for DSTATCOM allocation

Parameters	Summer	Autumn	Winter	Spring
Locations of DSTATCOM @bus	20	19	19	19
Size of DSTATCOM (KVAr)	1000	799	979	858
Maximum loading in %	100	59	71	63
TPL without DSTATCOM kW	357.8774	111.8167	166.7586	128.7262
TPL with DSTATCOM kW	291.9013	92.6509	136.9035	106.3435
% Reduction in TPL	18.44	17.14	17.9	17.38
Vmin (p.u) without DSTATCOM	0.8508	0.9174	0.8989	0.9113
Vmin (p.u) with DSTATCOM	0.8905	0.9448	0.9331	0.9409
Annual energy loss without DSTATCOM (KWh)	3,135,006.024	979,514.292	1,460,805.336	1,127,898.964
Annual energy loss with DSTATCOM (KWh)	2,557,055.388	811,621.884	1,199,274.66	931,569.06
AES in KWh	577,950.636	167,892.408	261,530.676	196,072.452
ACS in $	34,677.034	10,073.54	15,691.84	11,764.35

- Highest annual energy cost saving for multiple PV-DG is 50111.96 $ in summer, and for single PV + DSTATCOM is 52286.37 $ in summer.

The analysis carried out will help network operator to plan distribution system with Renewable Energy Sources.

Table 7 Season wise optimal solution using FPA with PV-DG and DSTATCOM allocation

Parameters	Summer	Autumn	Winter	Spring
Locations of PV-DG, DSTATCOM @bus	94, 20	28, 19	28, 19	28, 19
Size of PV-DG, DSTATCOM	166.7 kW, 1000 KVAr	162.40 KW, 799 KVAr	166.7 kW, 979 KVAr	165.10 kW, 858 KVAr
Maximum loading in %	100	59	71	63
TPL without PV-DG, DSTATCOM kW	357.8774	111.8167	166.7586	128.7262

(continued)

Table 7 (continued)

Parameters	Summer	Autumn	Winter	Spring
TPL with PV-DG, DSTATCOM kW	258.398	76.7842	116.3068	88.8376
% Reduction in TPL	27.79	31.33	30.25	30.98
Vmin (p.u) without PV-DG, DSTATCOM	0.8508	0.9174	0.8989	0.9113
Vmin (p.u) with PV-DG, DSTATCOM	0.9004	0.9534	0.9422	0.9498
Annual energy loss without PV-DG, DSTATCOM (KWh)	3,135,006.1	979,514.3	1,460,805.3	1,127,898.9
Annual energy loss with PV-DG, DSTATCOM (KWh)	2,263,566.4	672,629.6	1,018,847.5	778,217.4
AES in KWh	871,439.5	306,884.7	441,957.7	349,424.2
ACS in $	52,286.4	18,413.1	26,517.5	20,965.5

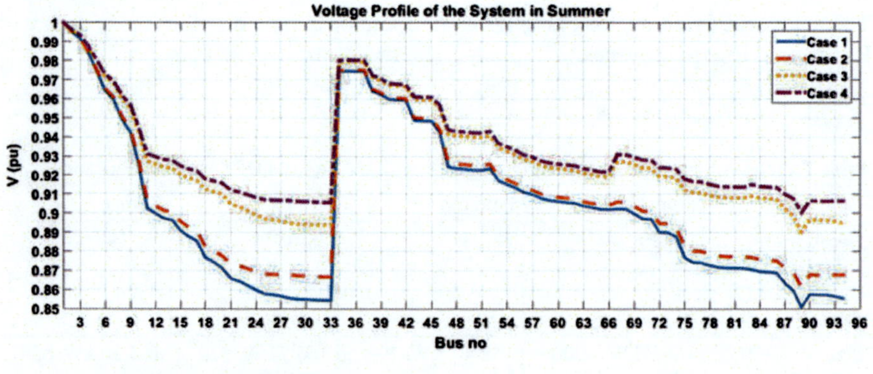

Fig. 4 Voltage profile of the system for summer

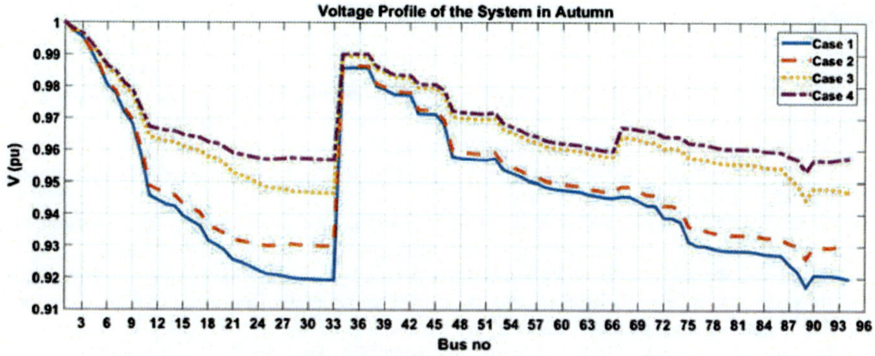

Fig. 5 Voltage profile of the system for autumn

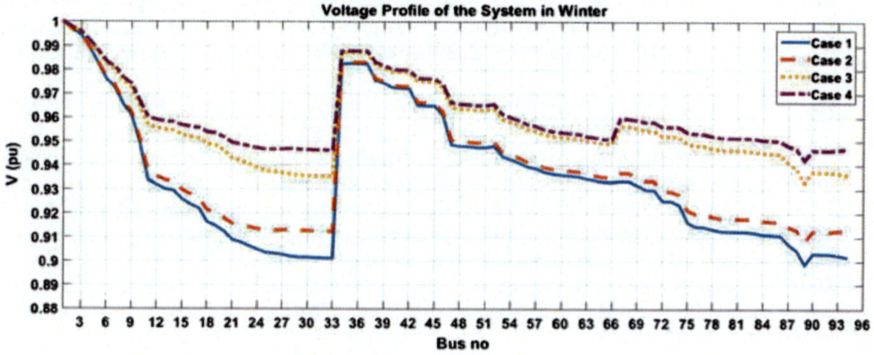

Fig. 6 Voltage profile of the system for winter

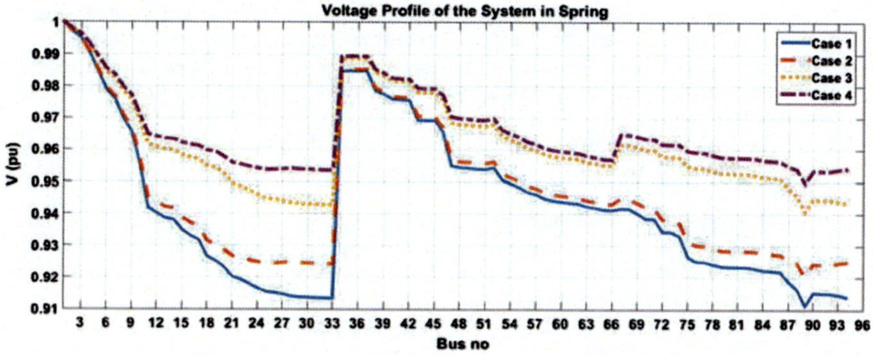

Fig. 7 Voltage profile of the system for spring

References

1. Mokryani G, Siano P, Piccolo A, Chen Z (2013) Improving fault ride through capability of variable speed wind turbines in distribution networks. IEEE Syst J 7(4):713–722
2. Banosa R, Manzano-Agugliaro F, Montoya FG, Gila C, Alcayde A, Gómez J (2011) Optimization methods applied to renewable and sustainable energy: a review. Renew Sustain Energy Rev 15(4):1753–1766
3. Kirschen D, Strbac G (2004) Fundamentals of power system economics. Wiley, Hoboken, NJ, USA
4. Palma-Behnke R, Cerda JLA, Vargas L, Jofre A (2005) A distribution company energy acquisition market model with the integration of distributed generation and load curtailment options. IEEE Trans Power Syst 20(4):1718–1727
5. Borghetti A, Bosetti M, Grillo S, Massucco S, Nucci CA, Paolone M, Silvestro F (2010) Short-term scheduling and control of active distribution systems with high penetration of renewable resources. IEEE Syst J 4(3):313–322
6. Radosavljevic J (2018) Metaheuristic optimization in power engineering, 1st edn. IET, London
7. Eid A, Abdel-Akher M (2019) Power loss reduction using adaptive PSO in unbalanced distribution networks. In: 2019 21st international middle east power systems conference (MEPCON). IEEE, pp 675–680
8. Gupta AR, Kumar A (2019) Annual energy savings with multiple DG and D-STATCOM allocation using PSO in DNO operated distribution network. In: Malik H, Srivastava S, Sood Y, Ahmad A (eds) Applications of artificial intelligence techniques in engineering. Advances in intelligent systems and computing, vol 698. Springer, Singapore
9. Kefayat M, Lashkar Ara A, Nabavi Niaki SA (2015) A hybridof ant colony optimization and artificial bee colony algorithm for probabilistic optimal placement and sizing of distributed energy resources. Energy Convers Manag 92:149–161
10. Das CK, Bass O, Kothapalli G, Mahmoud TS, Habibi D (2018) Optimal placement of distributed energy storage systems indistribution networks using artificial bee colony algorithm. Appl Energy 232(July):212–228
11. Evangelopoulos VS, Georgilakis PS (2014) Optimal distributed generation placement under uncertainties based on point estimate method embedded genetic algorithm. IET Gener Transm Distrib
12. Qin AK, Huang VL, Suganthan PN (2009) Differential evolution algorithm with strategy adaptation for global numerical optimization. IEEE Trans Evol Comput 13(2):398–417
13. Saxena NK, Kumar A (2021) Estimation of dynamic compensation for renewable-based hybrid DG in radial distribution system using least error iterative method. Iran J Sci Technol Trans Electr Eng 45:15–28
14. Devabalaji KR, Ravi K (2016) Optimal size and siting of multiple DG and DSTATCOM in radial distribution system using bacterial foraging optimization algorithm. Ain Shams Eng. J. 7(3):959–971
15. Sanjay R, Jayabarathi T, Raghunathan T, Ramesh V, Mithulananthan N (2017) Optimal allocation of distributed generation using hybrid grey Wolf optimizer. IEEE Access 5:14807–14818
16. Yang X-S (2012) Flower pollination algorithm for global optimization. In: Unconventional computation and natural computation, vol 9. Springer, Berlin, p 240
17. Alyasseri ZA, Khader AT, Al-Betar MA, Awadallah MA, Yang XS (2018) Variants of the flower pollination algorithm: a review. Springer, Berlin
18. Malik MZ, Kumar M, Soomro AM, Baloch MH, Farhan M, Gul M, Kaloi GS (2020) Strategic planning of renewable distributed generation in radial distribution system using advanced MOPSO method. Energy Rep 6:2872–2886
19. Teng JH (2000) A network-topology based three-phase load flow for distribution systems. Proc Natl Sci Council ROC (A) 24(4):259–264
20. Panchauri RK (Foreword) From sunlight to electricity: a practical handbook on solar photovoltaic applications, 2nd edn. New Delhi, TERI

21. Hung DQ, Mithulananthan N, Lee KY (2014) Determining PV penetration for distribution systems with time-varying load models. Power Syst IEEE Trans Energy Convers 29(6):3048–3057
22. Karaboga D (2005) An idea based on honey bee swarm for numerical optimization. Computer Engineering Dept, Erciyes University, Turkey. Technical report TR06
23. Pavlyukevich I (2007) L´evy flights, non-local search and simulated annealing. J Comput Phys 226:1830–1844

Modelling and Analysis of Vector Controlled Doubly Fed Induction Generator (DFIG)

Tanzeel Imtiyaz, Farhad Ilahi Bakhsh, and Anjali Jain

Abstract In the past few years, major increase in the exploitation and development of wind energy-based renewable energy systems have occurred. In wind energy systems, the power whether active or reactive coming from a wind turbine needs to be controlled and a doubly fed induction generator (DFIG) provides a viable solution for its control. DFIG is an induction machine with variable velocity having the stator connected to the grid directly and connection of the rotor with the grid is through back-to-back power electronics converters. In this paper, complete analysis and control strategy of DFIG is presented. The control strategy used is field-oriented vector control (VC) inspired from vector control of dc drives. VC makes independent control of torque and speed of a DFIG possible and hence giving the freedom to control the power of a DFIG-based wind energy system. The modelling and analysis of field-oriented VC DFIG is done in MATLAB/Simulink. From the analysis, it is found that by tuning the PI controller parameters properly, it is possible to control the speed and torque of DFIG independently.

Keywords Wind energy · Doubly fed induction generator (DFIG) · Vector control (VC) · Simulation

Nomenclature

$U_{r,s(dq)}$	Voltage of rotor and stator winding
$I_{r,s(dq)}$	Current of rotor and stator winding

T. Imtiyaz (✉) · A. Jain
Electrical Engineering Department, Amity University, Noida, Uttar Pradesh, India

A. Jain
e-mail: ajain5@amity.edu

F. I. Bakhsh
Electrical Engineering Department, National Institute of Technology Srinagar, Hazratbal, Jammu and Kashmir, India

A. R. Gupta et al. (eds.), *Power Electronics and High Voltage in Smart Grid*,
Lecture Notes in Electrical Engineering 817,
https://doi.org/10.1007/978-981-16-7393-1_21

$\phi_{r,s(dq)}$	Flux linkage of rotor and stator winding
$R_{r,s}$	Resistance of rotor and stator winding
$L_{s,r}$	Inductance of stator and rotor winding
L_m	Mutual inductance
p	No of poles pairs
T_e	Electromagnetic torque
$P_{r,s}$	Active rotor and stator power
$Q_{r,s}$	Reactive rotor and stator power
ϕ_s	Stator flux linkage

1 Introduction

In the last few decades, energy and environmental crisis came up as a major problem because of the increase in the demand of electrical power and depletion of natural resources. This has drawn a lot of attention to renewable sources of energy like that of solar and wind energy [1, 2]. The production of energy from wind has grown magnificently because of its renewability and cleanliness. In the year 2020, the power production from wind sources is about 10–12% of the total power of the world and it is anticipated to be two-fold or even more by 2040 [3, 4].

No doubt wind energy is becoming more famous with every passing day, it has several disadvantages such as efficiency. For any regular wind turbine, the average capacity factor is around 33% and it can go down to 22.5% because of wind energy's intermittent nature. Thus, transforming existing wind energy into electrical energy is a very vital step [5]. One more disadvantage is that the reactive power which the generator produces or consumes has to be kept within limits to preserve the power quality of the system [5, 6].

There are basically two categories in which wind turbines are divided i.e., constant and variable speed. The variable wind turbines are more commonly adapted because of various advantages like less mechanical stress on the turbine shaft, maximum power generated, and improved power quality [7]. The variable wind turbines can use synchronous generator (SG) with full-scale power electronics converters or DFIG with partial scale power electronics converters for generation purpose [8]. The DFIGs have less power losses and converters are relatively cheap than SG, hence they are largely adapted for high power applications. The use of small power electronics converters for control mechanism is the principal cause for DFIGs popularity in wind energy applications [9, 10].

Controlling the DFIG is very vital as the input (wind energy) to the generator is varying all the time and the generator should be able to operate in harmony with any variation in wind speed. The DFIG consists of the rotor side converter (RSC) controlling the stator power, and the grid side converter (GSC) that adjusts the voltage of dc bus and helps in controlling the absorbed reactive power of the grid [11].

Aydin et al. [5], presented various control strategies for DFIG, concluding that field-oriented VC is the best one and by proper control algorithm independent control of the reactive power and active power and adjusting the voltage dc bus of a DFIG is possible. Mohammadi et al. [3], have done a combined direct power and vector control strategy of DFIG-based wind energy systems and concluded that a combined control strategy gives lower power ripples and high dynamic response.

Li and Haskew [12], did analysis on decoupled d–q VC for a DFIG. In most of the wind power applications, control is done for the GSC. The active power is controlled by direct axis component of current and quadrature axis component of current helps in controlling the reactive power. Thus, in DFIG, the most familiar control technique is VC because of advantages like increased efficiency, lower switching frequency of converter, precise steady-state performance, and lower power ripples [13].

This paper is divided into six sections. Section 2 presents the doubly fed induction generator model. Section 3 gives the proposed control strategy. Section 4 implements the simulation model of VC of DFIG. Section 5 shows the results obtained using vector control strategy. At last, Sect. 6 concludes the paper.

2 Doubly Fed Induction Generator (DFIG)

DFIGs are quite famous for wind energy applications as they have the capability to deliver maximum power at variable wind speeds. DFIG is advantageous in high power applications by providing extra output power than the rated power and reducing the overall cost [14]. It also increases the overall system efficiency.

Figure 1 represents the DFIG connected to wind energy system block diagram. At constant amplitude and frequency, stator is fed with three-phase voltages from the grid and stator magnetic field is generated. Grid gives supply to the rotor but at different amplitude and frequency so as to operate the machine in different conditions.

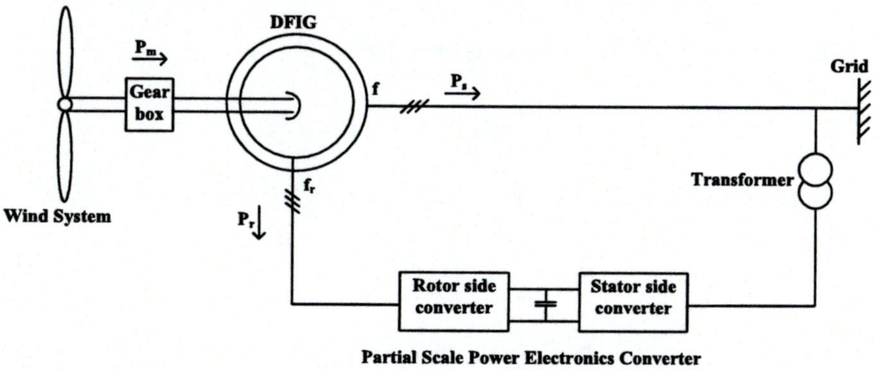

Fig. 1 DFIG connected to wind energy system block diagram

A back-to-back three-phase power converter is connected that permits bidirectional power flow [15, 16].

2.1 DFIG Modelling

To develop DFIG model, space vector theory is applied to the fundamental electrical equations of DFIG [17]. The DFIG's mathematical model in d–q frame designed for stator side and rotor side is given by the Eqs. (1)–(8), respectively.

$$Usd = RsIsd - \phi sq\omega s + \frac{d\phi sd}{dt} \tag{1}$$

$$Usq = RsIsq - \phi sd\omega s + \frac{d\phi sq}{dt} \tag{2}$$

$$\phi sd = LmIrd + LsIsd \tag{3}$$

$$\phi sq = LmIrq + LsIsq \tag{4}$$

$$Urd = RrIrd - \phi rq\omega r + \frac{d\phi rd}{dt} \tag{5}$$

$$Urq = RrIrq - \phi rd\omega r + \frac{d\phi rq}{dt} \tag{6}$$

$$\phi rd = LmIsd + LrIrd \tag{7}$$

$$\phi rq = LmIsq + LrIrq \tag{8}$$

The DFIG electromagnetic torque is given by

$$Te = p\frac{Lm}{Ls}(\phi sq.Ird - \phi sd.Irq) \tag{9}$$

The stator and rotor reactive and active power is specified as under

$$Qs = \frac{3}{2}(UsqIsd - UsdIsq) \tag{10}$$

$$Ps = \frac{3}{2}(UsdIsd + UsqIsq) \tag{11}$$

$$Qr = \frac{3}{2}(UrqIrd - UrdIrq) \tag{12}$$

$$Pr = \frac{3}{2}(UrdIrd + UrqIrq) \tag{13}$$

2.2 *Operating Principle*

In the DFIG, the stator windings are connected to the grid directly while the rotor windings are connected to the grid via slip rings or brushes and power electronics converters are present to have a control on the currents of stator and rotor. The set of brushes or slip-rings that are connected to the rotor are there to inject or absorb the power from winding of the rotor. By controlling the currents of rotor, the power of a DFIG is controlled independently [11].

3 Vector Control (VC)

To control any electrical machine, field-oriented VC is one of the most common techniques used. It was first developed by Hasse and Blaschke [18] and the basic idea behind it is independent control of speed and torque in ac drives just like that in dc drives. In VC, the phase currents are first transformed to αß reference frame and after that to d–q reference frame by Clarke and Park transformation [19].

From Eq. (9), it is clear that there is coupling between stator and rotor fluxes and currents making the control difficult. The basic principle of VC is alignment of the flux of DFIG to one of the axis, d or q. In this paper, the space vector of stator flux is aligned with direct axis. d–q frame of reference rotating synchronously is shown in Fig. 2. Due to this choice of alignment, the speed and torque are proportional to the direct axis rotor current component and quadrature axis rotor current component controls the reactive power.

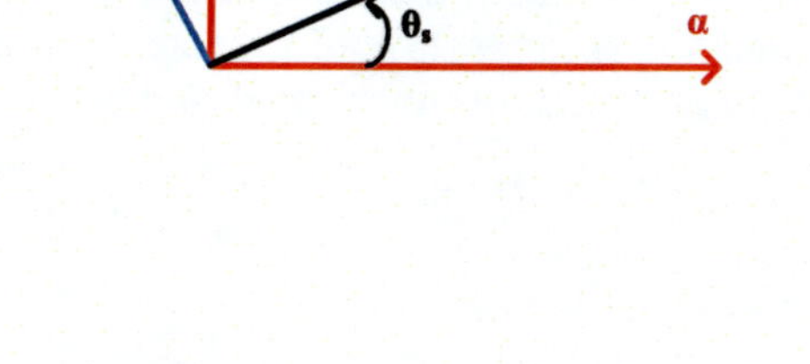

Fig. 2 d–q frame of reference rotating synchronously

The rotor voltages in d–q reference frame are shown as under

$$Urd = RrIrd + \sigma Lr\frac{dIrd}{dt} - \omega r\sigma LrIrq + \frac{Lm}{Ls}\frac{d\psi s}{dt} \tag{14}$$

$$Urq = RrIrq + \sigma Lr\frac{dIrq}{dt} + \omega r\sigma LrIrd + \omega r\frac{Lm}{Ls}\frac{d\psi s}{dt} \tag{15}$$

where,

$$\sigma = 1 - \frac{Lm^2}{LsLr} \tag{16}$$

.

With the space vector of stator flux aligned along direct axis, the electromagnetic torque equation gets modified as

$$Te = -\frac{3}{2}p\frac{Lm}{Ls}\psi sIrq \tag{17}$$

The modified reactive power equation is given as

$$Qs = -\frac{3}{2}\frac{Lm}{Ls}\omega s\psi s(Ird - \frac{\psi s}{Lm}) \tag{18}$$

Thus, it is clear from Eqs. (17) and (18) that d-axis and q-axis components of current help in controlling the torque and reactive power. Figure 3 represents the VC of DFIG. The figure very well illustrates that control is performed for d–q components of rotor current by use of one PI controller for each current component. The currents to be controlled are first transformed to d–q coordinate system and then they are transformed to αß coordinate system making independent control possible. Furthermore, two more PI controllers are used for speed and reactive power control of DFIG.

4 Simulation Model of Vector Control of DFIG

Figure 4 shows the simulation model of VC of DFIG. VC is implemented on the rotor-side of DFIG based on cascade multi-loop control. The simulation model is consisting of three-phase voltage source block acting as the grid, one V-I measurement block for measuring the stator currents and voltages and another V-I measurement block used to measure rotor currents, DFIG block, universal bridge block, PWM modulation block, and finally, the VC block. The input to the DFIG block is the load torque which is taken as 25% of the rated torque. The outputs from the DFIG block is the rotor angle, the rotor speed, and the electromagnetic torque present in the block

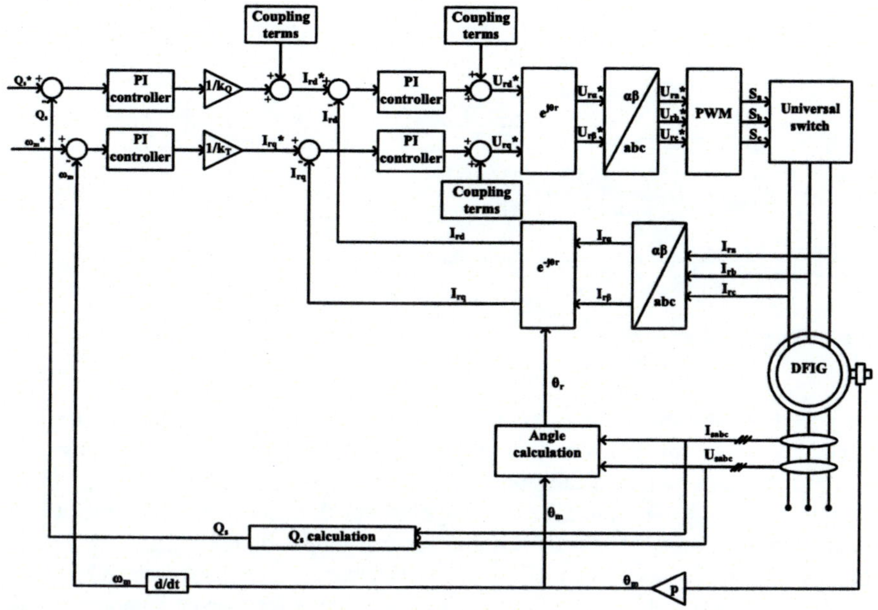

Fig. 3 Vector control of DFIG

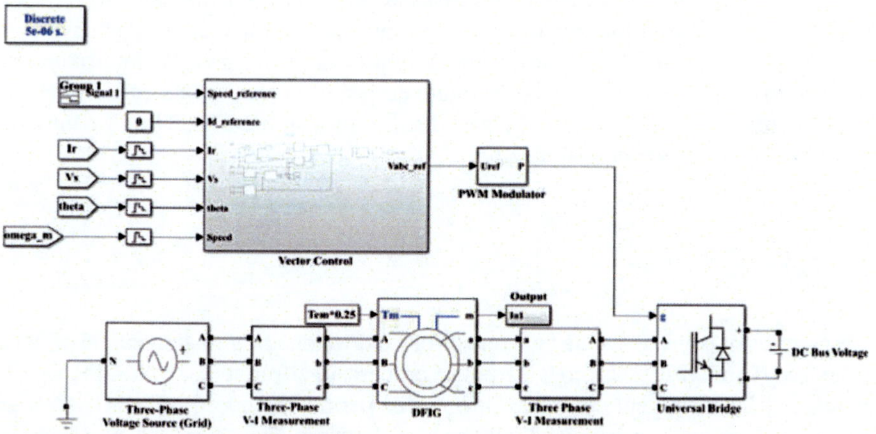

Fig. 4 Simulation model of VC of DFIG

named as output. The universal bridge is short-circuited with a simple dc voltage. In this simulation, PWM modulation block is used to control the power electronics converter (universal bridge block). At last, the vector control block is simulated for the entire system control.

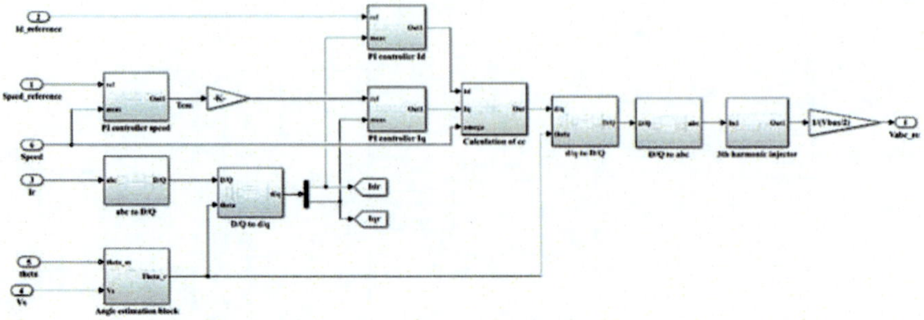

Fig. 5 Vector control subsystem

Vector control subsystem is represented in Fig. 5. The inputs to the subsystem are speed reference, direct axis current reference, rotor current, stator voltage, mechanical angle and speed, and the output is the stator voltage references. The PWM modulation block uses normalized triangular waveform ones so the output of VC subsystem must also be normalized and for that output is multiplied with a reciprocal of half of dc bus voltage of the power electronics converter. A third harmonic injection block is used in order to extract more voltage for a given dc boost voltage of power electronics converter and this block actually helps to increase 15% of the output voltage for a given dc boost voltage. Then αß to abc and d/q to αß conversion is done with the help of fundamental matrices of Inverse Clarke and Inverse Park transformation [20]. Similarly, for rotor current, abc to αß and αß to d/q conversion is done by Clarke and Park transformation matrices [20]. Three PI controllers are adapted to control the direct and quadrature currents and speed. The quadrature axis current component will control the speed and torque of a DFIG. The PI controller values are tuned to get desired results.

5 Results

This section of paper gives the final results of the developed simulation model of field-oriented VC of DFIG. Through the entire process, i.e., from t = 0 to 6 s, the supply voltage of DFIG is kept constant. During this period, the speed of DFIG is varied while load torque is maintained constant as shown in Figs. 6 and 7, respectively. This effect is studied on rotor currents, stator currents, voltage d–q components, and current d–q components as presented in Figs. 8, 9, 10, 11, 12 and 13, respectively. To explain the effect of change in speed on different quantities of DFIG, all the graphs are studied in three-time intervals, i.e., from t = 0 to 2 s, from t = 2 to 4 s, and t = 4 to 6 s.

Figure 6 illustrates the speed of DFIG. It is evident that the DFIG is started at zero speed at t = 0 s and then the speed is linearly increased to 65% of rated speed at t = 0.5 s. From t = 0.5 s to t = 2 s, speed is kept constant at 65% of rated speed and then

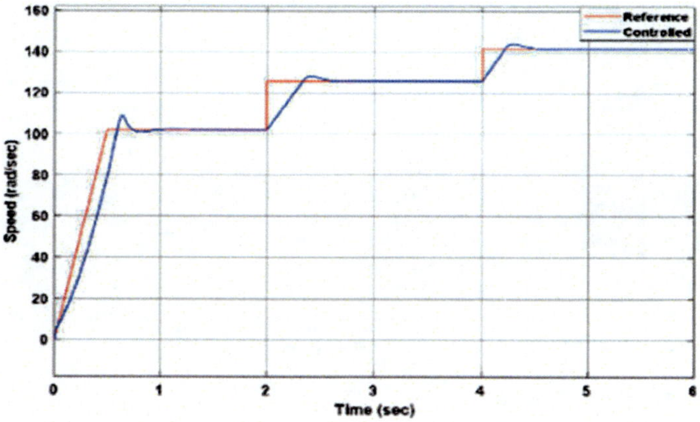

Fig. 6 Speed of DFIG

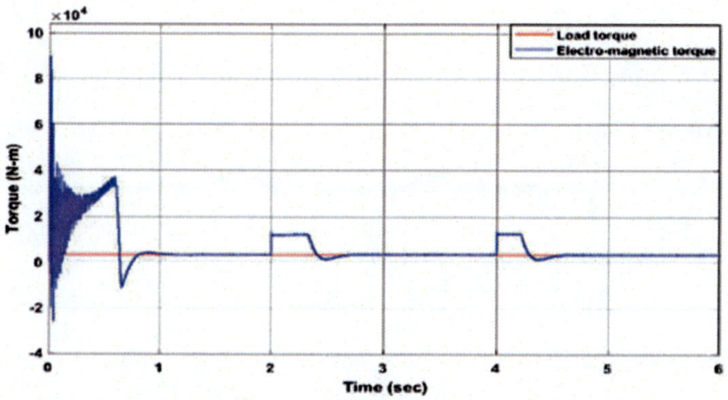

Fig. 7 Load torque and electromagnetic torque of DFIG

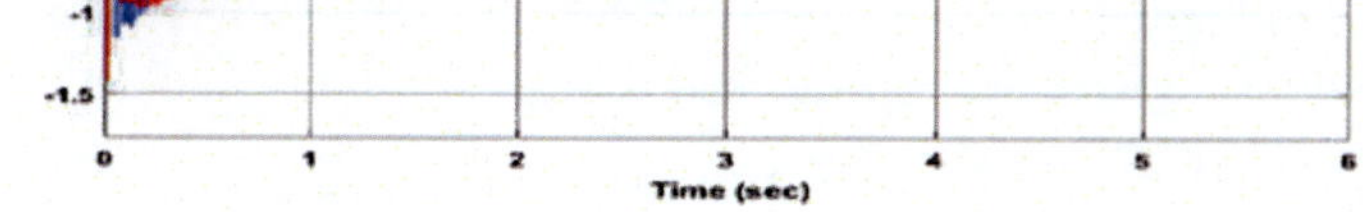

Fig. 8 Currents of stator windings

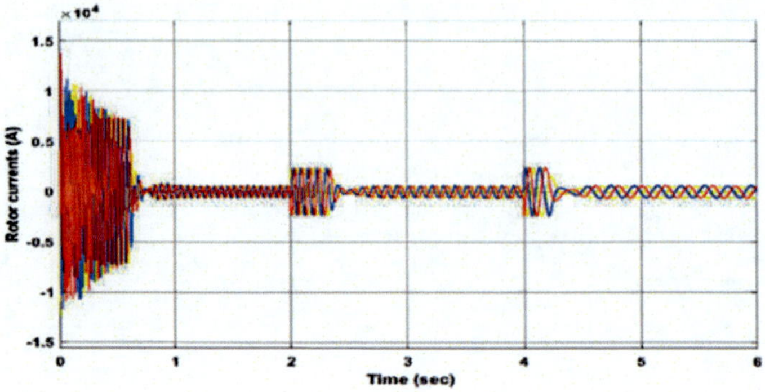

Fig. 9 Currents of rotor windings

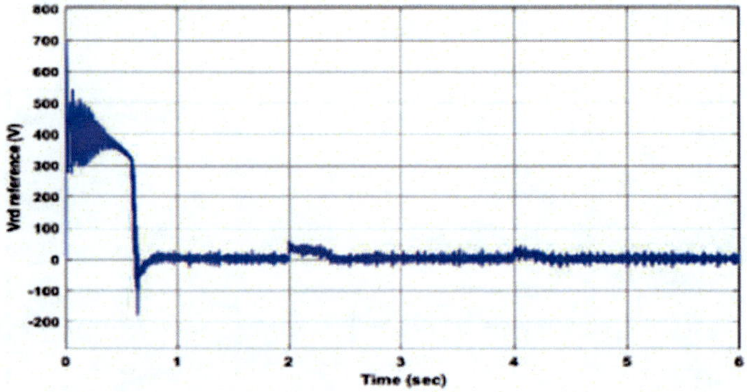

Fig. 10 Direct component of reference rotor voltage

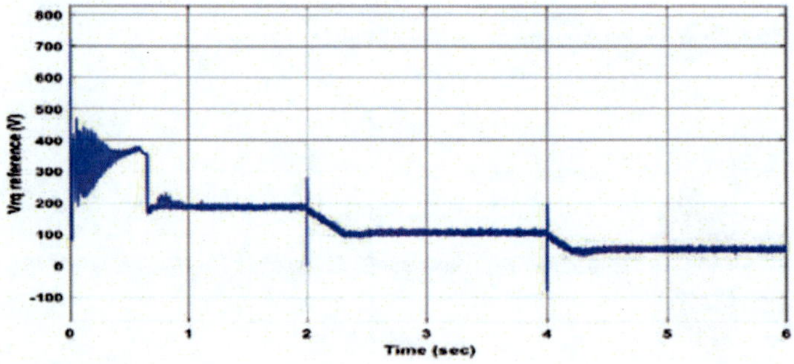

Fig. 11 Quadrature component of reference rotor voltage

at t = 2 s, increase in speed is from 65 to 80% of rated speed and then kept constant to this value till t = 4 s. The speed of DFIG is further amplified from 80 to 90% of the rated speed at t = 4 s and then maintained constant to this value till t = 6 s. Figure 7 gives the load torque and electromagnetic torque of DFIG. The load torque is kept constant at 25% (3183 N-m) of the rated DFIG torque and the electromagnetic torque is of varying nature. Electromagnetic torque is varying because of change in speed which causes variation in electromagnetic torque and after sometime, this torque comes back to its original value that is equal to the load torque.

It is evident from Figs. 8, 9, 10, 11 and 12 that in the first time interval from t = 0 to 2 s, the quantities are high during sub-transient state (t = 0–0.6 s) and then the quantities are comparatively low in transient state (t = 0.6–1 s) from sub-transient state. These quantities finally settle down to steady-state value at around t = 1 s and remain constant to their respective values till t = 2 s.

There is further increase in the speed of DFIG to 80% of its rated speed at t = 2 s. Due to increase in speed:

1. The DFIG electromagnetic torque is initially increased to 12,732 N-m and then becomes equal to the load torque (3183 N-m) at t = 2.6 s and remains at this value till t = 4 s.
2. Stator currents, rotor currents, reference rotor voltage d-axis and q-axis components values increase in sub-transient state (t = 2–2.3 s) and then decrease in the transient state (t = 2.3–2.6 s). These quantities reach their respective steady state values at t = 2.6 s and remain at that value till t = 4 s.
3. Rotor current q-axis component value undershoots in sub-transient state (t = 2–2.3 s) and then overshoots in transient state (t = 2.3 s to t = 2.6 s). It reaches steady-state value at t = 2.6 s and then remain in it till t = 4 s.

At t = 4 s, there is further increase in the speed of DFIG to 90% of its rated speed. Due to this increase in speed:

1. The DFIG electromagnetic torque is initially increased to 12,732 N-m and then at t = 4.6 s becomes equal to the load torque (3183 N-m) and remains at this value till t = 6 s.
2. Values of stator currents, rotor currents, reference rotor voltage d-axis and q-axis components are higher in sub-transient state (t = 4–4.2 s) than their values in the transient state (t = 4.2–4.6 s). These quantities reach their respective steady state values at t = 4.6 s and remain constant till t = 4 s.
3. Rotor current q-axis component value undershoots in sub-transient state (t = 4–4.2 s) and then overshoots in transient state (t = 4.2–4.6 s). At t = 4.6 s, it reaches steady state value and is same till t = 6 s.

It is clear from Fig. 13 that the rotor current d-axis component is having some oscillations at the start and at t = 0.6 s it starts to follow its reference signal.

It is clear from Figs. 6, 7, and 12, that by proper tuning of PI controller parameters the controlled quantities follow the reference quantities in a quite good manner. Thus, by using the proposed control strategy the speed of the DFIG can be controlled while

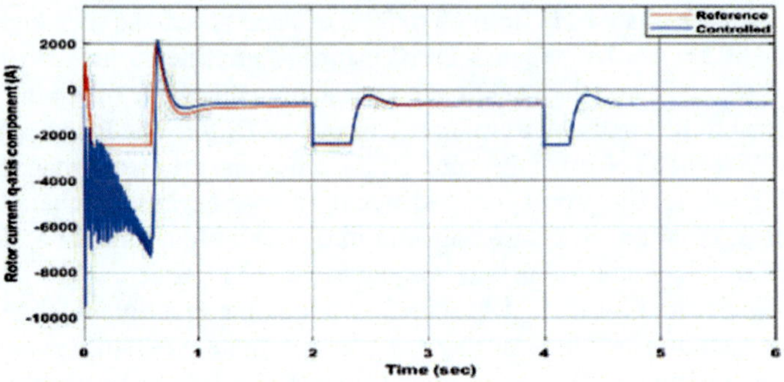

Fig. 12 Rotor current q-axis component

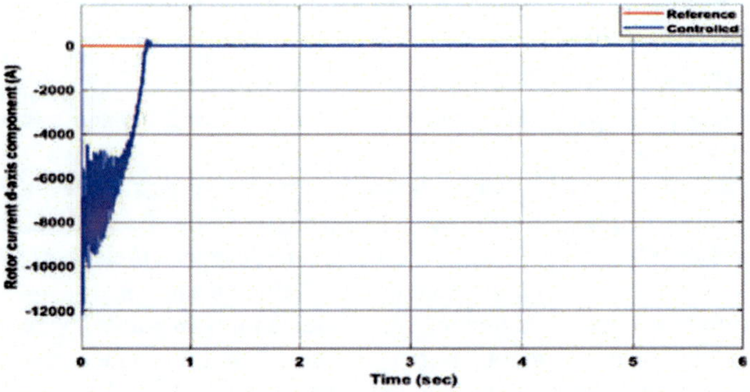

Fig. 13 Rotor current d-axis component

keeping its electromagnetic torque constant. Hence, the speed and torque of DFIG are independently controlled.

6 Conclusion

This paper presents the implementation of the field-oriented VC of DFIG. For its analysis, first a set of equations has been modelled in MATLAB/Simulink. Then the control for developed model is done. The control strategy shows that by doing some mathematical operations the three-phase currents and voltages of a DFIG can be transformed into two components (d–q) which makes the control easy. Due to which, controlling the speed and electromagnetic torque of DFIG is possible independently. The obtained results of simulation model depict that using the presented control strategy one can regulate the speed and torque of DFIG independently with decent dynamic performance by proper tuning of PI controllers (Table 1).

Table 1 Simulation parameters

Parameters	Value
Rated frequency	50 Hz
Rated current of stator	1.76 kA
Rated voltage of stator	0.69 kV
Rated torque	12,732 N-m
Rated power of stator	2 MW
Rated rotational speed	1500 rpm
Resistance of rotor and stator winding	0.0029 Ω and 0.0026 Ω
Mutual inductance	0.0025 H
Leakage inductance	0.000087 H
Stator and rotor winding inductance	0.0026 H
No of pole pairs	2
Inertia constant	127
Damping constant	0.001
Stator flux linkage estimation	1.79
DC bus voltage	325.3 V
Switching frequency of converter	4×10^4 Hz

References

1. Malik H, Iqbal A, Joshi P, Agrawal S, Bakhsh FI (2020) Meta heuristic and evolutionary computation: algorithms and applications. In: Studies in computational intelligence, vol 1096. Springer, Singapore. https://doi.org/10.1007/978-981-15-7571-6
2. Al-Dousari A, Al-Nassar W, Al-Hemoud A, Alsaleh A, Ramadan A, Al-Dousari N, Ahmed M (2019) Solar and wind energy: challenges and solutions in desert regions. Energy 176:184–194. https://doi.org/10.1016/j.energy.2019.03.180
3. . Mohammadi J, Zadeh SV, Afsharnia S, Daryabeigi E (2014) A combined vector and direct power control for DFIG-based wind turbines. IEEE Trans Sustain Energy 5(3):767–775. https://doi.org/10.1109/tste.2014.2301675
4. Bakhsh FI, Shees MM, Asghar MSJ (2014) Performance of wound rotor induction generators with the combination of input voltage and slip power control. In: Russian Electr Eng 85(6):403–417. https://doi.org/10.3103/s1068371214060042
5. Aydin E, Polat A, Ergene LT (2016) Vector control of DFIG in wind power applications. In: IEEE international conference on renewable energy research and applications (ICRERA), pp 478–483. https://doi.org/10.1109/icrera.2016.7884383
6. Bakhsh FI, Khatod DK (2016) A new synchronous generator-based wind energy conversion system feeding an isolated load through variable frequency transformer. Renew Energy 86:106–116. https://doi.org/10.1016/j.renene.2015.07.093
7. Menezes EJN, Araújo AM, da Silva NSB (2018) A review on wind turbine control and its associated methods. J Cleaner Product 174:945–953. https://doi.org/10.1016/j.jclepro.2017.10.297
8. Mohseni M, Islam SM, Masoum MA (2010) Enhanced hysteresis-based current regulators in vector control of DFIG wind turbines. In: IEEE Trans Power Electron 26(1):223–234. https://doi.org/10.1109/tpel.2010.2058816

9. Singh P, Kaur A (2014) Power control of doubly fed induction generator (DFIG) using back -to-back converters (PWM technique). In: IEEE international conference on advanced communications, control and computing technologies, Ramanathapuram, pp 73–77. https://doi.org/10.1109/icaccct.2014.7019202
10. Bouderbala M, Bossoufi B, Lagrioui A, Taoussi M, Aroussi HA, Ihedrane Y (2018) Direct and indirect vector control of a doubly fed induction generator based in a wind energy conversion system. Int J Electr Comput Eng (IJECE) 9(3):1531–1540. https://doi.org/10.11591/ijece.v9i3.pp1531-1540
11. Sheikh H, Bakhsh FI (2020) Maximum power extraction and monitoring from wind power generation system using intelligent controllers. In: Soft computing in condition monitoring and diagnostics of electrical and mechanical systems. Springer, Singapore, pp 345–362. https://doi.org/10.1007/978-981-15-1532-3_15
12. Li S, Haskew TA (2007) Analysis of decoupled d–q vector control in DFIG back-to-back PWM converter. In: IEEE power engineering society general meeting, 1–7. https://doi.org/10.1109/pes.2007.385461
13. Poza J, Oyarbide E, Sarasola I, Rodriguez M (2009) Vector control design and experimental evaluation for the brushless doubly fed machine. IET Electr Power Appl 3(4):247–256. https://doi.org/10.1049/iet-epa.2008.0090
14. Zeeshan A, Sinha SK (2016) Control of DFIG under unbalanced grid voltage conditions: A literature review. In: International conference on control, computing, communication and materials (ICCCCM), pp 1–6. https://doi.org/10.1109/iccccm.2016.7918253
15. Xu L, Cartwright P (2006) Direct active and reactive power control of DFIG for wind energy generation. IEEE Trans Energy Convers 21(3):750–758. https://doi.org/10.1109/tec.2006.875472
16. Cardenas R, Pena R, Alepuz S, Asher G (2013) Overview of control systems for the operation of DFIGs in wind energy applications. In: IEEE Trans Ind Electron 60(7):2776–2798. https://doi.org/10.1109/iecon.2013.6699116
17. Rub HA, Malinowski M, Haddad KA (2014) Power electronics for renewable energy systems, transportation and industrial applications. Wiley. https://doi.org/10.1002/9781118755525
18. Loucif M, Boumediene A, Mechernene A (2013) Backstepping control of double fed induction generator driven by wind turbine. In: 3rd international conference on systems and control, pp 153–158. https://doi.org/10.1109/icosc.2013.6750851
19. Rourk CJO, Qasim MM, Overlin MR, Kirtley JL (2019) A geometric interpretation of reference frames and transformations: dq0, clarke, and park. In: IEEE Trans Energy Conv 34(4):2070–2083. https://doi.org/10.1109/tec.2019.2941175
20. Chattopadhyay S, Mitra M, Sengupta S (2011) Clarke and park transform. In: Electric power quality. Springer, Dordrecht, pp 89–96. https://doi.org/10.1007/978-94-007-0635-4_12

Mitigation of Voltage Disturbances in Photovoltaic Fed Grid System Using Cascaded Soft Computing Controller

Kasa Sudheer, Suresh Penagaluru, and N. Prabaharan

Abstract Significant growth in global technology demands more energy from utility. Renewable integration is feasible option to serve growing energy demand in near future. Stochastic nature of renewable sources and nonlinear loads hinder over all power quality of the system. Felicitous control techniques and optimum utilization of renewable sources provide an ancillary service in improving power quality (PQ) of the system. Paper presents comparative analysis of soft computing-based control strategies for proposed PV connected Dynamic voltage Restorer to alleviate abnormalities in voltages on distribution feeder. Performance analysis of proposed control strategies is carried out in MATLAB/SIMULINK platform. Harmonic reduction and suppression of voltage disturbances are analyzed for the proposed parallel feeder distribution system.

Keywords Particle swarm optimization · ANFIS controller · Power quality · Power conditioning devices · PI controller

1 Introduction

The main hurdles for designing an efficient power system for the current modern world are power quality (PQ) and pollution-free power. The most facing issues in residence and industrial loads regarding PQ are voltage disturbances like voltage sag, power interruption, and voltage swell [1, 2]. The usual causes for voltage sag are induction motors' high starting current, lightening strokes, inrush currents, and

K. Sudheer (✉) · S. Penagaluru
SV College of Engineering, Tirupati, India

S. Penagaluru
e-mail: suresh.penagaluru@hmail.com

N. Prabaharan
SASTRA Deemed University, Thanjavur, Tamil Nadu, India
e-mail: prabaharan@eee.sastra.edu.in

A. R. Gupta et al. (eds.), *Power Electronics and High Voltage in Smart Grid*,
Lecture Notes in Electrical Engineering 817,
https://doi.org/10.1007/978-981-16-7393-1_22

short circuit faults [3]. The No linear loads cause harmonics in voltage and currents which affect the consumer loads [4].

DVR is the most cost-effective power conditioning device connected in series with the line to mitigate voltage imperfections such as sag/swell and interruption [5]. In literature, PV fed DVR topologies [6, 7] proven added advantages over energy storage-based DVR topologies [8].

DVR consists of a voltage source inverter connected between grid and sensitive voltage, which injects a voltage through injection transformer to compensate the voltage imperfection in sensitive load [9]. IEEE 1159–1995 and IEEE-519–1992 describe the standards and definition of voltage sag/swell. The duration of voltage swell/sag is 10 ms to 1 min [10, 11]. Block Diagram of proposed DVR model is shown in Fig. 1. Key objectives of proposed model are to alleviate sag and interruption along with harmonic reduction in Load voltage.

Paper is organized as Sect. 1, gives the literature information about DVR, Sect. 2 explains the proposed system and its controlling action, Sect. 3 presents supporting simulated results for the proposed system, and section for draws the keys highlights of proposed system.

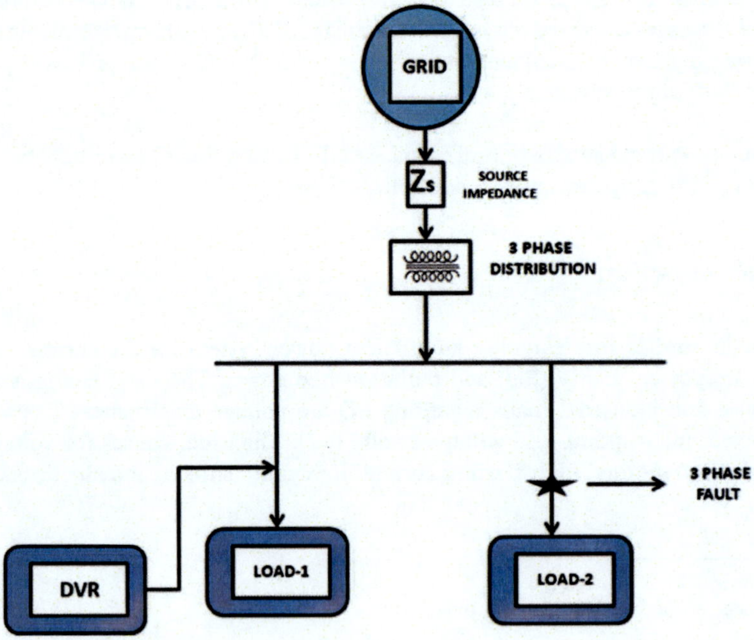

Fig. 1 Block Diagram

2 Description of Proposed Model

Proposed single-line diagram is presented in Fig. 1. Considered 3 phase 3 wire system consists of two adjacent distribution feeders connected with sensitive loads. Fault is considered on Line-2 which causes voltage disturbance on Line-1. DVR is placed at Line-1 through the voltage injection transformer to compensate the disturbances in sensitive load-1 voltage. Voltage source inverter, Injection transformer, and PV unit are the internal parts of DVR in the presented model. Detailed proposed model is shown in Fig. 2. Control technique description and working of proposed PV fed DVR are explained as follows.

2.1 PV Unit

Here, the photovoltaic unit uses P&O based MPPT technique, due to its predominant qualities over incremental conductance method [12, 13]. The connected PV provides the real power required by the DVR to inject the compensating voltage [14]. The governing equations of PV unit are given by,

$$P_P = V_P * I_P \tag{1}$$

Solar cell terminal voltage & currents given by

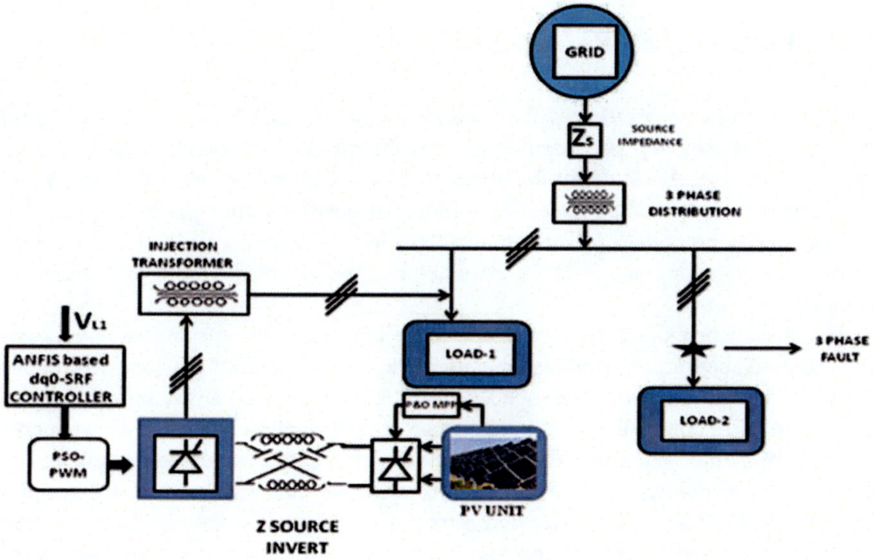

Fig. 2 Proposed System

Table 1 Modes of ZSI

S4	S3	S2	S1	Mode
1	0	0	1	Active
0	1	1	0	
0	1	0	1	Zero
1	0	1	0	
0 or 1	0 or 1	1	1	Shoot through

$$I_p = I_{sc} - I_{sat}\exp[\frac{q}{AkT}(V_p + I_p R_{se}) - 1] - \frac{V_p + I_{sc} R_{se}}{R_{sh}} \quad (2)$$

$$V_p = \frac{AkT}{q}\ln\{\frac{I_{sc}}{I_p} + 1\} \quad (3)$$

2.2 *Impedance Source Inverter (ZSI)*

It comprises of lattice structured impedance circuit. In comparision with traditional inverters like voltage & current source, ZSI holds Buck-Boost capabilities. They improvize reliability as ZSI's are immune for EMI noise (Table 1).

2.3 *Working of DVR Controller*

Due to the fault occurred on Line-2 cause's voltage disturbance on Line-1 Load voltage. DVR uses in phase compensation technique to compensate disturbances in load side voltage. DVR controller uses ANFIS-dq0-Synchronous reference frame technique to generate the reference currents for desired gating pulses to inject the compensated component for maintaining stable load voltage. PSO aids the PWM process by selecting the best PI parameters. The detailed control technique is described in Fig. 3.

In dq0-synchronous reference frame technique, initially sensed source voltages transformed abc frame to dq0 frame using park's transformation. These dq0 components are compared with reference values (d = 1 & q = 0) and then response is given to ANFIS controller. With inverse park transformation tuned dq0 components inturn transformed back to abc frame. These desired abc frame signals are fed to Pulse width modulator to generate the consequential gate pulses for IGBT-based 3 leg VSI. The LC filter used after VSI eliminates the harmonics from the inverter to the line. With the voltage injection transformer required compensating voltage is fed

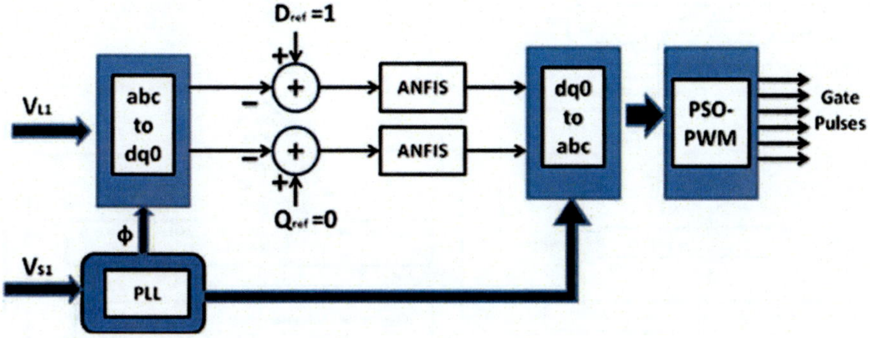

Fig. 3 ANFIS-dq0-SRF-based DVR Controller

to the line during abnormality in the load voltage. In this model, the dqo components error signal is tuned with different control strategies to study the harmonic suppression in the load voltage. They are traditional PI controller, Fuzzy controller, and ANFIS controller.

2.4 PSO (Particle Swarm Optimization)

Figure 4 presents PSO technique used for selecting PI parameters Kp & Ki values in PWM controller. PSO process is depicted below (Table 2).

- Particle fitness value is computed for every sample
- GBP(global best position) & fitness values are updated
- Update particle velocity & position.

2.5 ANFIS

ANFIS is a Multilayer network that makes use of both Fuzzy inference mechanism [15, 16] and Neural Networks adaptive learning[17] features. The Neuro-Fuzzy combination has proven its ability in tuning the error signal [18]. ANFIS controller effectively deals with the dq components smoothening for achieving desired reference dq components. Over PI controller ANFIS has predominant nonlinear characteristics. ANFIS structure and rules considered in the model are shown in Figs. 5 and 6.

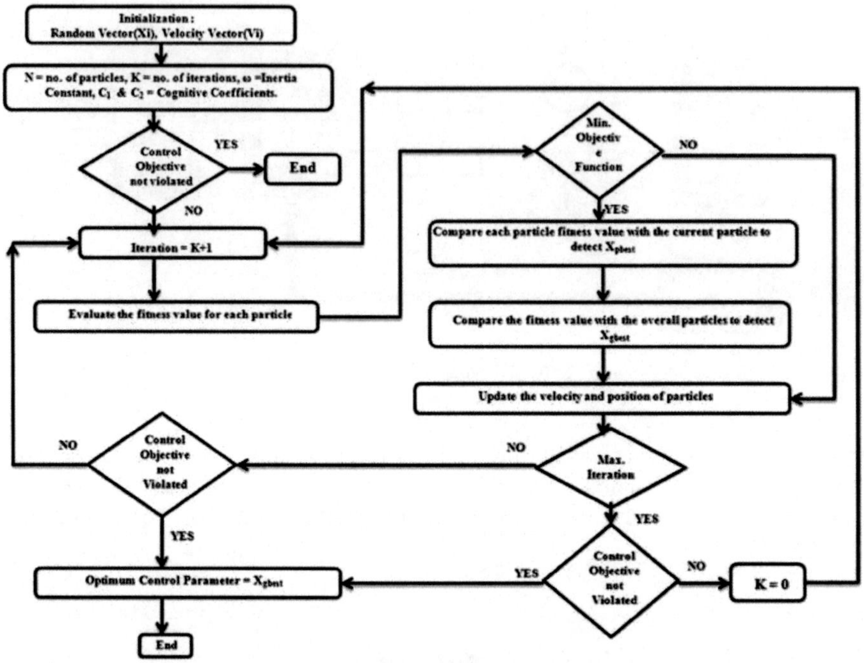

Fig. 4 Flow chart of PSO Algorithm

Table 2 PSO Initialization

Parameter	Value
Inertia, w	0.7
C_1, C_2	1,0
Fitness	1
V_{max}, V_{min}	1, −1
Swarm size	40
Kp, Ki	1.085, 0.732

3 Simulation Results

The proposed PV fed ANFIS-dqo-SRF based DVR controlled three-phase system SIMULINK model is shown in Fig. 7 (Table 3).

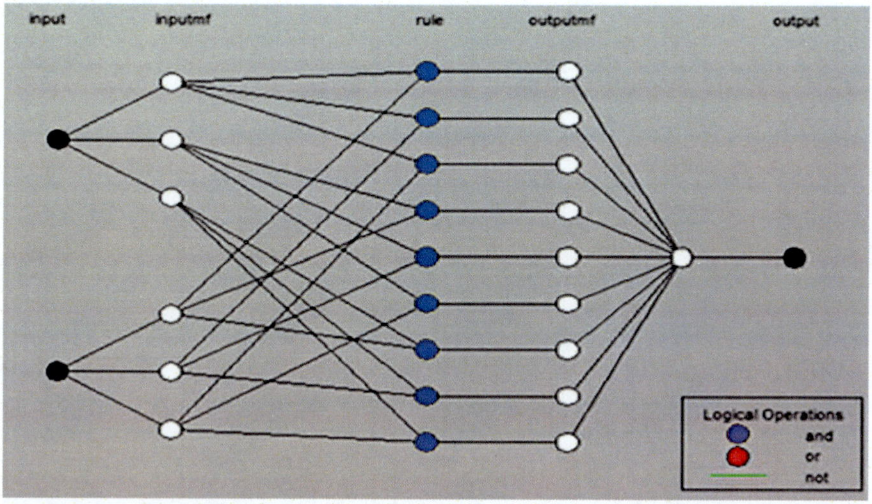

Fig. 5 Structure of ANFIS controller

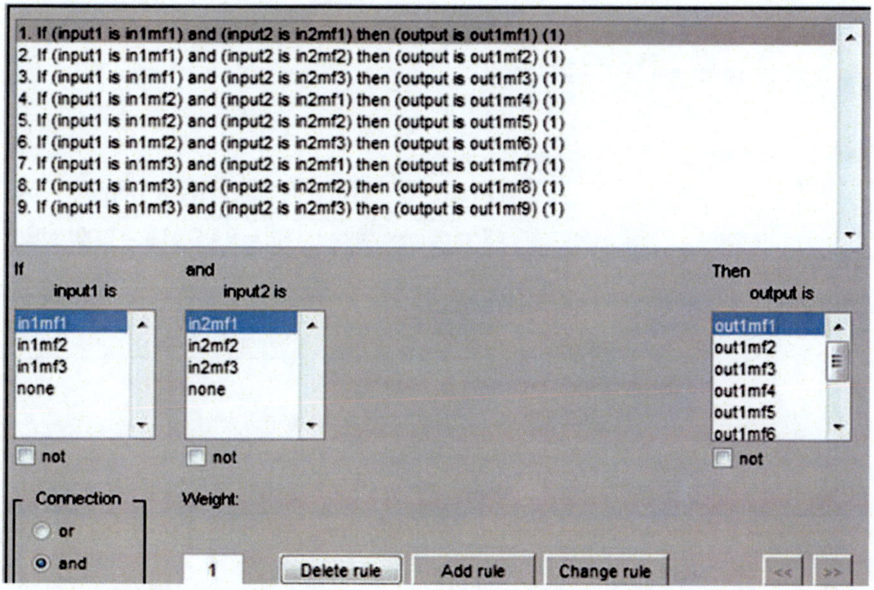

Fig. 6 Rules used in ANFIS controller

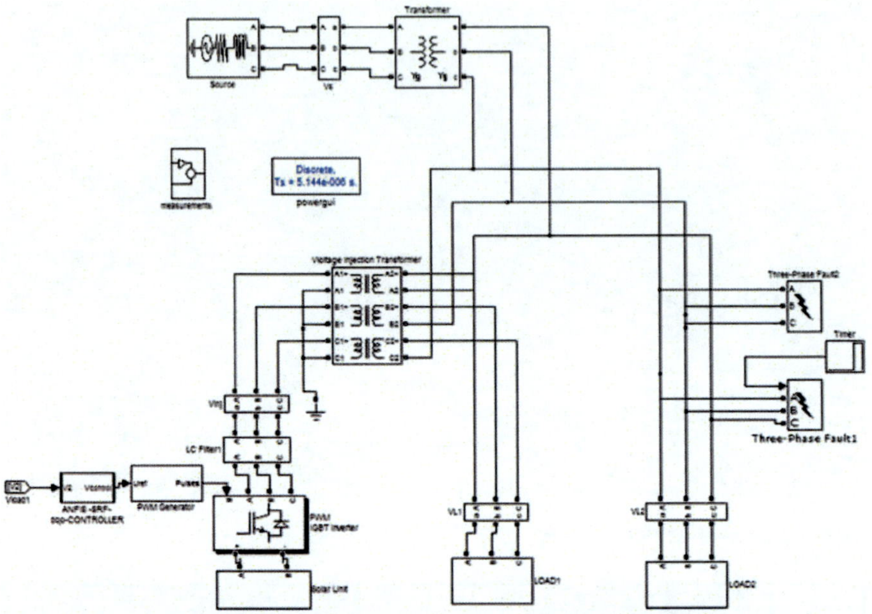

Fig. 7 Proposed System SIMULINK model

Table 3 Parameter description

S.no	Parameters	Ratings
1	Supply	3-phase. 415 V, 50 Hz
2	Source impedance	$R_S = 0.1\ \Omega$, $L_S = 0.001$ mH
3	Z network	L = 1 mH, C = 1 μF
4	Solar cell	35 V.6A
5	Load	8 ohms, 12 mH

3.1 Mitigation of Voltage Imperfections

The performance of proposed model is studied under three different fault cases with three different control techniques. The proposed three controllers work for the three fault conditions.

Case 1: Here, three-phase to ground fault is created between t = 0.02 and 0.09 s. This causes a 25% sag in Load-1 voltage. Interruption is created between t = 0.12 and 0.16 s. The proposed ANFIS-dq0-SRF-based DVR controller compensates the Load-1 voltage imperfections by injecting the compensating voltage. Figure 8 shows the Source voltage, Load-1 voltage, injected voltage, and Load-2 voltage. The power required is delivered from the PV unit.

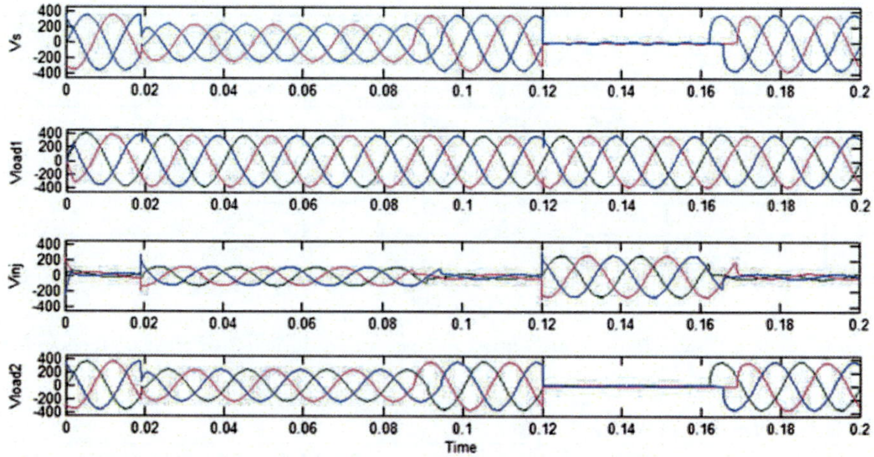

Fig. 8 Source, Load-1, Injected, and Load-2 voltages with AFIS-dqo-SRF-based DVR controller

Case 2: Here double line to ground fault is created between t = 0.02 and 0.09 s. This causes a 25% sag in Load-1 voltage. Interruption is created between t = 0.12 and 0.16 s. The proposed ANFIS-dq0-SRF-based DVR controller compensates the Load-1 voltage imperfections by injecting the compensating voltage. Figure 9 shows the Source voltage, Load-1 voltage, injected voltage, and Load-2 voltage The power required is delivered from the PV unit.

Case 3: Here line to ground fault is created between t = 0.02 and 0.09 s. This causes a 25% sag in Load-1 voltage. Interruption is created between t = 0.12 and 0.16 s. The proposed ANFIS-dq0-SRF-based DVR controller compensates the Load-1 voltage imperfections by injecting the compensating voltage. Figure 10 shows the Source voltage, Load-1 voltage, injected voltage, and Load-2 voltage The power required is delivered from the PV unit.

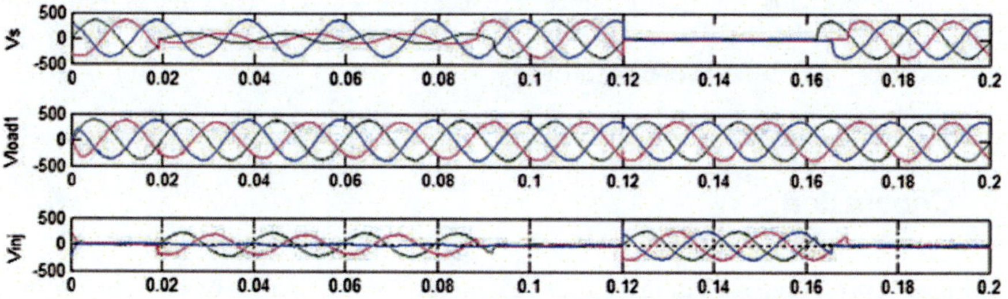

Fig. 9 Source, Load-1, Injected, and Load-2 voltages with AFIS-dqo-SRF-based DVR controller

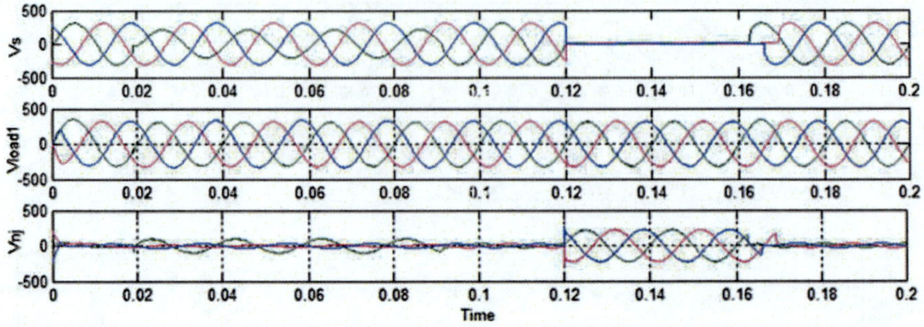

Fig. 10 Source, Load-1, Injected, and Load-2 voltages with AFIS-dqo-SRF-based DVR controller

Table 4 Comparison of voltage % THD

Parameters	Without ZSI-DVR	With ANFIS based ZSI-DVR	With PSO-ANFIS based ZSI-DVR
%THD			
V_{La}	7.5	2.78	0.91
V_{Lb}	7.89	2.19	0.88
V_{Lc}	8.1	2.54	0.89

3.2 *Harmonics Mitigation*

Load voltage harmonic suppression is studied with ANFIS-dqo-SRF controller and PSO-tuned DVR (Table 4).

Case1: %THD of Load Voltage with ANFIS based ZSI-DVR controller is presented in Fig. 11.

Case2: %THD of Load Voltage with PSO-ANFIS-based ZSI-DVR controller is presented in Fig. 12.

Table 2 presents the performance ZSI-based PV-fed DVR control action in suppressing the harmonics in load voltage. PSO and ANFIS techniques aid the performance of DQ–DVR control strategy.

4 Conclusion

The model presented here in this paper has proven the dominance of soft-computing-based DVR effectively alleviates the disturbance in voltages and harmonics suppressed well within the limit. ZSI-based PV-fed DVR effectively operated under three different fault conditions. Comparison shows that PSO & ANFIS-based SRF controller reduces %THD to a considerable value than traditional PI controller. ZSI-based PV-fed DVR serves real power demanded under grid outage conditions.

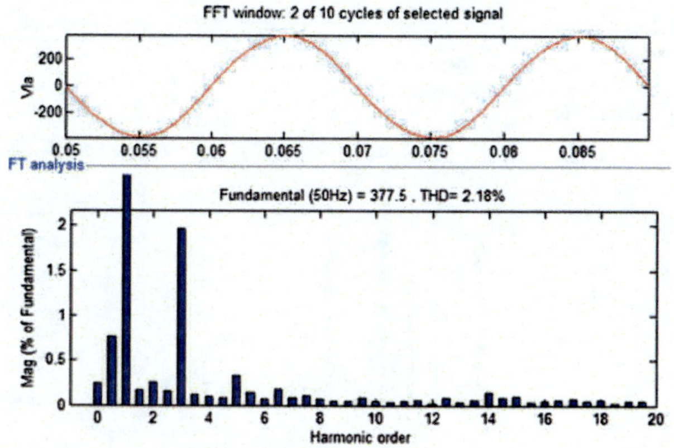

THD = 2.18

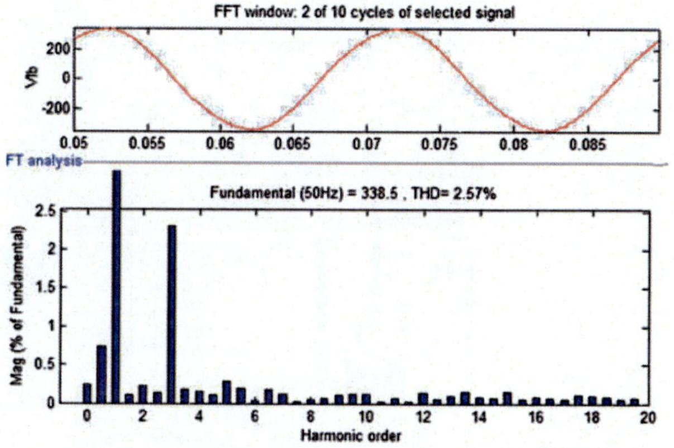

THD = 2.67

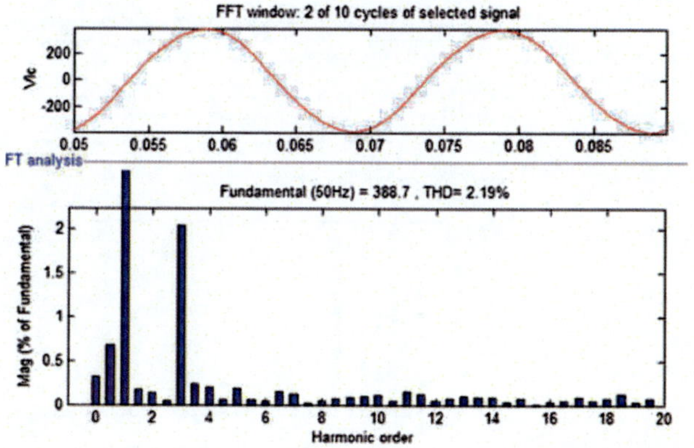

THD = 2.19

Fig. 11 Source, Load-1, Injected, and Load-2 voltages with AFIS-dqo-SRF-based DVR controller

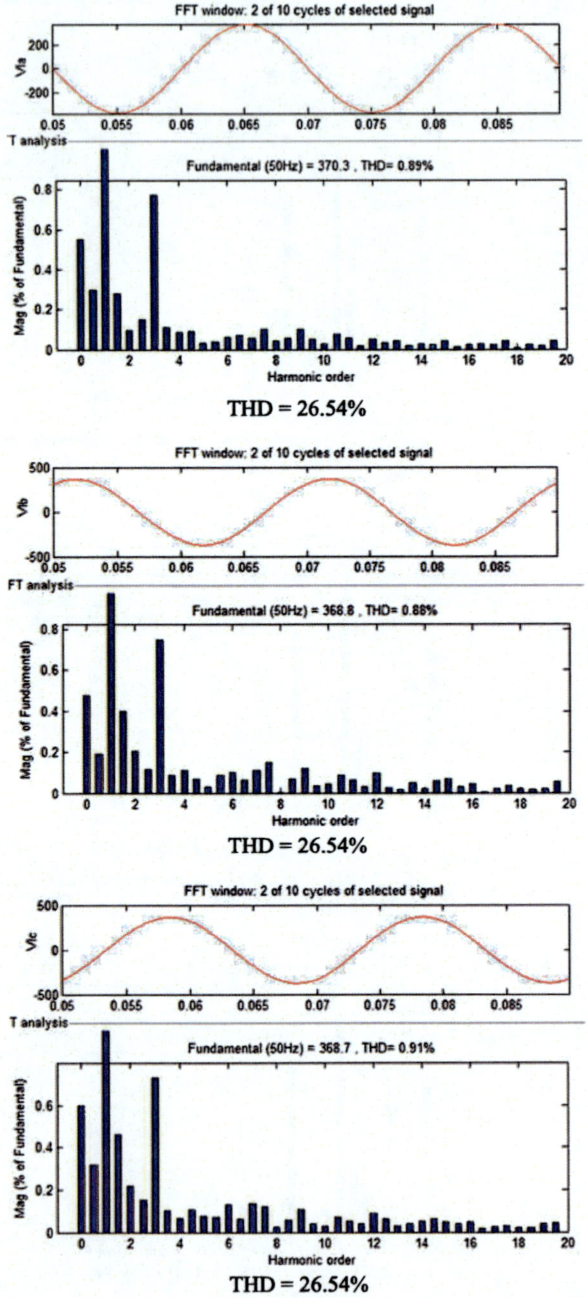

Fig. 12 Source, Load-1, Injected, and Load-2 voltages with AFIS-dqo-SRF-based DVR controller

- Sag and Voltage interruption alleviated from load voltage.
- PSO and ANFIS techniques aid the performance of dq control strategy
- Harmonics effectively compensated and achieved %THD is within limits.

References

1. Kasa S, Ramanathan P, Ramasamy S, Kothari DP (2016) Effective grid interfaced renewable sources with power quality improvement using dynamic active power filter. Int J Electr Power Energy Syst 82(10):150–160
2. Ramasamy S, Sudheer K (2018) Enhancement of power quality in renewable source integrated three phase system with ICC controller. J Green Eng 8(1):71–88
3. Wahab SW, Yusof AM (2006) Voltage sag and mitigation using dynamic voltage restorer (DVR) system. Elektrika 8(2):32–37
4. Zobaa AF, Abdel Aleem SHE (2014) A new approach for harmonic distortion minimization in power systems supplying nonlinear loads. IEEE Trans Ind Inform 10(2):1401–1412
5. Youssef K (2001) Industrial power quality problems electricity distribution, IEE Conf. Pub1 482(2):5
6. Kasa S, Ramasamy S (2016) Photovoltaic fed dynamic voltage restorer with voltage disturbance mitigation capability using ANFIS controller. Int J Renew Energy Res 6(3):825–832
7. Ramasamy M, Thangavel S (2013) Experimental verification of PV based dynamic voltage restorer (PV-DVR) with significant energy conservation. Int J Electr Power Energy Syst, Elsevier 49:296–307
8. Prakash Y, Sankar S (2014) Power quality improvement using DVR in power system. In: IEEE conference power and energy systems: towards sustainable energy
9. Huang Y, Xu J, Sun Y, Huang Y (2016) Modeling and control design of dynamic voltage restorer in microgrids based on a novel composite controller. J Electr Eng Technol 11(6):1645–1655
10. Ribeiro H, Marques H, Borges BV (2012) Characterizing and monitoring voltage transients as problem to sensitive loads. Int J Electr Power Energy Syst 43:1305–1317
11. Honrubia-Escribano A, Gomez-Lazaro E, Molina-Garcia A, Fuentes JA (2012) Influence of voltage dips on industrial equipment: analysis and assessment. Int J Electr Power Energy Syst 41:87–95
12. Penagaluru S, Manohar TG (2018) Review on renewable source integrated topologies with power quality enhancing strategies. Int J Renewable Energy Res 8(4):2350–2366
13. Sudheer K, Sudha R (2018) Enhancement of power quality in multi feeder three phase system with photovoltaic fed ANFIS-unified multi converter controller. In: MATEC web of conferences 2018, vol 225. EDP Sciences, p 03015
14. Ramasamy M, Thangavel S (2013) Photovoltaic based dynamic voltage restorer with power saver capability using PI controller. Int J Electr Power Energy Syst, Elsevier 49:296–307
15. Kasa S, Ramasamy S, Ramanathan P (2015) Hybrid fuzzy-ZN PID control based Grid Interfaced distribution level renewable energy source with power quality. ICCPCT, IEEE, pp1–7
16. Sudheer K, Ramasamy S (2017) Mitigating voltage imperfections with photovoltaic fed ANFIS based ZSI-DVR in three phase system. Int J Renewable Energy Res 7(4):2103–2110
17. Aali S, Nazarpour D (2011) Voltage quality improvement with neural network-based interline dynamic voltage restorer. J Electr Eng Technol 6(6):769–775
18. Roger Jang J-S, Sun C-T, Mizutani E (1997) Neuro-fuzzy and soft computing—a computational approach to learning and machine intelligence. Prentice-Hall

Induction Machine Characteristics Control in Field Weakening Region for Propulsion Application

G. K. Nisha and Z. V. Lakaparampil

Abstract Starting and climbing at low speeds require high torque and almost power constant at high speeds entail for electric vehicle applications. In DC series motor, the characteristic for electric vehicle application is achieved through its construction and by variable DC voltage. Induction motor is considered an ideal candidate for electric vehicle application due to its features such as reliability, ruggedness, lightweight, low cost, and less maintenance. This paper put forward a novel approach to obtain better torque-speed characteristics of Induction machine as nearly as that of DC series machine for electric vehicle application. A control strategy is built-up for Space Vector Modulated (SVM) inverter-fed Field Oriented Control (FOC) of induction machine in Field Weakening (FW) province. For electric vehicle applications at high speed, it is enviable to hold on to utmost torque capacity in FW region. Modification in design has to be carried out to change the behavior of Induction motor along with sliding mode controller to suit propulsion requirements using MATLAB/Simulink.

Keywords Space vector modulation · Field oriented control · Sliding mode Control · Field weakening

1 Introduction

AC induction motors are used in modern mechanized countries to convert in excess of half the total electrical energy used to mechanical energy. They are used to drive compressors, pumps, conveyors, machine tools, cranes, etc., and are so accepted due to its easeness, reliability, less maintenance, and reduced cost. Various PWM algorithms are compared by the concept of harmonic distortion. Highest voltage with the low harmonic distortion is the most important aspire of the modulation

G. K. Nisha (✉)
Department of EEE, Mar Baselios College of Engineering and Technology, Trivandrum, India
e-mail: nisha.gk@mbcet.ac.in

Z. V. Lakaparampil
Department of EEE, Amal Jyothi Engineering College, Kottayam, Kerala, India
e-mail: principal@amaljyothi.ac.in

A. R. Gupta et al. (eds.), *Power Electronics and High Voltage in Smart Grid*, Lecture Notes in Electrical Engineering 817,
https://doi.org/10.1007/978-981-16-7393-1_23

index [1, 2]. These advantages make it popular for inverter control and hence the research is still active especially in the switching sequence of SVM. Various SVM sequences are investigated in [3] and its application in AC drives [4]. More degrees of freedom in space vector approaches and diverse vector series found on the rate of the modulation index have been presented in [5–8]. SVM application in erratic speed electric drive is explained in [9].

Sensorless speed control has become a full-grown technology during the past decade both in electrical drives and in active control for wide speed range applications [10]. Sensorless AC drives have exclusion of sensor cable, enhanced noise immunity, reduced hardware complexity, lower cost, less maintenance constraints, and higher robustness [11]. Sliding Mode Observers (SMO) are renowned as an impending control tactic owing to less commotion, diminished order, and simple execution [12, 13]. FW can be sorted out as deviation of stator flux in reciprocal ratio to speed of rotor and feed-forward reference flux production on machine equations. In foremost approach [14], the machine gets over-excited by the best possible steadiness between the magnetizing and the torque generating current elements. In the second approach [15] can afford exact outcome if magnetic saturation is well thought-out with identified machine factors of adequate precision. The next advance is depicted in [16]. Most MRAS speed observers make use of a simple preset gain linear PI regulator which includes nonlinearities contributed by the inverter and may plunge the performance stage due to incessant disparity in machine parameters and working states. Spotlight has been on alternative techniques such as sliding mode control for compensating the speed alteration signal to calculate speed in FW region.

This paper is sorted out as follows: In the initial part, FOC of induction machine is evaluated. In the subsequent section, realization of SVM-based inverter, MRAS Sliding mode, and the FW are discussed. Subsequently, the design modifications are projected. Lastly, discussions on simulation results are presented.

2 Field Oriented Control

To achieve good dynamics, high precision, and steady-state response the closed-loop feedback controls: FOC, DTC, and Predictive Control are to be implemented. For acquiring good dynamic performance compared to DC series motor, FOC primarily relies on the mathematical modeling of AC machine which decouple its torque from flux control. Figures 1 and 2 show the diagram of vector control and phasor diagram in stationary and revolving reference frame, respectively, of an induction motor.

Fig. 1 Field oriented control

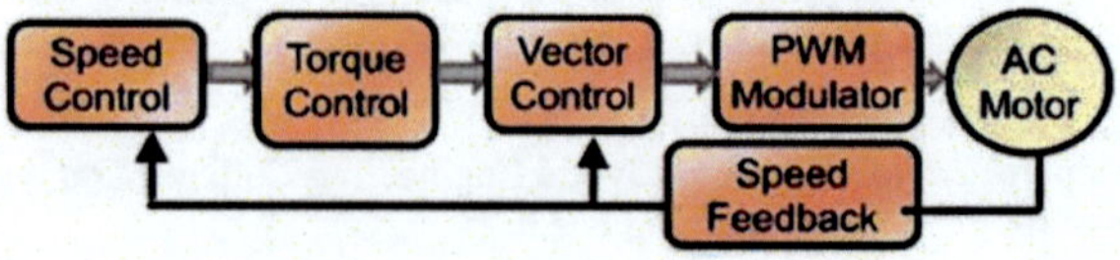

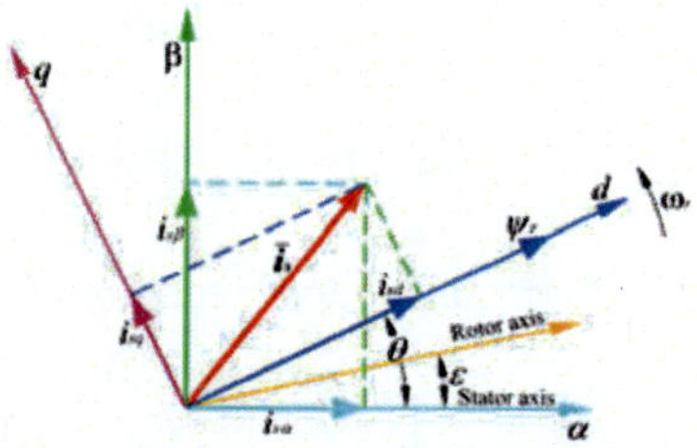

Fig. 2 Reference frame in stationary and rotating

Voltage and current complex space vector is resolved into components of d and q as:

$$\underset{s}{\vec{V}}(t)e^{-j\theta} = \left|\underset{s}{\vec{V}}\right| e^{-j(\gamma_1-\theta)} = \left(V_{sd} + jV_{sq}\right) \tag{1}$$

$$R_s \underset{s}{\vec{i}}\, e^{-j\theta} = R_s \underset{s}{\vec{i}}\, e^{-j(\gamma-\theta)} = R_s\left(i_{sd} + ji_{sq}\right) \tag{2}$$

$$R_r \underset{r}{\vec{i}}\, e^{-j(\varepsilon-\theta)} = R_r\left(i_{rd} + ji_{rq}\right) \tag{3}$$

$$\frac{d\omega_r}{dt} = \frac{T_d(t) - T_L(t)}{J} = \frac{1}{J}\left(\frac{2}{3}\frac{P}{2}\frac{L_m}{L_r}\left(\psi_{rd}i_{sq} - \psi_{rq}i_{sd}\right) - T_L\right) \tag{4}$$

where, stator and rotor resistance-Rs and Rr; poles-P, the moment of inertia-J; electromagnetic torque-T_d; rotor speed-ω_r; stator currents-i_{sd}, i_{sq}; rotor flux linkages-ψ_{rd}, ψ_{rq}.

Equivalent dynamic circuit for d-q-axis is in Fig. 3. The dynamics of magnetizing current accountable for the magnetizing flux generation involves a large time constant.

The stator and rotor rotational voltage term is expressed as:

$$\psi_{sd}\omega_e = (L_s i_{sd} + L_m i_{rd})\omega_e \tag{5}$$

$$\psi_{sq}\omega_e = \left(L_s i_{sq} + L_m i_{rq}\right)\omega_e \tag{6}$$

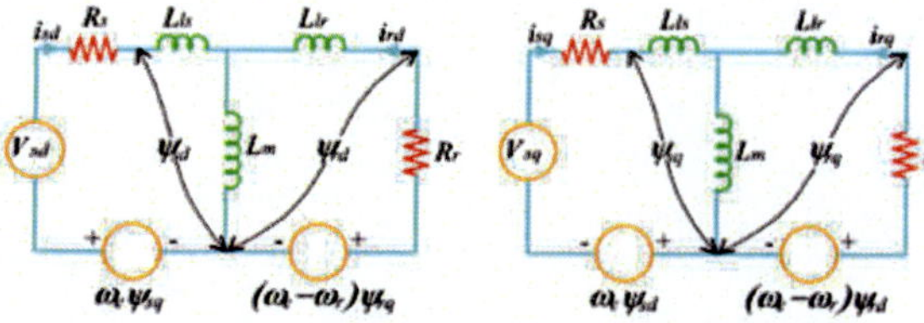

Fig. 3 d-q axis dynamic equivalent circuit

$$\psi_{rd}(\omega_e - \omega_r) = (L_r i_{rd} + L_m i_{sd})(\omega_e - \omega_r) \tag{7}$$

$$\psi_{rq}(\omega_e - \omega_r) = \left(L_r i_{rq} + L_m i_{sq}\right)(\omega_e - \omega_r) \tag{8}$$

Matrix form representation of dynamic voltage as:

$$\begin{bmatrix} Vsd \\ Vsq \\ 0 \\ 0 \end{bmatrix} = \begin{bmatrix} R_s + L_s p & -\omega_e Ls & L_m p & -L_m \omega_e \\ \omega_e L_s & R_s + L_s p & L_m \omega_e & L_m p \\ L_m p & -(\omega_e - \omega_r)L_m & R_r + L_r p & -(\omega_e - \omega_r)L_r \\ (\omega_e - \omega_r)L_m & L_m p & (\omega_e - \omega_r)L_r & R_r + L_r p \end{bmatrix} \begin{bmatrix} i_{sd} \\ i_{sq} \\ i_{rd} \\ i_{rq} \end{bmatrix} \tag{9}$$

Electromagnetic torque of the machine becomes:

$$T_d = \frac{3}{2}\frac{P}{2}\frac{L_m}{(1+\sigma_r)} i_{mr} i_{sq} \tag{10}$$

2.1 SVM Based Inverter

Reference vector can be blended by putrid into components along six active voltage vectors and magnitude of each vector is equal to (2/3)V_{DC}. Block diagram of SVM pulse generation is in Fig. 4.

Particular fundamental angular position for each stationary vector is shown in Fig. 5. In linear modulation range, the reference voltage follows a circular trajectory.

A typical seven-segment sector switching time is in Fig. 6.

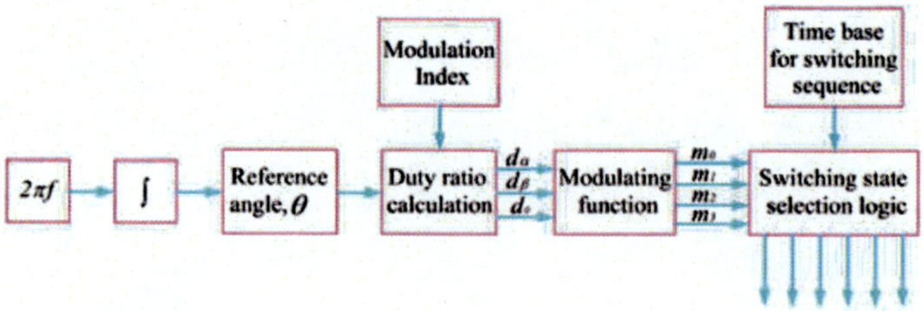

Fig. 4 SVM pulse generation

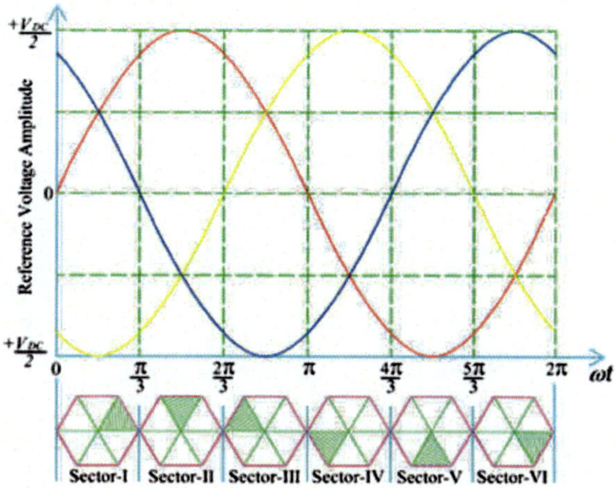

Fig. 5 Phasor angular position of inverter in fundamental cycle

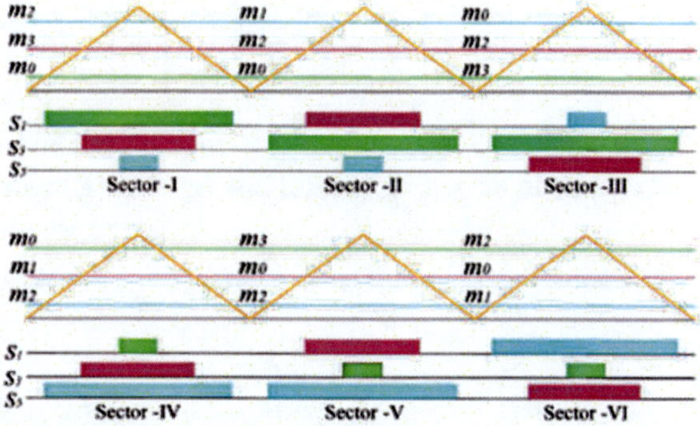

Fig. 6 Switching time for each sector

3 MRAS-Sliding Mode Control

An open-loop linear time-invariant and a nonlinear feedback time-varying subsystem constitute a nonlinear feedback system. Stator voltage and current put into the linear time-invariant system is represented as u, and the output speed-tuning signal as y as shown in Fig. 7. The variable ω is the output of the nonlinear time-invariant system and u = -ω..

$$\eta_{(0,t)} = \int_0^t y^T \omega dt \geq -\gamma^2 \tag{11}$$

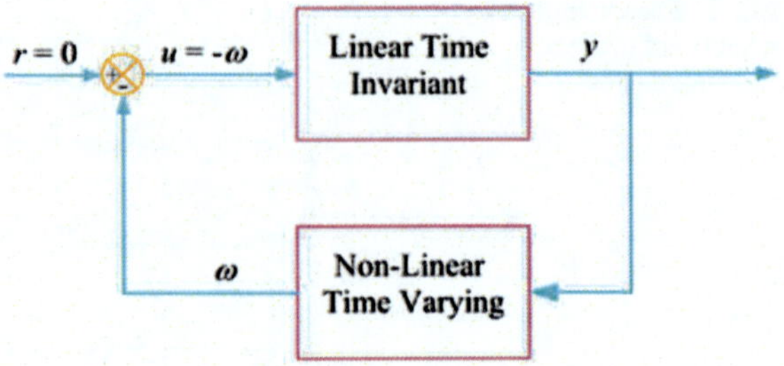

Fig. 7 Switching time for each sector

where y: Input.

ω: Feedback block output.

γ^2: Finite positive constant.

For tuning the adaptive model the disparity amid the two predictable vectors shall be fed to an amendment mechanism. The adaptive rates of rotor field elements as:

$$p\begin{bmatrix} \hat{\psi}_{r\alpha} \\ \hat{\psi}_{r\beta} \end{bmatrix} = \frac{L_m}{\tau_r}\begin{bmatrix} i_{s\alpha} \\ i_{s\beta} \end{bmatrix} + \begin{bmatrix} -\frac{1}{\tau_r} & -\hat{\omega}_r \\ \hat{\omega}_r & -\frac{1}{\tau_r} \end{bmatrix}\begin{bmatrix} \hat{\psi}_{r\alpha} \\ \hat{\psi}_{r\beta} \end{bmatrix} \tag{12}$$

The state model in stationary reference structure with state variables as stator flux and current in designing control law for sliding mode controller expressed as:

$$\dot{i}_{s\alpha} = -\frac{R_s}{\sigma L_s} i_{s\alpha} - \omega_r i_{s\beta} - \frac{R_r}{\sigma L_r} i_{s\alpha} + \frac{R_r}{\sigma L_s L_r}\psi_{s\alpha} + \frac{\omega_r}{\sigma L_s}\psi_{s\beta} + \frac{1}{\sigma L_s} V_{s\alpha} \tag{13}$$

$$\dot{i}_{s\beta} = -\frac{R_s}{\sigma L_s} i_{s\beta} - \omega_r i_{s\alpha} - \frac{R_r}{\sigma L_r} i_{s\beta} + \frac{R_r}{\sigma L_s L_r}\psi_{s\beta} + \frac{\omega_r}{\sigma L_s}\psi_{s\alpha} + \frac{1}{\sigma L_s} V_{s\beta} \tag{14}$$

where, $\sigma = 1 - \frac{L_m^2}{L_s L_r}$

$$\dot{\psi}_{s\alpha} = V_{s\alpha} - i_{s\alpha} R_s$$

$$\dot{\psi}_{s\beta} = V_{s\beta} - i_{s\beta} R_s$$

The control law for the sliding mode controller should be stable and the discontinuous control law can be derived as:

$$V(S) = \frac{1}{2} S_\psi^2 \tag{15}$$

4 Field Weakening Operation

For allowing maneuver in erratic speed induction machine drives at high speeds for electric vehicle applications FW is used where the rotor field is abridged beneath its rated value. As operating speed increases, dimension of the voltage constraint ellipse diminishes as shown in Fig. 8. In the current plane, torque is depicted as an inverse curve.

For space vector modulation, the maximum phase voltage V_{sm} is attained in linear control range as $V_{DC}/\sqrt{3}$. Reference current, Ism, should gratify (17) despite the reference frame.

$$I_{sm}^2 \geq i_{sd}^2 + i_{sq}^2 \tag{16}$$

D-axis current is the intersection point of circle and ellipse, which should be put as rated to put off the saturation. Highest accessible torque is decisive by the utmost q-axis current. Highest torque is always the same, known as constant torque region. It terminates at base speed ω_b given as:

$$\omega_b = \frac{V_{sm}}{L_s}\sqrt{\frac{1}{[\sigma^2(I_{sm}^2 - i_{sd}^2) + i_{sd}^2]}} \tag{17}$$

Field weakening region I begins from base speed and extended to ω_1, here the rated current in d-axis is always greater than reference current as in Fig. 9 and is infered as:

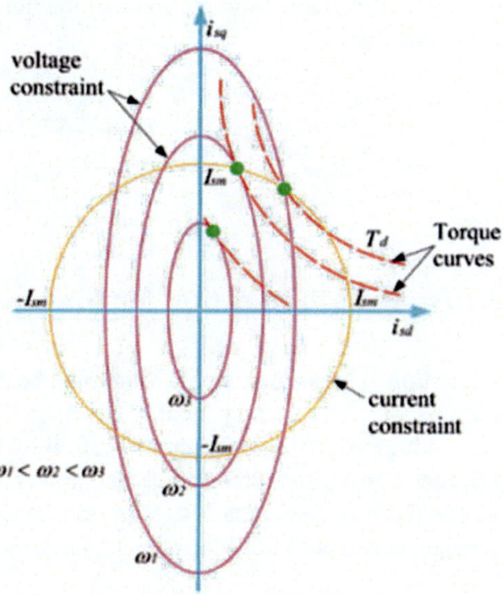

Fig. 8 Voltage and current constraint at different operating speed

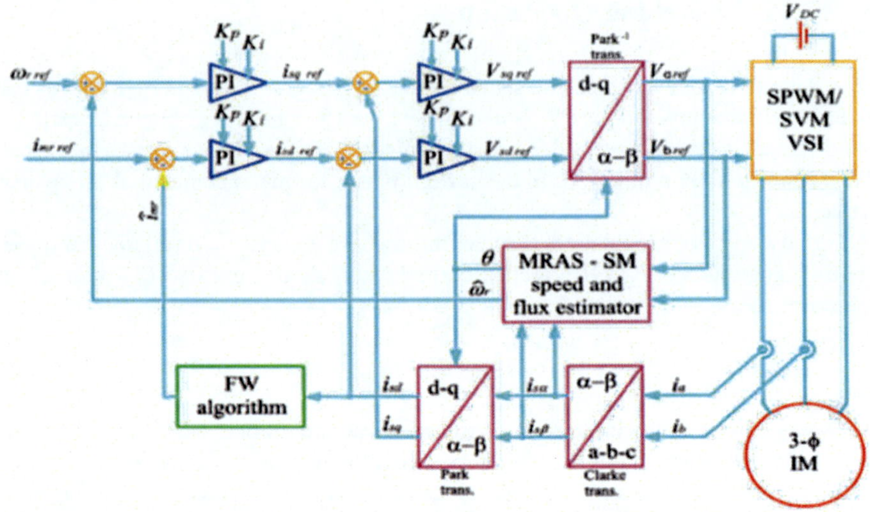

Fig. 9 Induction machine in FW region

$$
i_{sd} = \sqrt{\frac{\left(\frac{V_{sm}}{\omega_e}\right)^2 - (\sigma L_s I_{sm})^2}{L_s^2 - (\sigma L_s)^2}} \tag{18}
$$

If the operating frequency further increases, then the ellipse shrinks and the ellipse impending to be entirely within the circle, this region is termed as field weakening region II or characteristic province of the induction machine, where this region ends is surmised as:

$$
\omega_1 = \frac{V_{sm}}{I_{sm}} \sqrt{\frac{1 + \sigma^2}{2(\sigma L_s)^2}} \tag{19}
$$

5 Design Modification for Maximum Torque Production

Excitation stage of flux weakening must be aptly accustomed to turn out the utmost torque that motor could probably build up. This pioneering approach proposed is to succeed appropriate design modification by varying the rated power factor so as to encounter at utmost rated torque and power. For designing a squirrel cage induction motor, the main specifications taken are rated power output, power factor, frequency, voltage, speed, and full load current. Dimensions of stator and rotor are resoluted by

Table 1 Machine parameters

Parameters	Values	Unit
Line voltage	220	Volt
Number of poles	4	
Rated torque	195	N-m
Synchronous speed	1500	Rpm
Frequency	50	Hz

independent variables like stack length, stator outer diameter, stator wire size. Table 1 shows the machine parameter.

Design spreadsheets are built-up for squirrel cage motor for 30 kW and design options are design-1, 2, 3, and 4 created by tumbling the values of power factors equal to 0.9, 0.8, 0.7, and 0.6, correspondingly. The desired values attained from spreadsheet are tabulated and presented in Table 2.

6 Simulation

The Simulink/MATLAB model of sensorless SVM inverter fed FOC induction machine extended from base speed to FW region shown in Fig. 9. For comparing performance of Design-1, 2, 3, and 4 simulations are carried out using the above model.

Figure 10 shows the speed responses for reference speeds: 140, 100, and 50 rad/s for sensorless FOC drive with MRAS-SM. At $t = 1$ s, the load torque applied as, $T_L = 5$ Nm for all the speeds, the predictable speed lock close with the actual speed and following the reference in all speed. MRAS-SM shows excellent performance when reference speed is at 50 rad/s and provides a good result in low speed range.

Stator winding current variation is shown in Fig. 11. Motor torque has to balance for friction and the load torque in steady-state.

The vibrant activities of field weakening algorithm is assessed by pertaining chain of numerous step transforms of reference speed signal among 0–4 p.u. in four quadrants as in Fig. 12. Results prove that MRAS-SM observer with SVM inverter approximates the rotor speedwell and holds enhanced performance in following the speed command.

In this simulation, the speed responses when the motor initiates from no-load condition as in Fig. 13. The results show that the rotor speed raises from Design-1, 2, 3, and 4 with a drop in power factor. The highest possible speed at 8 s is increased from Design-1, 2, 3, and 4 as given in Table 3.

Figure 14 shows electromagnetic torque (p.u.) variation as regards to rotor speed (p.u.) for different options. Torque capacity is enhanced in the FW region as the power factor reduces is confirmed in Table 4.

Table 2 Machine parameters for 30 kW

Parameters	Design				
	1	2	3	4	Unit
Rotor					
Air gap length	0.72	0.70	0.68	0.66	mm
Diameter of Rotor	294	279	269	259	mm
Slot pitch	24.3	24.4	23.6	22.7	mm
Rotor bar current	335	366	396	426	A
C/s area of rotor bar	67	73	79	85	mm^2
Provided c/s area of rotor bar	70	77	84	88	mm^2
Resistance of each bar	84.0	73.6	65.0	60.9	Ω
Copper loss in bars	358	374	387	421	watts
End ring current	1014	1106	1198	1290	A
C/s area of rotor ring	203	221	240	258	mm^2
Provided c/s area of rotor ring	204	228	247	260	mm^2
Copper loss in end rings	166.5	163.9	168.3	177.9	watts
Total copper loss	524.9	537.8	555.7	598.5	watts
Full load slip	0.02	0.02	0.02	0.02	
Resistance of rotor refers to stator	0.07	0.07	0.07	0.08	Ω
Depth of rotor core	51	48	46	45	mm
No Load Current:					
Magnetizing MMF/pole	967	1058	1051	805	A
Magnetizing current	66.9	64.5	63.2	56.8	A
Magnetizing reactance	5.60	5.58	6.09	8.56	Ω
Core loss (CL)	1288	1220	1086	988	watts
Friction&Windage loss(FWL)	300	300	300	300	watts
No load loss	1588	1520	1386	1288	watts
Loss component	2.41	2.30	2.10	1.95	A
No load current (NLC)	66.9	64.5	63.2	56.8	A
Short circuit current					
Slot leakage reactance	0.11	0.13	0.15	0.16	Ω

(continued)

Disparity of power (p.u.) with regard to rotor speed (p.u.) is shown in Fig. 15. In FW-II range, power increases from Design-1, 2, 3, and 4 which entails that power boost occurs with a fall in power factor.

Table 5 shows that raise in power for Design-4 judges against Design-1 having rotor speed of 2.0 p.u and 3.0 p.u., respectively.

Table 2 (continued)

Parameters	Design				
	1	2	3	4	Unit
Overhang leakage reactance	0.17	0.19	0.22	0.24	Ω
Zigzag leakage reactance	0.11	0.11	0.12	0.17	Ω
Total leakage reactance	0.39	0.43	0.49	0.57	Ω
Performance at full load					
Total losses	3573	3183	2846	2636	watts
Efficiency	89.4	90.4	91.3	91.9	%

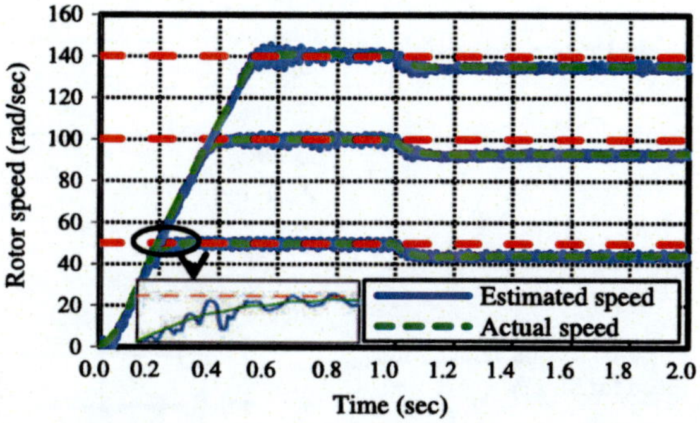

Fig. 10 Rotor speed versus time for MRAS-SM

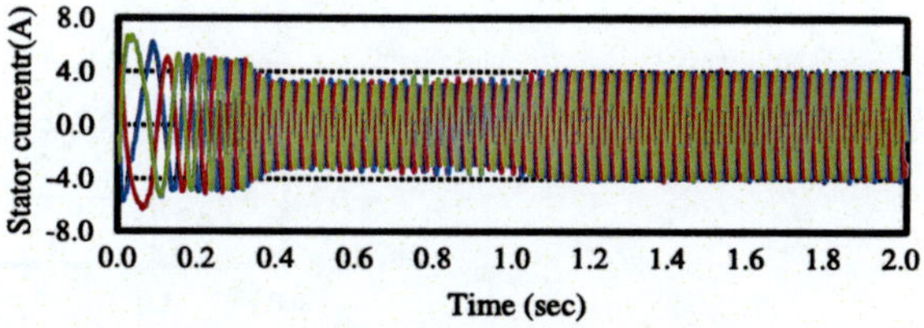

Fig. 11 Stator current for MRAS-SM for ω_{ref} = 100 rad/s

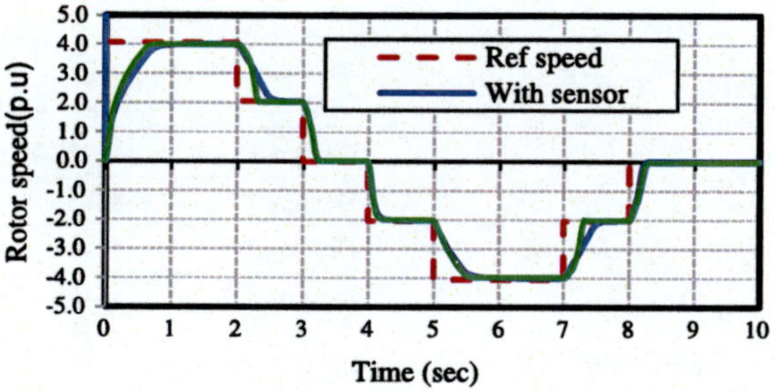

Fig. 12 Rotor speed versus time

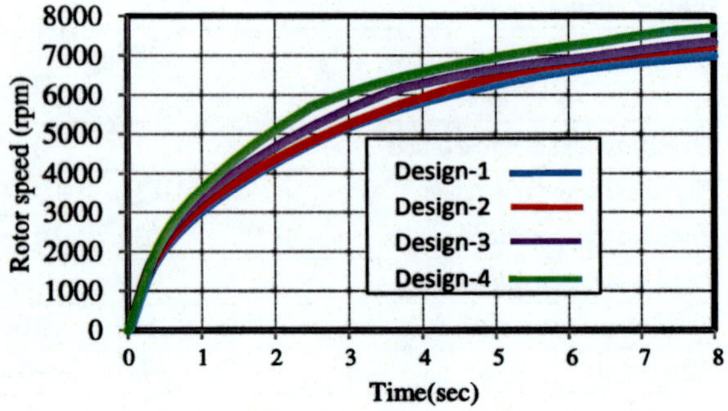

Fig. 13 Rotor speed versus time

Table 3 Speed response (at 8 secs)

Options	ω_r (rpm)
Design-1	6973
Design-2	7225
Design-3	7335
Design-4	7687

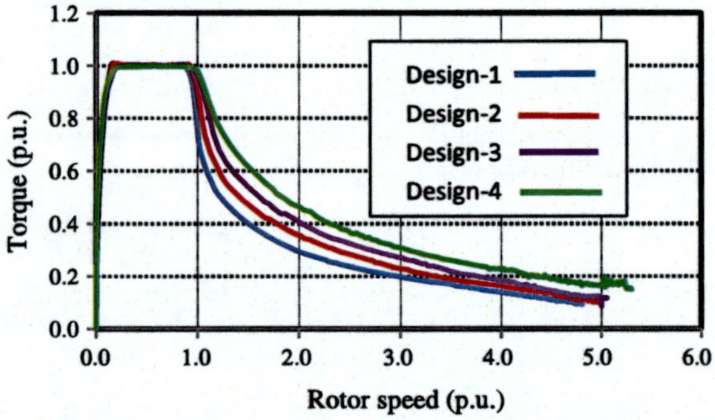

Fig. 14 Torque versus rotor speed for 30 kW

Table 4 Torque capability (p.u.)

Options	$\omega_r = 2.0$ p.u	$\omega_r = 3.0$ p.u
Design-1	0.292	0.196
Design-2	0.353	0.229
Design-3	0.405	0.271
Design-4	0.464	0.307

The rotor flux locus with SVM inverter is shown in Fig. 16. The speed evaluation by the rotor flux MRAS with SM observer guaranteed excellent precision in all sorts of speed control in four quadrants.

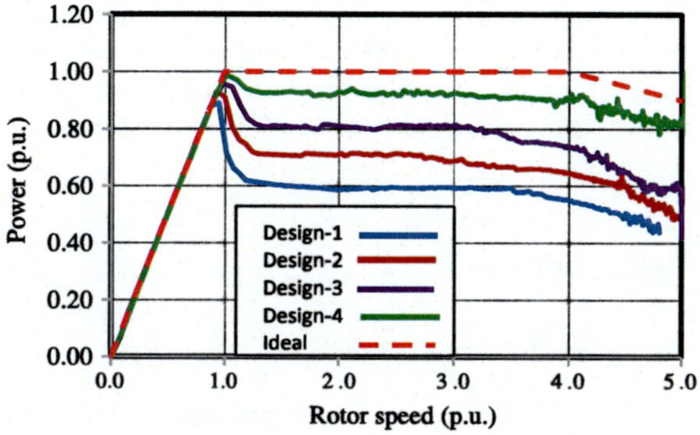

Fig. 15 Power versus rotor speed for 30 kW

Table 5 Maximum power (p.u.)

Design options	$\omega_r = 2.0$ p.u	$\omega_r = 3.0$ p.u
	Case-1	Case-1
Design-1	0.580	0.588
Design-2	0.706	0.687
Design-3	0.810	0.813
Design-4	0.928	0.921

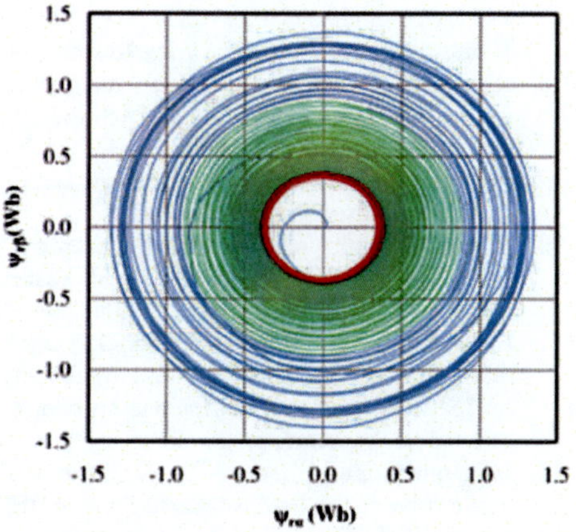

Fig. 16 Rotor flux locus with SVM inverter

7 Conclusion

In this paper, sensorless operation of field-oriented controlled induction motor with model reference adaptive system based on sliding mode control in flux deteriorating is designed. Simulation results ensure that the sliding mode-based adaptive approach gives better performance. The innovative contribution in this paper is an inventive method of varying machine parameters to get better torque capacity in FW region.

Simulation models for SVM inverter fed field-oriented induction machine is adopted to drive various design options for evaluating the torque-speed behavior. The highest achievable rotor speed and torque capacity of induction machine in FW region is enhanced by diminishing power factor of the machine from the rated condition.

References

1. Abu Rub H, Iqbal A, Guzinski J (2012) High performance control of AC Drives. Wiley, New York, pp 375–388
2. Nisha GK, Ushakumari S, Lakaparampil ZV (2012) CFT based optimal PWM strategy for three phase inverter. In: Proceedings of IEEE international conference on power, control and embedded systems, ICPCES 2012, pp 1–6
3. Nisha GK, Ushakumari S, Lakaparampil ZV (2012) Online harmonic elimination of SVPWM for three phase inverter and a systematic method for practical implementation. IAENG Int J Comput Sci 39(2), 220–230
4. Beig AR, Narayanan G, Ranganathan VT (2007) Modified SVPWM algorithm for three level VSI with synchronized and symmetrical wave forms. IEEE Trans Ind Electron 54(1):486–494
5. Nisha GK, Lakaparampil ZV, Ushakumari S (2013) FFT analysis for field oriented control of SPWM and SVPWM inverter fed induction machine with and without sensor. Int J Adv Electr Eng 2(4):151–160
6. Mirafzal B, Skibinski GL, Tallam RM (2009) Determination of parameters in the universal induction motor model. IEEE Trans Ind Appl 45(1):142–151
7. Nisha GK, Ushakumari S, Lakaparampil ZV (2012) Method to eliminate harmonics in PWM: a study for single phase and three phase. In: International conference on Emerging Technology, Trends on Advanced Engineering Research, Kollam, India, pp 598–604
8. Nisha GK, Lakaparampil ZV, Ushakumari S (2012) Sensorless vector control of SVPWM fed induction machine using MRAS – sliding mode. In: Proceedings of IEEE international conference on green technology, ICGT 2012, pp 29–36
9. Huang MS (2003) Improved field-weakening control for IFO induction motor. IEEE Trans Aerosp Electron Syst 39(2):647–658
10. Nisha GK, Lakaparampil ZV, Ushakumari S (2013) Performance study of field oriented controlled induction machine in field weakening using SPWM and SVM fed Inverters. Int Rev Model Simul 6(3):741–752
11. Nisha GK, Lakaparampil ZV, Ushakumari S (2014) Four-quadrant operation of sensorless FOC induction machine in field weakening region using MRAS-sliding mode observer. In: IEEE international conference on control communication and computing (ICCC'2013), Thiruvananthapuram, pp 33–38
12. Mengoni M, Zari L, Tani A, Serra G, Casadei D (2008) Stator flux vector control of induction motor drive in the field weakening region. IEEE Trans Power Electron 23(2):941–949
13. Nisha GK, Lakaparampil ZV, Ushakumari S (2013) Sensorless field oriented control of SVM inverter fed induction machine in field weakening region using sliding mode observer. In: The proceedings of 7th world congress on engineering (WCE'2013). Imperial College, London, pp 1174–1181
14. Tarvirdilu-Asl R, Nalakath S, Xia Z, Sun Y, Wiseman J, Emadi A (2020) Improved online optimization-based optimal tracking control method for induction motor drives. IEEE Trans Power Electron 35(10):10654–10672

15. Nisha GK, Lakaparampil ZV, Ushakumari S (2017) Torque capability improvement of sensorless FOC induction machine in field weakening for propulsion purposes. J Electr Syst Inf Technol 4(1):173–184
16. Dong Z, Yu Y, Li W, Wang B, Xu D (2018) Flux-weakening control for induction motor in voltage extension region: torque analysis and dynamic performance improvement. IEEE Trans Ind Electron 65(5):3740–3751

A Novel Hybrid GMPPT Scheme Based on P&O-MM with Reduced Output Power Oscillations Under PSC for PV System

Muralidhar Nayak Bhukya, P. T. Krishna Sai, Manish Kumar, Shobha Rani Depuru, and T. Sudhakar Babu

Abstract Traditional Perturb and Observe (P&O) controllers are preferred over metaheuristic algorithms during uniform irradiance conditions but fail to replicate the same performance during Partial Shaded Conditions (PSC). Hence, a novel hybrid GMPP Tracking (GMPPT) scheme based on Perturb and Observe and Mean Method (PO-MM), which gives effective performance under any weather condition with reduced output power oscillations, is proposed. During PSC, the P&O controller remains at the first obtained peak of the P–V characteristics. Therefore, the rest of the characteristics are examined by the Mean Method to attain exact GMPP.

Keywords Perturb and observe · Mean method · Power oscillations

1 Introduction

Nowadays pollution-free electric power generation is the main agenda of many nations. This is the main motive behind the extensive use of solar energy as source to generate electric power via Photovoltaic (PV) system. Apart from this, PV system

M. N. Bhukya · M. Kumar (✉)
Department of Electrical Engineering, School of Engineering and Technology, Central University of Haryana, Jant-Pali, Haryana 123031, India
e-mail: manish.kumar@cuh.ac.in

M. N. Bhukya
e-mail: murlidhar.nayak@cuh.ac.in

P. T. K. Sai
Department of Electrical and Electronics Engineering, Dhanekula Institute of Engineering and Technology, Gangur, Andhra Pradesh 521131, India

S. R. Depuru
Department of Electrical and Electronics Engineering, Institute of Aeronautical Engineering, Hyderabad 500043, India

T. S. Babu
Institute of Power Engineering, Universiti Tenaga National, 43000 Kajang, Malaysia

A. R. Gupta et al. (eds.), *Power Electronics and High Voltage in Smart Grid*, Lecture Notes in Electrical Engineering 817,
https://doi.org/10.1007/978-981-16-7393-1_24

is vulnerable to non-uniform distribution of irradiance on the panels. Irrespective of various weather conditions, various tracking techniques are developed to extract and integrate solar power. The application of traditional tracking schemes such as Perturb and Observe (P&O) [1], Incremental Conductance (INC), Hill Climbing (HC) method, Fractional Open Circuit Voltage (FOCV) and Fractional Short Circuit Current (FSCC) are limited to uniform weather conditions whereas the intelligent and hybrid (intelligent and traditional) tracking schemes are venerable to limited search space.

Ibrahim et al. [2] investigated the effectiveness of modified PSO algorithm under shaded conditions. The number of oscillations in the steady state conditions are reduced to the possible extend compared to artificial neural networks. The essential duty cycle to the converter is generated based on the change in velocity function with respect to environmental conditions. Amin et al. [3] presented a two-stage global tracking schemes based on the intelligent sampling procedure. Initially, P–V characteristics are sampled and the samples which are not in the global region are eliminated such that the search scale is reduced. In continuation using fractional open circuit voltage tracking algorithm the global MPP is attained under any weather conditions. The proposed technique is validated through several simulation and experimental results. Eftichios et al. [4] increased the power generation from the solar array by ensuring GMPPT under shaded condition through updating the PWM control signal given to the power converter. The real-time dynamic PWM signals are generated based on duty cycle (δ) and switching frequency.

Mustafa et al. [5] proposed and presented a simple and innovative strategy for energy storage devices using a buck converter. To produce sufficient amount of hydrogen the electrolysis needs low potential and high current discharge. PV system with P&O algorithm and PI controller is used as source to the converter. Majad et al. [6] improved the tracking time of the PV system to a percentage in between 10 and 30% using a novel Harris Hawk Optimization (HHO) algorithm. The comparative study and key contributions are carried out using P&O, DFOA, CS, and GWO in terms of computational procedure and efficiency. Chandrasekaran et al. [7] experimentally developed a 2-kW PV system to implement the proposed Sine Cosine algorithm to detect partial shading and avoid local mean paths. The out performance is demonstrated using simulation as well as experimental results (Fig. 1).

As discussed earlier, tracking schemes based on metaheuristic algorithms such as PSO [8, 9], Ant Colony Optimization (ACO), Simulated Annealing (SA), Genetic Algorithm (GA), Differential Evolution (DE), Bee Algorithm (BE), Tabu Search (TS), Harmony Search (HS), Fireworks, social mimic optimization algorithm (SMO), and artificial ecosystem-based optimization are developed to path GMPP are complex, expensive, and moreover changes its Gbest value for every trial run. Apart from this, the traditional P&O controller is preferred over metaheuristic algorithms. But the traditional algorithm fails to perform effectively during PSC. Based on the above statements a novel hybrid GMPPT scheme is developed based on P&O and MM which gives effective performance under any weather condition with reduced output power oscillations [10–12]. During Partial Shaded Conditions (PSC), the P&O

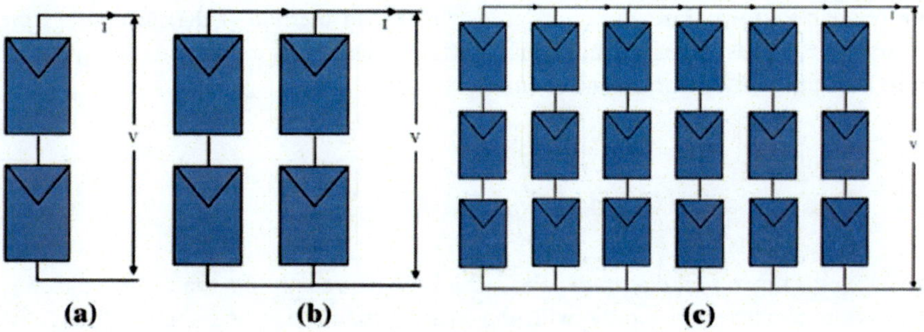

Fig. 1 PV configurations: **a** 2S; **b** 2S2P; and **c** 3S6P

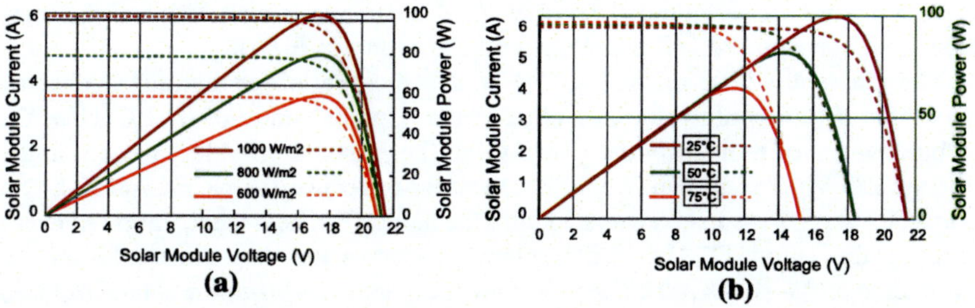

Fig. 2 PV system characteristics at constant: **a** Temperature of 25 °C; **b** Irradiance of 1,000 W/m^2

controller remains at the first obtained peak of the P–V characteristics [13–15]. Therefore, the rest of the characteristics are examined by the MM to attain exact GMPP [16]. Figure 2 depicts P–V and I-V curves at different irradiance and temperatures. Typically, PV cell possess low voltage and its mathematical equation is

$$I_{PV} = I_{Ph} - I_D\left[\exp\left(\frac{q(V_{PV} + I_{PV}Rs)}{N_S A B_K T}\right) - 1\right] - \frac{V_{PV} + I_{PV}R_S N_S}{N_S R_{Sh}} \quad (1)$$

2 Proposed Hybrid GMPPT Scheme Based PO-MM

The P&O scheme has technical issues to be addressed such as oscillations around MPP and inability to perform under PSC. After attaining MPP, the P&O scheme perturb in both forward and reverse direction and fail to remain at a certain point. Therefore, the confusion involved in the operation creates oscillations around MPP and reduces net output power. At the same time, the logical confusion limits the algorithm to perform only under uniform irradiance conditions. Hence, in this paper, the regular P&O scheme is empowered with Mean method (MM) to address the limitations. The working of the hybrid GMPPT scheme is explained clearly under both PSC and uniform irradiance conditions.

2.1 Under Partial Shaded Conditions

It is a well-known fact that under PSC, the P–V curve exhibits multiple peaks, and P&O fails to choose the peak with the highest magnitude. As shown in Fig. 3a, the P&O scheme starts reverse perturbation immediately after reaching the first local peak of the P–V curve. This reverse and forward perturbation not only creates oscillations but also depowers the P&O algorithm to track the global power path. Therefore, as the direction of perturbation reverses, MM technique is operated, and the local MPP obtained from P&O is addressed as X, as shown in Fig. 3b.

The classical Mean method (MM) is a technique which optimizes towards maximum or minimum value in a closed interval with a ratio of 0.41%. Here the technique is used to identify the global value (i.e., Maximum Value) present in the interval [X, V_{OC}] as shown in Fig. 3b. In order to reduce the closed interval, two points, $\alpha 1$ and $\alpha 2$ are to be generated with the help of the below equations. At every iteration, the interval [X, V_{OC}] will be converged, and a new closed interval is framed as either [$\alpha 1$, V_{OC}] or [X, $\alpha 2$]. The procedure is repeated with the new obtained closed interval until it converges at the point $\alpha 1 = \alpha 2$. At this particular point voltage at $\alpha 1$ and $\alpha 2$ are averaged to obtain the maximum value in the initial chosen closed interval. Let the MPP obtained from MM technique be addressed as 'Y'. And finally, the magnitude of X and Y are compared to decide the GMPP of the PV system.

$$\left.\begin{aligned} \alpha_1 &= X + 0.41(V_{OC} - X) \\ \alpha_2 &= V_{OC} - 0.41(V_{OC} - X) \end{aligned}\right\} \quad (2)$$

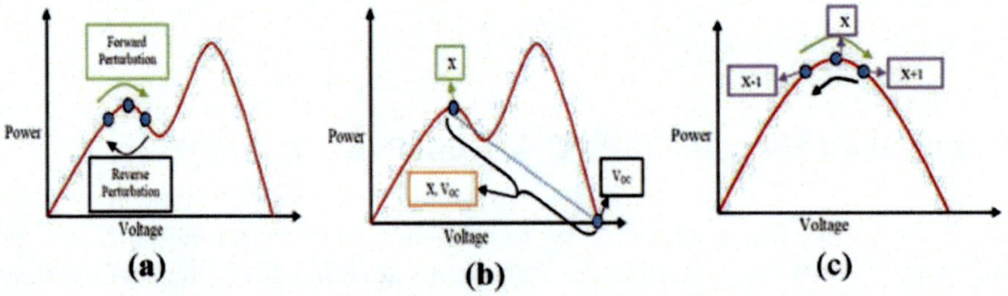

Fig. 3 **a** Perturbation using P&O; **b** Choosing closed interval for MM technique and **c** Closed interval under uniform irradiance

2.2 Under Uniform Irradiance Conditions

As shown in Fig. 3c, under uniform irradiance, the P&O scheme perturbs in forward and reverse directions, and these perturbations create oscillations in the output parameters. Hence, in order to control oscillations around MPP, two points are chosen X−1 and X + 1, as shown in Fig. 3c. These two points are framed as a closed interval [X−1, X + 1] and given to MM, such that the technique iterates in the closed interval and converges to the point X. Until there is a change in irradiance, the hybrid scheme forces the PV system to operate at point X. The proposed scheme is independent of weather conditions, i.e., for any weather condition, the MM technique examines from the point X to V_{OC} and thereafter decides whether the system is experiencing PSC or uniform irradiance.

2.3 Step by Step Procedure

1. Initially P&O scheme is operated under any weather condition.
2. MM technique is initiated as soon as the P&O direction of perturbation is reversed.
3. The MPP obtained from P&O is addressed as 'X'.
4. An interval is chosen with the point X and V_{OC}.
5. The closed interval obtained is [X, V_{OC}].
6. Two points $\alpha 1$ and $\alpha 2$ are chosen in the closed interval with the help of Eq. 5.
7. As soon as the MM starts iterating the interval will be converged to [X, $\alpha 2$] and [$\alpha 1$, V_{OC}].
8. With every new interval, the procedure is repeated until the condition obtained $\alpha 1 = \alpha 2$.
9. At this condition, MPP is obtained.
10. Let the maximum value in the interval is 'Y', which is the MPP, obtained from MM.
11. Compare the points X and Y

    - If X < Y, Y is the GMPP and the PV system is under PSC
    - If X > Y, X is the MPP and the PV system is under uniform irradiance.
    - Under uniform irradiance, a closed interval is chosen as [X−1, X + 1] and given to MM technique.
    - MM converges the interval [X−1, X + 1] to X and forces the PV system to operate at X.

12. For every change in irradiance, repeat the procedure.

3 Result Analysis

Initially, a comparative study is carried out between P&O, modified P&O, and proposed GMPPT scheme using 2S configuration. Figure 4a and b depicts irradiance and high temperature on the solar panel from 9:00 a.m. to 5:00 p.m., and the wattage of each panel is 50 W. From this figure, it is clear that from 12:30 p.m. to 3:00 p.m. the solar panel has experienced maximum irradiance and temperature. As the irradiance and temperature intensity changes for every second, a significant change is renowned for every half an hour. Figure 4(c) represents power obtained using three dissimilar MPPT schemes along with actual power. As the day approaches 5:00 p.m., the numerical value of irradiance and temperature is very low at this particular low value, the modified and traditional P&O schemes have low power extraction compared to the proposed GMPPT scheme. The ratio of actual power with respect to power extracted from different MPPT schemes is termed as efficiency and plotted in Fig. 4d.

Similarly, the performance of P&O, modified P&O, and proposed schemes using 2S2P configuration under constant temperature and constant irradiance conditions. Figure 5a and b depicts output power and efficiency of the considered configuration under constant temperature with variable irradiance. In the same way, a comparative study is carried on the same configuration under constant irradiance, and variable temperature and the results are plotted in Fig. 5c and d. From this test, the maximum tolerable limits are estimated individually for both the irradiance and temperature.

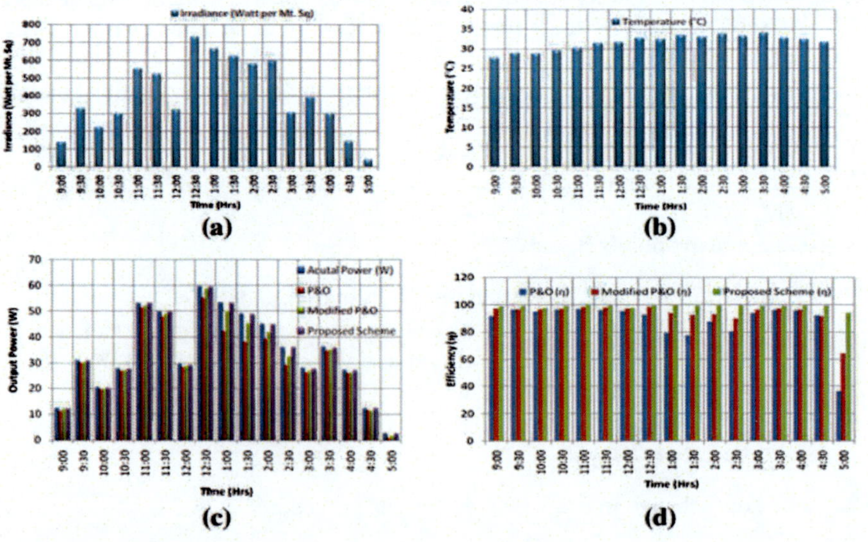

Fig. 4 **a** Irradiance (W/m^2); **b** Temperature for every half an hour change in time; **c** Power obtained from different schemes and **d** Efficiency for a 2S configuration

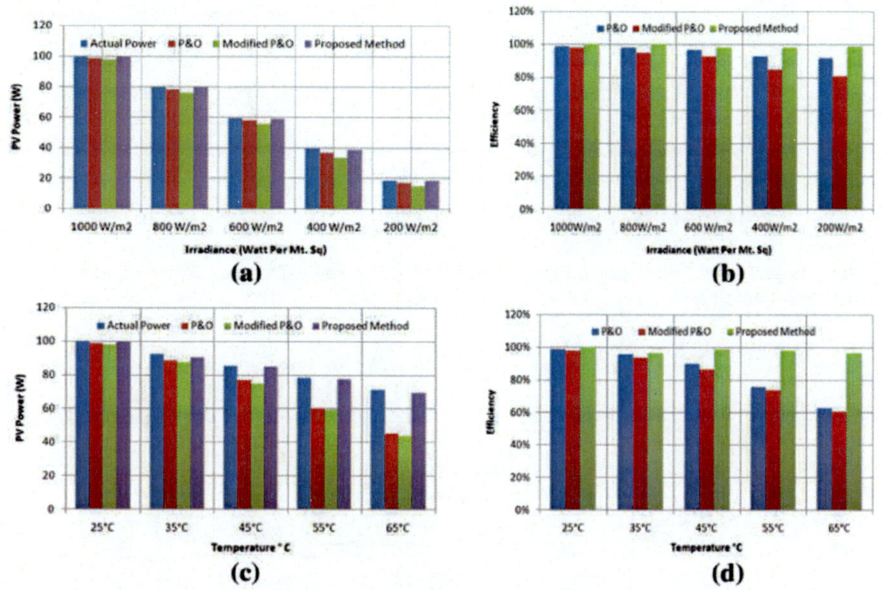

Fig. 5 Output power and efficiency of 2S2P configuration under constant temperature **a** Power; **b** Efficiency under constant irradiance; **c** Power; and **d** Efficiency

4 Experimental Validation

In order to validate the accuracy and reliability of the proposed GMPPT, a model is developed, as in Fig. 6, and the examination is conceded out using 3S6P configuration. Hall effect based ACS712 current sensor and TBV-LV voltage sensors are connected as an intermediate between solar configuration and converter. The proposed GMPPT scheme is programmed in MSP430FR4133 microcontroller developed by Texas Instruments. The essential gate pulses are generated to the switching devices using 7408IC. The over-current and under-voltage issues are supervised by 4027IC.

Fig. 6 The prototype used for Experimentation

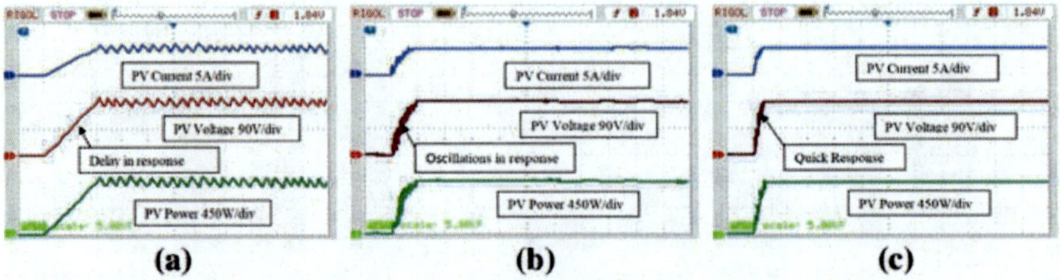

Fig. 7 Experimental results of **a** P&O; **b** Modified P&O, and **c** Proposed scheme under uniform irradiance condition

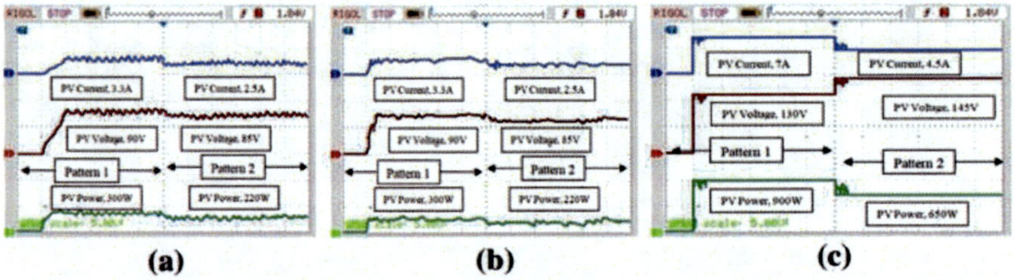

Fig. 8 Experimental results for pattern 1 and pattern 2 **a** P&O; **b** Modified P&O, and **c** Proposed scheme under uniform irradiance condition

Figure 7a–c depicts experimental results of the proposed scheme in comparison with P&O and modified P&O controllers using 3S6P configurations under uniform irradiance. From Fig. 7a, it is comprehensible that the traditional P&O exhibits oscillations and delay in response in the output parameters due to logical forward and reverse perturbations. Even in the modified P&O scheme, oscillations are observed whereas the proposed scheme extracts the solar power with zero oscillations.

The proposed scheme is tested under PSC, along with P&O and modified P&O controllers, and their corresponding results are plotted in Fig. 8a–c. The numerical values of shading patterns are given in Table 1. From Fig. 8a and b, it is observable that the traditional P&O and modified P&O controllers operate at first obtained local peak and fail to track the global path and that directly results in low power as shown in the figure. The proposed scheme has succeeded in extracting the maximum solar in both the shading pattern with zero oscillations, and during the shift in shading pattern, little oscillations are observed until the decision is completed.

5 Conclusion

This paper puts a Hybrid Global Maximum Power Point Tracking Scheme forwards based on Perturb and Observe and Mean method. Under PSC, the traditional controller operates at the primary obtained local peak of the P–V characteristics and, therefore, fails to pathway the global power path and hence a novel technique

Table 1 Shading patterns on the solar panel

S.no	Shading pattern 1	Shading pattern 2
1	1,000 W/m^2	500 W/m^2
2	800 W/m^2	400 W/m^2
3	600 W/m^2	300 W/m^2
4	400 W/m^2	200 W/m^2
5	200 W/m^2	100 W/m^2
6	1,000 W/m^2	500 W/m^2
7	800 W/m^2	400 W/m^2
8	600 W/m^2	300 W/m^2
9	400 W/m^2	200 W/m^2
10	200 W/m^2	100 W/m^2
11	1,000 W/m^2	500 W/m^2
12	800 W/m^2	400 W/m^2
13	600 W/m^2	300 W/m^2
14	400 W/m^2	200 W/m^2
15	200 W/m^2	100 W/m^2
16	1,000 W/m^2	500m^2

is attached to the existing scheme such that it overcomes the operational limitations. The presentation of the proposed scheme is experimentally verified with the traditional P&O and modified P&O controllers. It is benchmarked that the proposed scheme has extracted at most solar power under uniform or partial conditions with reduced output power oscillations.

References

1. Femia N, Petrone G, Spagnualo G, Vitelli M (2005) Optimization of perturb and observe maximum power point tracking method. IEEE Trans Power Electron 20:963–973
2. Ibrahim A-W, Shafik MB, Ding M, Sarhan MA, Fang Z, Alareqi AG, Almoqri T, AI-Rassas AM (2020) PV maximum power point tracking using modified particle swarm optimization under partial shaded condition. Chinese J Electr Eng 6(4):106–121
3. Ghasemi MA, Foroushani HM, Blaabjerg F (2020) Marginal power-based maximum power point tracking control of photovoltaic system under partial shaded condition. IEEE Trans Power Electron 35(6):5860–5872
4. Koutroulis E, Sason N, Georgiads V (2019) Combined tracking of the maximum power and maximum efficiency operating points for reel time maximization of the energy production of PV system. IEEE Trans Power Electron 34:8634–8645
5. Mustafa Ergin Sahin (2020) A photovoltaic powered electrolysis converter system with maximum power point tracking control. Int J Hydrogen Energy 45(6):9293–9304
6. Mansor M, Mirza AF, Ling Q (2020) Harris Hawk optimization based MPPT control for PV system under partial shaded conditions. J Cleaner Product 274:122857

7. Chandrasekaran K, Sankar S, Banumalar K (2020) Partial shading detection for PV arrays in a maximum power tracking system using the sine cosine algorithm. Energy Sustain Develop 55:105–121
8. Kota VR, Bhukya MN (2017) A novel linear tangents based P&O scheme for MPPT of a PV system. Renew Sustain Energy Rev 71:257–267
9. Kota VR, Bhukya MN (2016) A simple and efficient MPPT scheme for PV module using 2-dimensional lookup table. In: IEEE power and energy conference at Illinois (PECI), 2016, pp 1–7
10. Bhukya MN, Kota VR, Rani DS (2019) A simple, efficient and novel standalone photovoltaic inverter configuration with reduced harmonic distortion. IEEE Access 7(6287639), 43831–43845
11. Bhukya MN, Kota VR (2017) A new MPPT scheme based on trifurication of PV characteristic for photovoltaic power generation. Int J Pure Appl Math 114(10):439–447
12. Kota VR, Bhukya MN (2019) A novel global MPP tracking scheme based on shading pattern identification using artificial neural networks for photovoltaic power generation during partial shaded condition. IET Renew Power Gener 13(10):1647–1659
13. Babu TS, Rajasekar N, Sangeetha K (2015) Modified particle swarm optimization technique based maximum power point tracking for uniform and under partial shading condition. Appl Soft Comput 34:613–624
14. Ram JP, Babu TS, Rajasekar N (2017) A comprehensive review on solar PV maximum power point tracking techniques. Renew Sustain Energy Rev 67:826–847
15. Sangeetha K, Babu TS, Rajasekar N (2016) Fireworks algorithm-based maximum power point tracking for uniform irradiation as well as under partial shading condition. In: Artificial intelligence and evolutionary computations in engineering systems. Springer, New Delhi, pp 79–88
16. Bhukya MN, Kota VR (2019) A quick and effective MPPT scheme for solar power generation during dynamic weather and partial shaded conditions. Eng Sci Technol Int J 22(3):869–884

Biogeography-Based Optimization for Power Quality Improvement in HRES System

B. Srikanth Goud, B. Loveswara Rao, Aymen Flah, Mohit Bajaj, Naveen Kumar Sharma, and Ch. Rami Reddy

Abstract Grid Integration using Hybrid Renewable Energy Sources is rapidly growing its importance to meet the energy demand. This design creates mainly Power Quality issues (PQ) due to the existence of nonlinear loads. To solve the PQ problems and compensate for the issues, this paper proposes Biogeography-Based Optimization (BBO) controller with Unified Power Quality Conditioner (UPQC). The performance of the proposed method is compared with the existing methods like Genetic Algorithm (GA), Genetic Search Algorithm (GSA) with PI controller and obtained the best results in terms of total harmonic distortions (THD). The proposed system is implemented on MATLAB/Simulink platform.

Keywords PV · Wind · Battery · BBO · UPQC

1 Introduction

Sustainable power-source-based Distributed Generation (DG) is increasing to a greater significance because of the progress in innovation and environmental concerns because of extremely great interest for the capacity to the utility grid [1].

B. S. Goud · B. L. Rao
Koneru Lakshmaiah Education Foundation, Guntur 522502, India
e-mail: srikanth.b@anuraghyd.ac.in

A. Flah
National School On Engineering of Gabès, University of Gabès, 6072 Gabès, Tunisia

M. Bajaj (✉)
National Institute of Technology, Delhi 110040, India
e-mail: mohitbajaj@nitdelhi.ac.in

N. K. Sharma
I. K. Gujral Punjab Technical University, Jalandhar 144603, India
e-mail: dr.naveensharma@ptu.ac.in

Ch. R. Reddy
Malla Reddy Engineering College (Autonomous), Dhulapally 500100, India

A. R. Gupta et al. (eds.), *Power Electronics and High Voltage in Smart Grid*, Lecture Notes in Electrical Engineering 817,
https://doi.org/10.1007/978-981-16-7393-1_25

In order to manage various power electronics and to control innovation with greater adaptability, multiple optimization strategies have been designed when coordinating various alternate sources such as sun, wind, battery, etc. [2], which are taking on a growing job in meeting the enormous power needs [3]. Due to the limitations of the generation of power from non-renewable energy sources (NRES), DGs are more important because they provide end-users with more efficient, high quality and stable power that they constantly require [4]. RES has taken its position because NRES alone will not fulfil the requirements due to the rise in population and the growing demand for power production has become a demanding challenge. So, DGs of which the battery is widely used for solar, wind, [5]. The DG framework in the proposed paper includes a source linked to the DC connection of the BBO-driven interconnected inverter that is significantly controlled to take care of real power from the DG to the grid [6]. The suggested solution greatly remunerates harmonics and imbalances under disrupted voltage conditions if the load is connected to the non-linear or imbalanced point of typical coupling (PCC) and both [7]. We decided to mitigate the load reactive capacity [8], in comparison to the existing power from RES to the grid and grid-interfacing inverters. In the ongoing decade, the use of sustainable power source is rising; it has numerous points of interest. Be that as it may, while reconciliation of inexhaustible source to the electric framework, it will make some power quality issues. In this way, in the written works numerous procedures have been proposed, yet the issue isn't yet totally fathomed. For example, [9–15] have introduced a technique to improve the power quality while using a sustainable power source. Still, the power quality issue isn't corrected. Subsequently in this paper, an improvement-based control system is proposed to upgrade the power quality while joining a sustainable power source. The significant commitment of the proposed work is as follows. An HRES Grid-Connected System is developed with three force sources such as Wind, PV and Battery. Modelled BBO for controlling the framework and equalization, the power quality issues as far as current, voltage, real power, and reactive power. At that point, the proposed framework is developed and examined in MATLAB/Simulink. The rest of the paper is composed as follows: Sect. 2 proposed HRE framework with BBO, in Sect. 3, BBO and in Sect. 4, results and discussions.

2 Proposed HRES System with BBO

The proposed system is as shown in Fig. 1, which mainly consists of distribution generation consisting of PV-Wind connected to the inverter to interface the grid via DC-link capacitor [10–13]. By implementing BBO, the PV and Wind power are extracted. From BBO a reference DC link voltage is created. The reference DC link voltage is set to its default value at the time when solar energy is absent. The power quality issues of the HRES are considered which are mainly due to non-linear load on-grid side [16]. To overcome all the issues, the proposed system is designed with UPQC which compensates the problems by control techniques with the help of series and shunt controllers as shown in Fig. 1. The proposed HRES consists of PV and

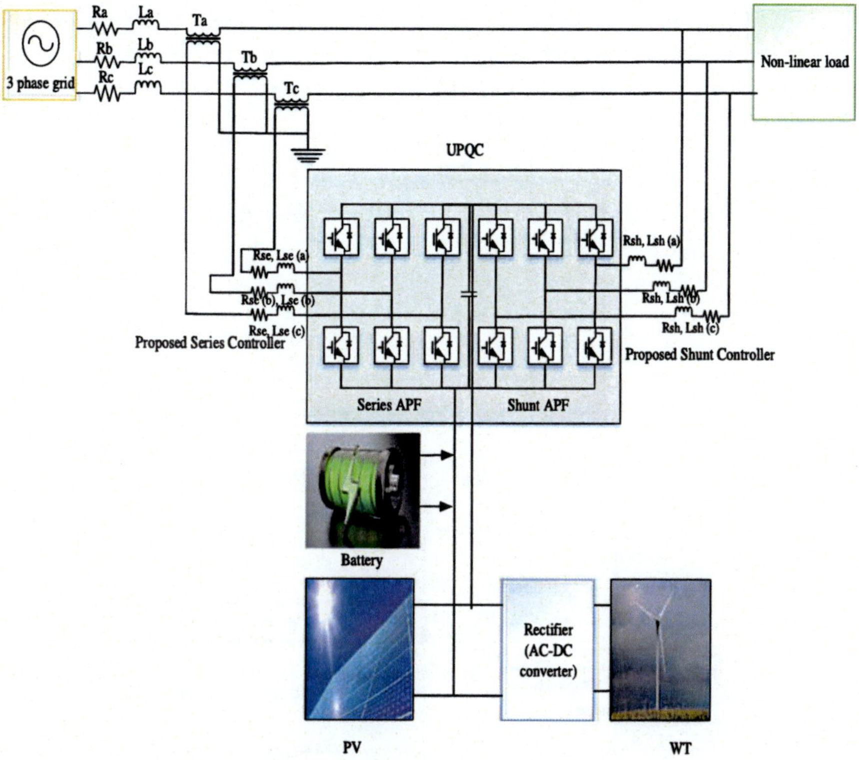

Fig. 1 Block diagram of grid connected HRES

Wind system connected to the grid. The battery is provided as an energy storing device and acts as energy management during system failure. The power quality problems occurred in the HRES system due to non-linear loads on the grid side [17–22]. To overcome such issues, the system is designed using UPQC which would be given the best performance characteristics to compensate for the PQ problems. Control operation of the system is done using optimization technique like BBO.

3 Biogeography-Based Optimization

BBO is a developmental enhancement that is again propelled by Swarm conduct in nature. Organic species and their exercises are watched. Migration and displacement are the qualities of any calculation. As a rule, the region has land, precipitation, vegetation, temperature and so on which demonstrates a high territory appropriateness list, so the species shifts from one island to the next. Reasonableness Index Variables which demonstrates the tenability. Species with a huge number demonstrates HIS is called migration and less shows low HIS is called movement. BBO advancement is the most recent way to deal with critical thinking [10]. The flow chart of BBO is presented in Fig. 2. The objective function of the proposed technique is expressed as Eq. (1).

$$J = \int_{0}^{t} t|e(t)|dt = \int_{0}^{t} t\left|v_{dc} - v_{dc(ref)}\right|dt \tag{1}$$

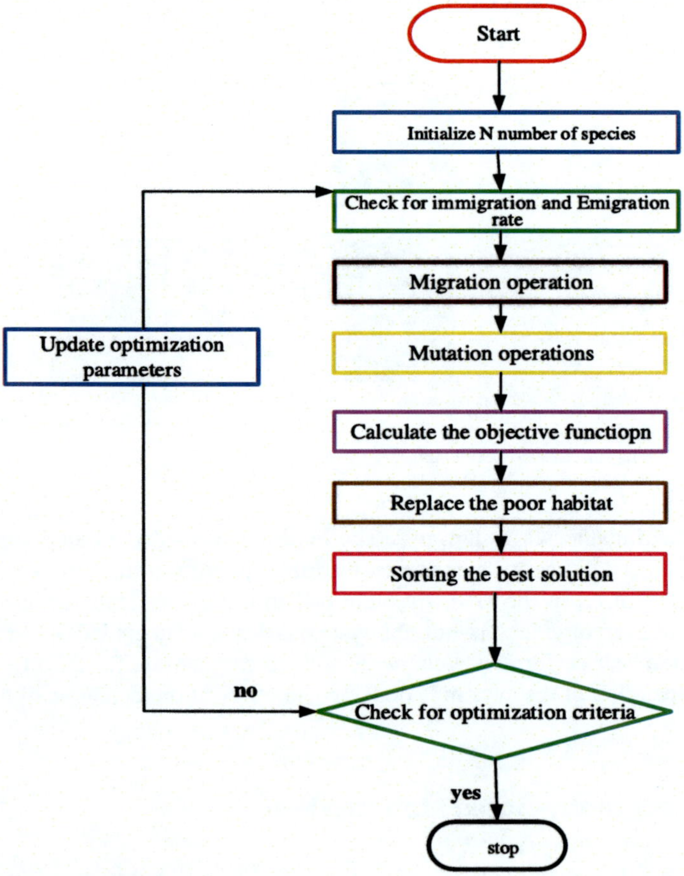

Fig. 2 Biogeography-based optimization

4 Results and Discussions

The PV and WT frameworks are considered as the wellspring of the framework. For this situation investigation, the PV and WT base boundaries of irradiance and wind speed are fixed as a steady level just as examination of the control structure attributes. The PV irradiance conditions are taken as 1,000 W/m^2. Identified with the irradiance level of PV, the vitality is created in a framework that is utilized to reward the heap request. The WT speed is taken as 12 m/s, in view of the speed the WT has produced the force. Thus, HRES has produced the necessary capacity to satisfy the heap need and make up for PQ issues. The battery likewise is associated with the framework which is empowered with just basic states of PV and WT vitality frameworks. The PV and WT boundaries with produced power are shown in Fig. 3. The real, reactive power and THD (Table 1) of different harmonics with BBO, GA, GSA and PI controller are depicted in Figs. 4, 5 and 6.

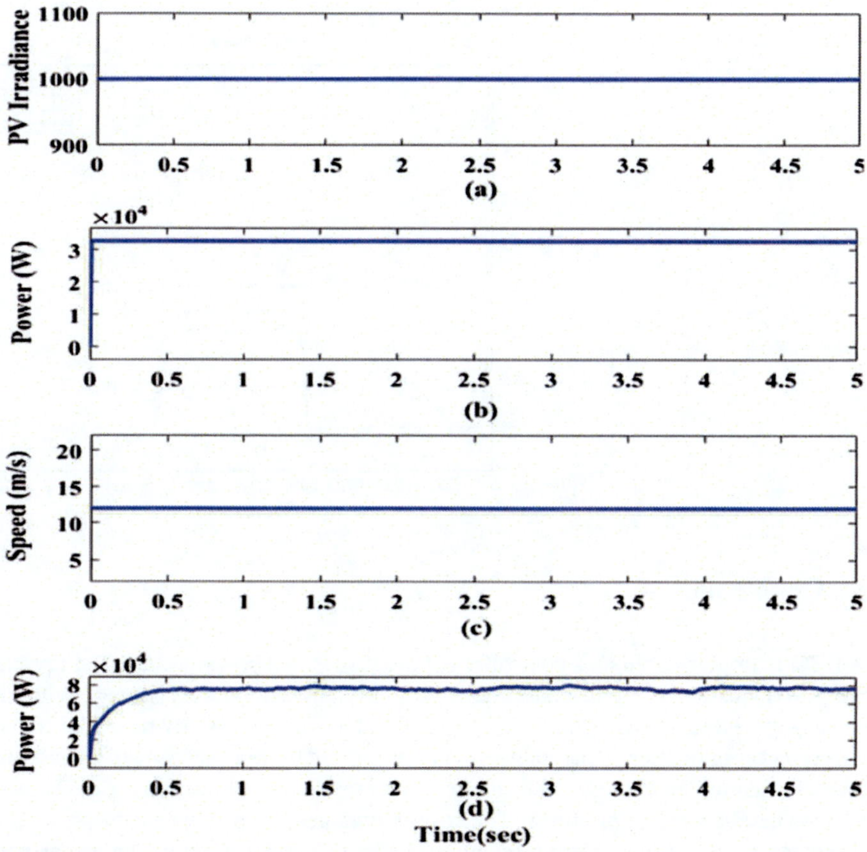

Fig. 3 Performance characteristics of HRES

Fig. 4 Active power supplied by HRES

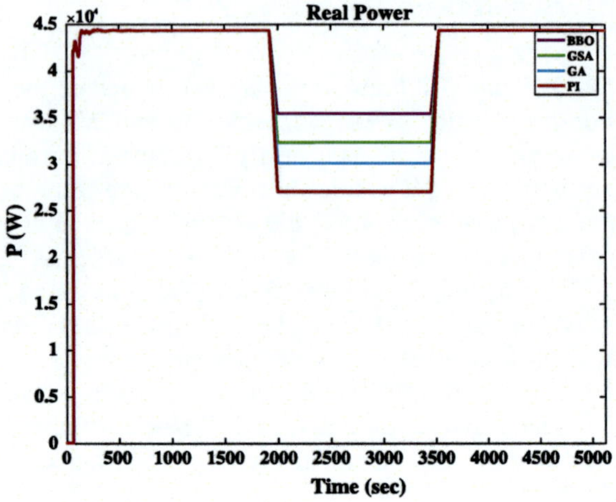

Fig. 5 Reactive power supplied by HRES

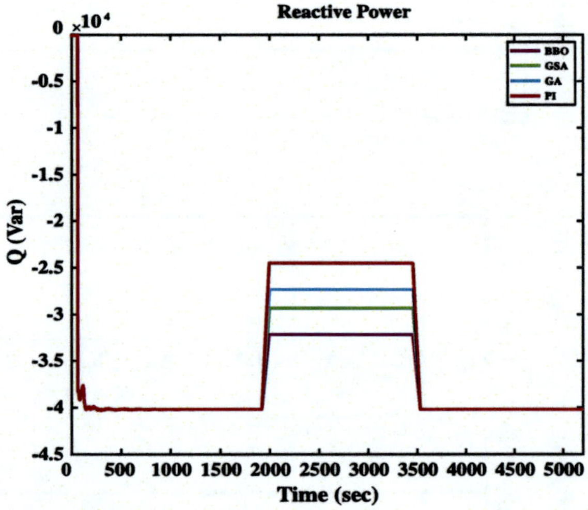

5 Conclusion

A BBO is implemented as a controller for the distributed power system to improve the power quality in a grid-connected hybrid system with battery storage without causing any effects to its normal operation of real power transfer. By using the BBO, the perfect combination of parameters is generated and the optimal control signals are predicted using PI technique. The results of the proposed optimization are developed in MATLAB/Simulink platform. The results obtained of the proposed system are compared with various existing techniques like GA, GSA with PI controller and concluded that BBO has produced best results in terms of THD.

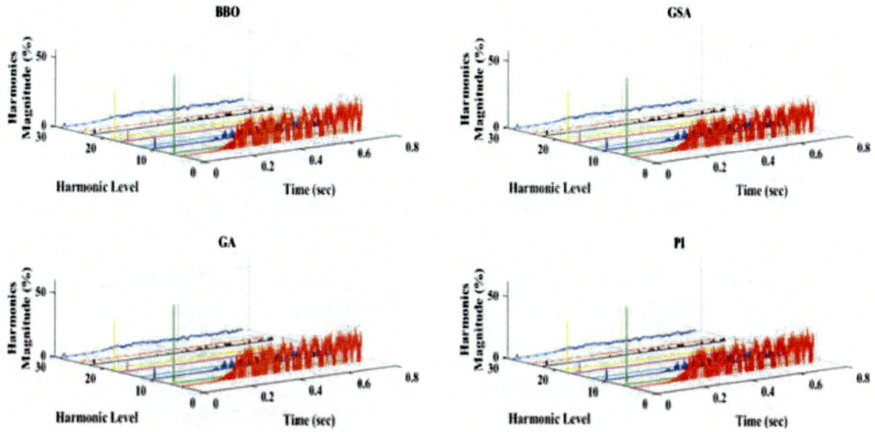

Fig. 6 Comparison analysis of THD's

Table 1 THD comparison of existing and proposed methods

Sl. no	Methods	Before UPQC								
		5	7	11	13	17	19	23	25	29
1	Proposed	51	38	36	25	32	51	32	40	26
2	GSA	65	34	28	36	51	42	38	42	32
3	GA	71	38	22	32	45	38	35	48	36
4	PI	82	35	25	12	48	41	38	45	32
		After UPQC								
		5	7	11	13	17	19	23	25	29
1	Proposed	25	8	12	15	21	9	12	8	12
2	GSA	35	15	17	18	22	12	15	11	15
3	GA	37	16	18	21	22	10	8	6	4
4	PI	35	12	13	17	21	12	7	5	1

1. References

1. Goud BS, Rao BL, Reddy CR (2020) Essentials for grid integration of hybrid renewable energy systems: a brief review. Int J Renew Energy Res (IJRER) 10(2):813–830
2. Bajaj M, Singh AK, Alowaidi M, Sharma NK, Sharma SK, Mishra S (2020) Power quality assessment of distorted distribution networks incorporating renewable distributed generation systems based on the analytic hierarchy process. IEEE Access 8:145713–145737
3. Goud BS, Rao BL (2019) PV-wind integrated grid with P&O and PSO MPPT techniques. Int J Recent Technol Eng (IJRTE) 8(1):3217–3224
4. Bajaj M, Singh AK (2020) Grid integrated renewable DG systems: a review of power quality challenges and state-of-the-art mitigation techniques. Int J Energy Res 44(1):26–69
5. Goud BS, Rao BL, Devi BN, Kumar KS, Keerthi N (2020) Review on optimization techniques employed in distribution generation. J Crit Rev 7(2):639–645

6. Bajaj M, Singh AK (2020) An analytic hierarchy process-based novel approach for benchmarking the power quality performance of grid-integrated renewable energy systems. Electr Eng 102(3):1153–1173
7. Goud BS, Rao BL (2019) Review of optimization techniques for integrated hybrid distribution generation. Int J Innovat Technol Explor Eng (IJITEE) 8(5):527–533
8. Bajaj M, Rana AS (2018) Harmonics and reactive power compensation of three phase induction motor drive by photovoltaic-based DSTATCOM. Smart Sci 6(4):319–329
9. Reddy CR, Goud BS, Reddy BN, Pratyusha M, Vijay Kumar CV, Rekha R (2020) Review of Islanding detection parameters in smart grids. In: 2020 8th international conference on smart grid (icSmartGrid), Paris, France, pp 78–89
10. Goud BS, Rao BL (2020) An intelligent technique for optimal power quality enhancement (OPQE) in an HRES grid-connected system: ESA technique. Int J Renew Energy Res (IJRER) 10(1):317–328
11. Bajaj M (2020) Design and simulation of hybrid DG system fed single-phase dynamic voltage restorer for smart grid application. Smart Sci 8(1):1–15
12. Bajaj M, Aggarwal S, Singh AK (2020) Power quality concerns with integration of RESs into the smart power grid and associated mitigation techniques. In: 2020 IEEE 9th power India international conference (PIICON), 2020, pp 1–6
13. Bajaj M, Singh AK (2019) Power quality challenges associated with distributed generation planning: a simulation-based demonstration. In: 2019 international conference on electrical, electronics and computer engineering (UPCON), 2019, pp 1–6
14. Madhusudan R, Reddy PL (2019) A rationalized controller based FACTS device for power quality enhancement in the distribution system. Int J Eng Adv Technol 8(5):275–284
15. Goud BS, Rao BL, Reddy CR (2021) An intelligent technique for optimal power quality reinforcement in a grid-connected HRES system: EVORFA technique. Int J Numer Model: Electron Netw, Dev Fields 34(2):e2833
16. Madaria PK, Bajaj M, Aggarwal S, Singh AK (2020) A grid-connected solar PV module with autonomous power management. In: 2020 IEEE 9th power India international conference (PIICON), 2020, pp 1–6
17. Goud BS, Varma PS, Rao BL, Reddy MSK, Pandian A, Reddy CR (2020) Cuckoo search optimization based MPPT for integrated DFIG-wind energy system. In: 2020 international conference on decision aid sciences and application (DASA), Sakheer, Bahrain, pp 636–639. https://doi.org/10.1109/DASA51403.2020.9317072
18. Reddy BN, Chandra Sekhar O, Ramamoorthy M (2018) Analysis and implementation of single-stage buck-boost buck converter for battery charging applications. J Adv Res Dyn Control Syst (JARDCS) 10(4):462–475
19. Rana AS, Bajaj M, Gairola S (2019) Optimal power flow solution in smart grid environment using SVC and TCSC. In: Advanced communication and control methods for future smartgrids, intechopen, pp 1–22
20. Nagi Reddy B, Bharathi M, Pratyusha M, Bhargavi KS, Srikanth Goud B (2020) Design of a novel isolated single switch AC/DC integrated converter for SMPS applications. Int J Emerg Trends Eng Res 8(4):1111–1119
21. Bajaj M, Kumar Singh A (2020) A streamlined approach for assessing the power quality in renewable energy systems. In: IEEE 17th India council international conference (INDICON), New Delhi, pp 1–6. https://doi.org/10.1109/INDICON49873.2020.9342487
22. Bajaj M, Kumar Singh A (2021) Hosting capacity enhancement of renewable-based distributed generation in harmonically polluted distribution systems using passive harmonic filtering. Sustain Energy Technol Assess 44:101030

Comparative Study of Closed Loop Controlled Parallel Connected Converter Methods for Minimization of Circulating Current

G. K. Nisha and Jeneena Babu

Abstract Distributed Energy Resources (DER) such as wind, photovoltaic systems, fuel cells, biomass etc. are used for energy management and production and the usage of renewable sources increases day by day. DERs are integrated into common DC or AC microgrid through power electronic converters. The main problems related with parallel connected DC-DC converters in DC microgrid are sharing of current between loads, circulating current between converters, regulation of output voltage etc. Droop control method is a common technique used for load current sharing and the limitations of conventional droop method are poor current sharing and drop in DC grid voltage. This paper studies the various reasons for circulating current and how it can be reduced with different switching methods. The results and observations from MATLAB/Simulink model are presented to substantiate the effects of closed loop parallel connected boost converters on circulating current using different methods.

Keywords Load sharing · Circulating current minimization · Converter

1 Introduction

Distributed energy sources like wind, photovoltaic systems, fuel cells etc. are connected with storage units like batteries and super capacitors together to form a microgrid and these systems improve the production and energy management. The major increase in demand for power tends to integrate renewable sources into the grid and the power electronic devices are interfaced with DC microgrid to improve the flexibility in conversion and to increase the power levels.

The control issues associated with microgrid are connection schemes between distributed energy sources and DC grid, load sharing between converters, maximum

G. K. Nisha (✉) · J. Babu
Mar Baselios College of Engineering and Technology, Trivandrum, Kerala, India
e-mail: nisha.gk@mbcet.ac.in

J. Babu
e-mail: jeneena.babu@mbcet.ac.in

A. R. Gupta et al. (eds.), *Power Electronics and High Voltage in Smart Grid*,
Lecture Notes in Electrical Engineering 817,
https://doi.org/10.1007/978-981-16-7393-1_26

power point tracking, voltage control among parallel converters and energy storage. The advantages of parallel connected converters include reliability, ease of maintenance, expandability of output power, efficiency etc. The challenges in paralleling DC sources are stability of the system needs a proper converter design and effective load sharing among Distributed Generators (DGs). Circulating current is obtained in parallel connected converters due to the difference in the output of the converters. The various reasons for the circulating current are asynchronous switching, variations in input, delay in the control loops etc. Various controllers like proportional, integral, derivative, PID, fuzzy etc. can be used to enhance the performance of the system and this paper performs the comparative study of different controller outputs of parallel connected closed loop controlled boost converters.

2 Parallel Operation of DC-DC Boost Converters

Boost converters are connected in parallel to meet higher power demands. The different factors which include the output of boost converters are input source given to the converters, the working of closed loop feedback loop used in the system, on and off-time of different switches etc. Based on the switching of parallel connected converters there are two modes of operation known as synchronous and asynchronous switching [2].

In synchronous switching, both the switches S_{W1} and S_{W2} in a converter system will be turned on or off at the same time which means no delay in switching action. Therefore, the duty signal given to parallel connected converters is synchronous in nature, whereas in asynchronous mode either one of the switch is closed at a time.

In Fig. 1a both the switches S_{W1} and S_{W2} are kept to be in closed position, therefore the diodes D_2 and D_1 are reverse biased and as a result the capacitance-voltage will be applied to the load. In Fig. 1b both the switches S_{W2} and S_{W1} are opened. The diodes D_2 and D_1 will conduct and capacitors will be in charging mode and the load current is directly supplied from the individual sources.

In Fig. 2a Switch S_{W2} is opened and switch S_{W1} is closed, D_1 is reverse biased,

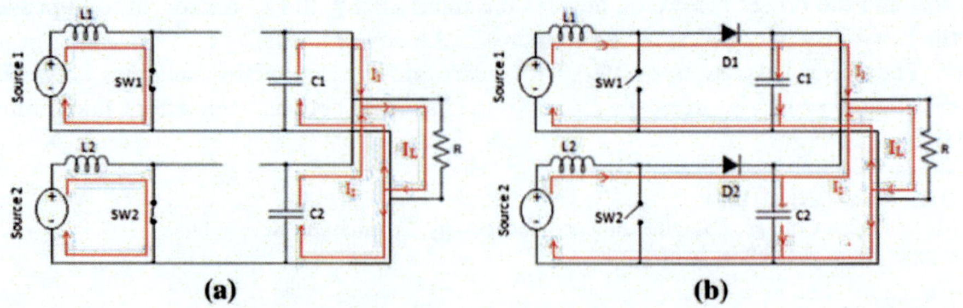

Fig. 1 Synchronous mode of operation

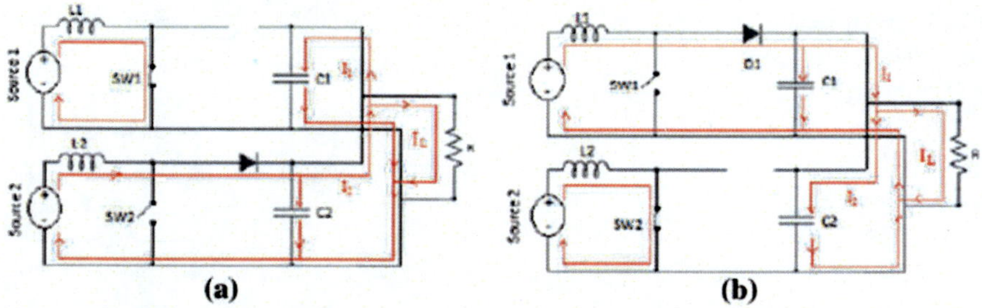

Fig. 2 Asynchronous mode of operation

D_2 is forward biased which leads to circulating current from converter II to charge capacitor. In Fig. 2b Switch S_{W1} is opened and switch S_{W2} is closed and D_2 is reverse biased and D_1 is forward biased, therefore circulating current from converter I to charge capacitor. Asynchronous switching arises due to delay in the feedback control loops of the system which may arise circulating currents. Overloading occurs because of a change in load sharing between converters due to circulating current.

3 Control Methods for Minimization of Circulating Currents in Parallel Connected Converters

The parallel connected closed loop controlled boost converters reduce the circulating current, and the different controllers used are proportional controller, integral controller, PI controller, PID controller, Neural Network, Fuzzy logic controller etc. To reduce the circulating currents, two methods used are series resistor and virtual output resistance [3]. In series resistor method a resistor is placed in series with distributed generator output and the resistor value is set using a potentiometer. Therefore, by adjusting the potentiometer values the output of all paralleled distributed generators are made almost identical. If the output voltage drop is large, high power loss occurs in the series resistor and therefore this method is not practically possible in real systems. In second method virtual resistance concept is used for identical outputs of distributed parallel connected generators.

Figure 3 shows the steady state equivalent circuit of parallel connected converters with load. V_{DC1}, V_{DC2}, I_1, I_2, R_1 and R_2 represents the output voltage, output current and resistance of the cables of converter 1 and converter 2, respectively. If $V_{DC2} > V_{DC1}$, I_{C21} is the circulating current flowing from converter to converter 1. By using Kirchoff's law,

$$V_{DC1} - I_1 R_1 - I_L R_L = 0 \tag{1}$$

$$V_{DC2} - I_2 R_2 - I_L R_L = 0 \tag{2}$$

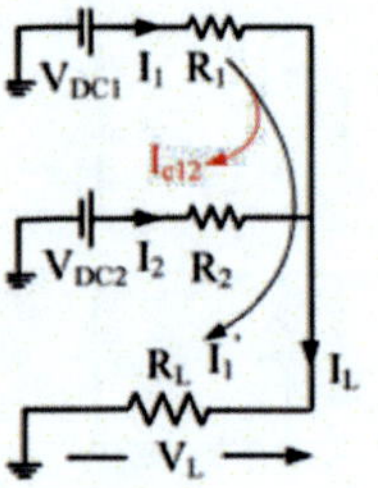

Fig. 3 Steady state equivalent circuit of two parallel connected converters

The output current of the converters,

$$I_1 = \frac{(R_2 + R_L)V_{DC1} - R_L V_{DC2}}{R_1 R_2 + R_1 R_L + R_2 R_L} \tag{3}$$

$$I_2 = \frac{(R_1 + R_L)V_{DC2} - R_L V_{DC1}}{R_1 R_2 + R_1 R_L + R_2 R_L} \tag{4}$$

The circulating current is

$$\begin{aligned} I_{C12} = -I_{C21} &= \frac{V_{DC1} - V_{DC2}}{R_1 + R_2} = \frac{I_1 - I_2}{2}(if \;\; R_1 = R_2) \\ &= \frac{I_1 R_1 - I_2 R_2}{R_1 + R_2}(R_1 \neq R_2) \end{aligned} \tag{5}$$

Figure 4 represents block diagram of parallel connected closed loop controlled converters. The reference voltage or set point is fixed. The PI controllers used consist of the outer voltage loop controller and inner current loop controller. These controllers provide a synchronous duty ratio input to the parallel connected DC-DC boost converters. The current signals for both converters are synchronized and the PWM signals thus generated are sent to the converters.

To remove the offset error, Fuzzy PI Controller is combined with the conventional PI controller to form a Hybrid Fuzzy PI Controller. The signal from the controller is sent to PWM generator to produce the PWM signal. The fuzzification process converts the input data into fuzzy sets and the inference system used is Mamdani which work on if-then rule. (e.g. If mf1 is input then output is mf1; if mf2 is input then output is mf2). Fig. 5a shows fuzzy logic controller and Fig. 5b shows the integration of PI controller with a fuzzy logic controller. The hybrid fuzzy PI Controller is a single input single output system.

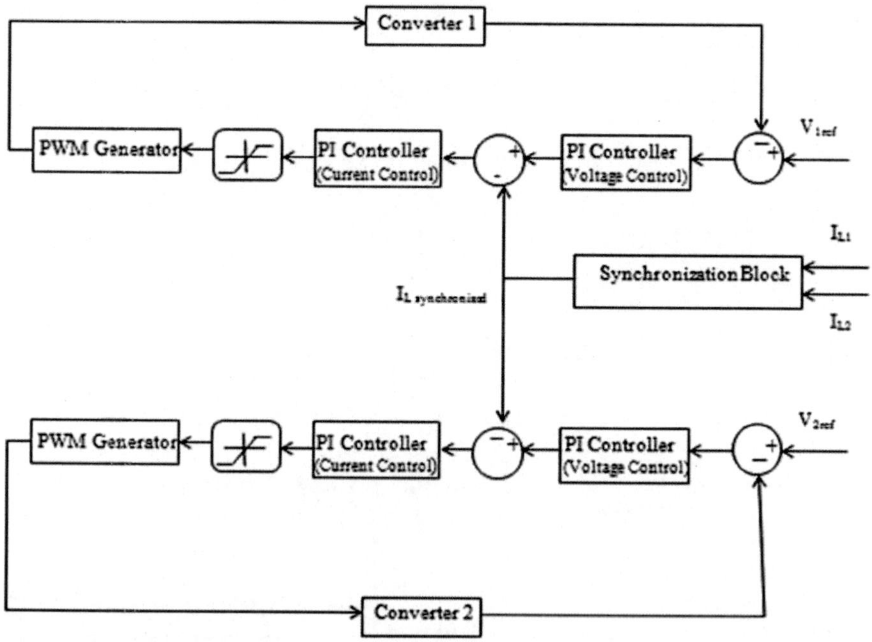

Fig. 4 Block diagram of closed loop parallel connected converters with synchronization block

4 Results and Discussions

This section presents simulation studies of different closed loop control strategies for minimizing circulating current in low voltage stand-alone DC microgrid in MATLAB/Simulink environment (Table 1).

4.1 Synchronous Switching

MATLAB/ Simulink model of synchronous switching of parallel connected converters is shown in Fig. 6. Two converters are connected in parallel and the power ratings are 96 W and 192 W. Switching frequency of 25 kHz is used and the duty ratio of two switches is synchronous in nature. The variation of output voltage and current of the converter is shown in Fig. 7.

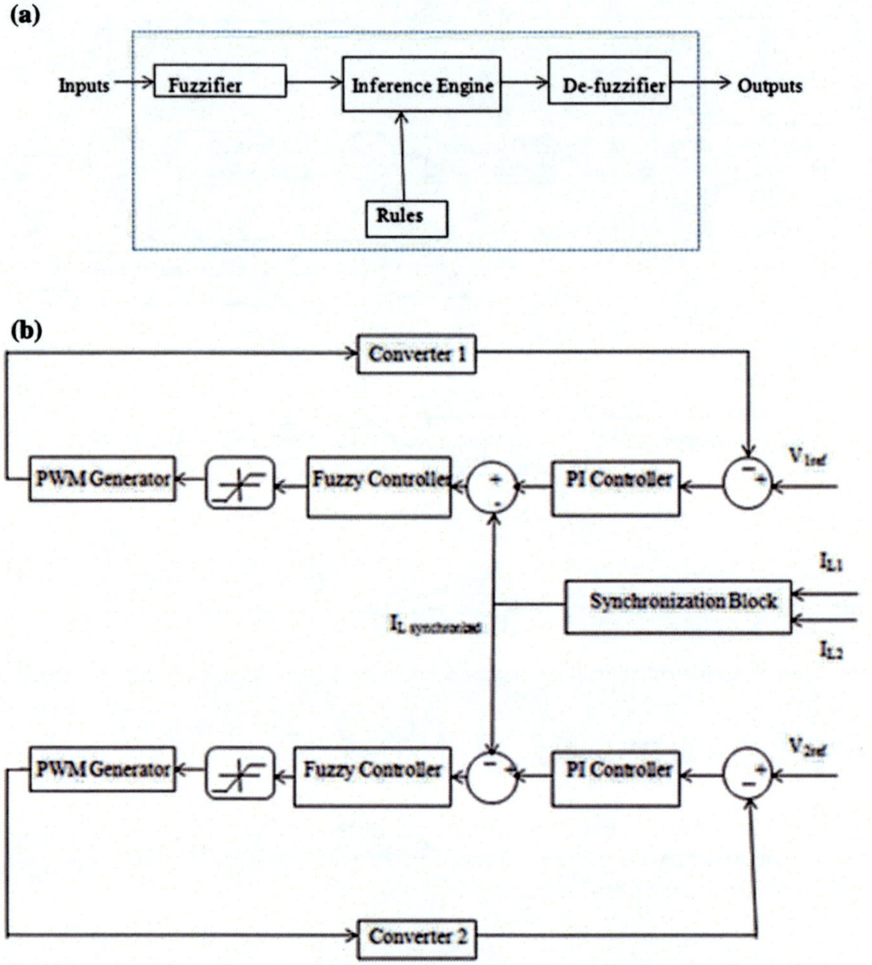

Fig. 5 **a** Fuzzy logic controller, **b** Block diagram of fuzzy controller integrated with PI controller

Table 1 Simulation parameters of parallel connected boost converters

Parameters	Converter 1	Converter 2
Input voltage (V)	23	23
Switching frequency (KHz)	25	25
Inductance (L)	0.023712	0.011856
Capacitance (F)	0.000086667	0.000173333
Power (W)	96	192

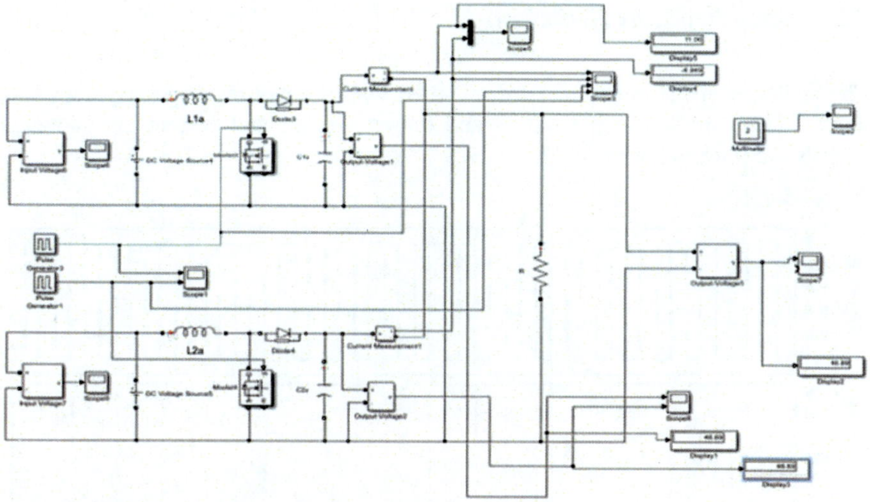

Fig. 6 Simulink model of synchronous switching

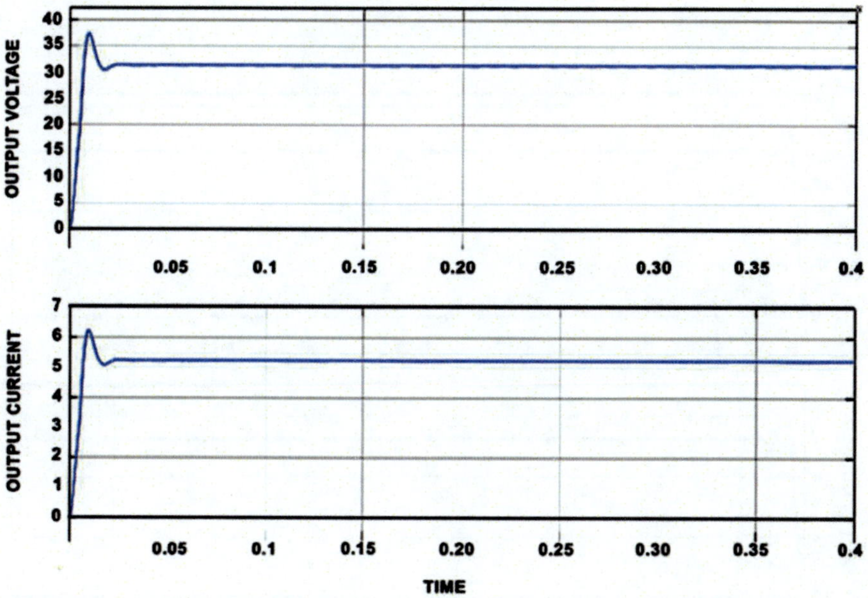

Fig. 7 Output voltage and Current waveform

4.2 Asynchronous Switching

The switches of the converters are operated with a delay results in asynchronous nature. The variation in the pulses and output voltage and current are shown in Figs. 8 and 9, respectively.

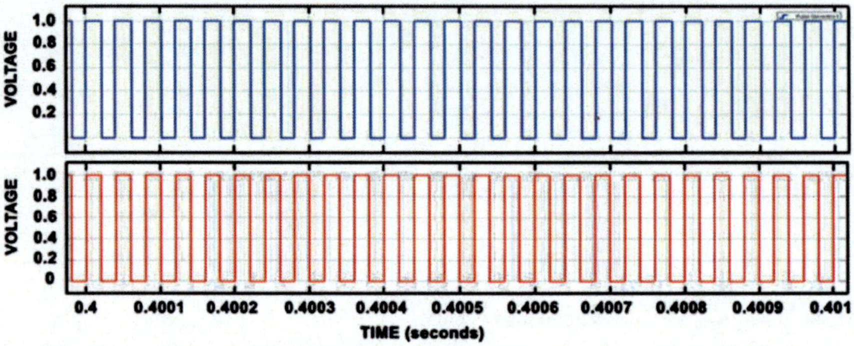

Fig. 8 Pulse obtained by asynchronous mode of operation

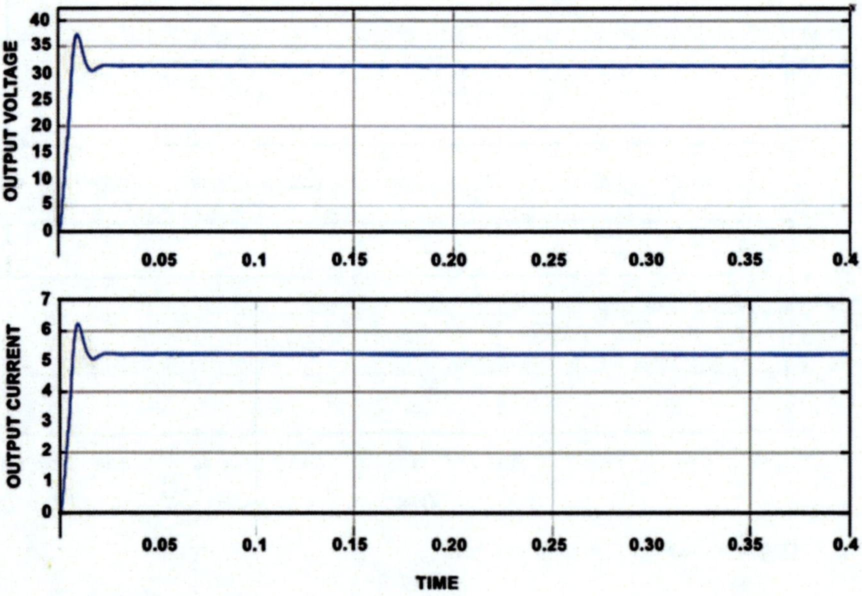

Fig. 9 Output voltage and Current waveform

4.3 Closed Loop Control of Two Parallel Connected Converters with PI Controllers

Two parallel connected converters are controlled with an inner current loop and outer voltage loop. The reference voltage is compared with the output of the boost converter and the error is given to PI controller and the current reference is compared with first PI controller output. The output from second controller is given as the PWM control for the switches. The output voltage and current of the system is shown in Fig. 10.

The pulses and output voltage and current obtained by a closed loop PI controller and synchronization block are shown in Figs. 11 and 12, respectively. The synchronization block is obtained by using the equations

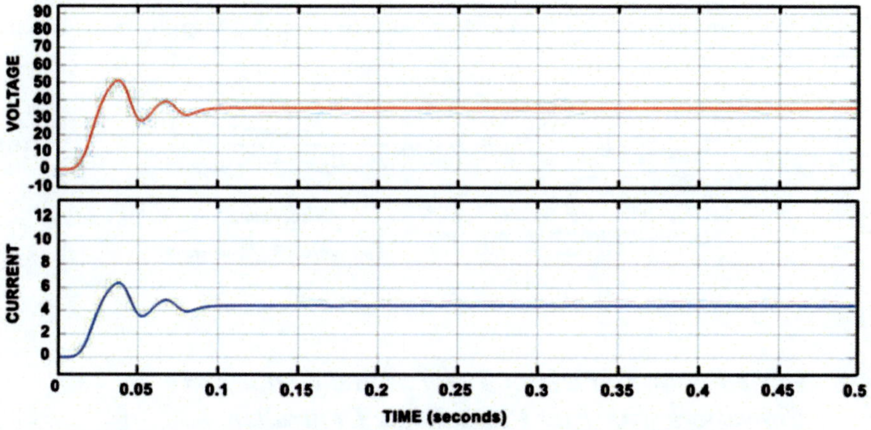

Fig. 10 Output voltage and Current waveform obtained by closed loop control

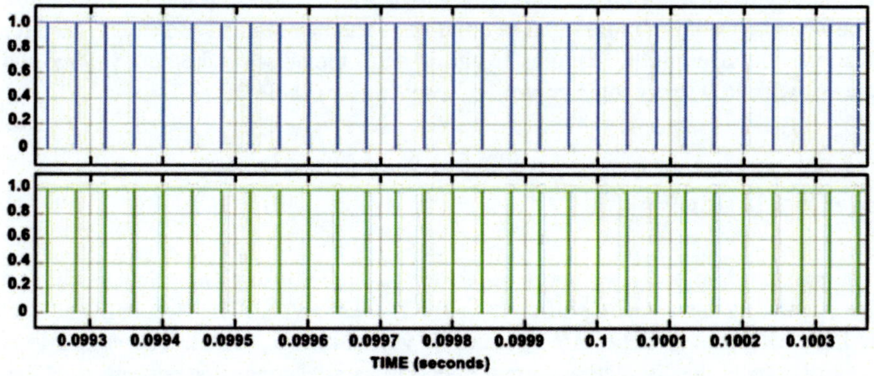

Fig. 11 Pulse obtained by closed loop controllers and synchronization block

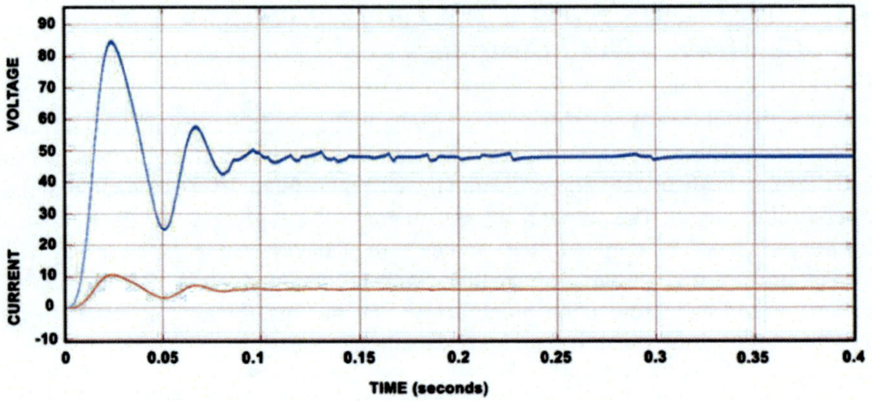

Fig. 12 Output voltage and Current waveform obtained by closed loop PI controllers and synchronization block

$$I_{L1\text{synchronised}} = (I_{L1} + I_{L2}) \times \frac{P_{\text{converter1}}}{P_{\text{converter1}} + P_{\text{converter2}}} \tag{6}$$

$$I_{L2\text{synchronised}} = (I_{L1} + I_{L2}) \times \frac{P_{\text{converter2}}}{P_{\text{converter1}} + P_{\text{converter2}}} \tag{7}$$

4.4 Closed Loop Control of Two Parallel Connected Converters with PID Fuzzy Logic Controller and Synchronization Block

PI controllers are replaced by hybrid fuzzy logic PI controllers in closed loop control. The advantages of both fuzzy and PI controllers can be obtained and can be used for system with uncertainties. Figures 13 and 14 show the pulses and output voltage and current with PID fuzzy logic controller, respectively (Table 2).

5 Conclusion

The parallel connection of boost converters is studied using Simulink model and the circulating current using different methods are analyzed. The circulating current is high when the switches are operated in asynchronous nature. The closed loop control reduces the circulating current effects in parallel connected converters. When compared to other systems, the hybrid fuzzy PID controllers with a synchronization block have minimum circulating current.

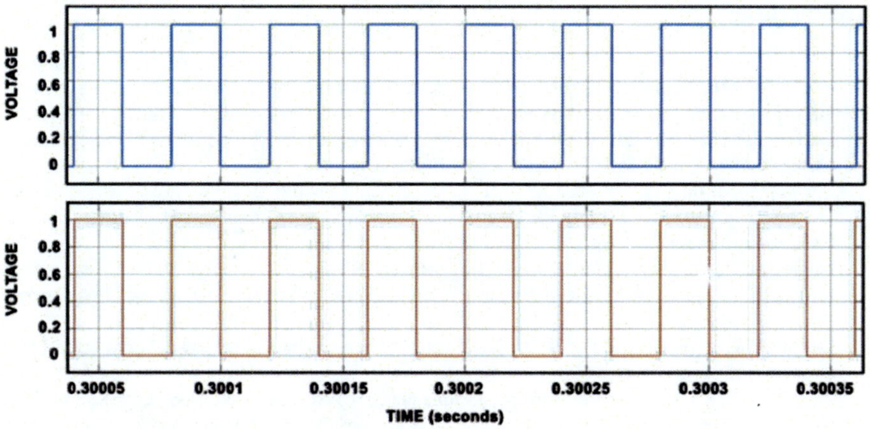

Fig. 13 Pulse obtained by fuzzy logic controller

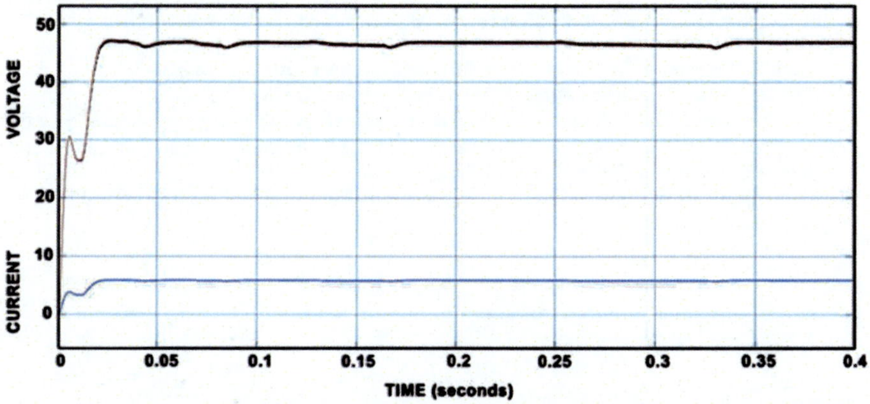

Fig. 14 Output voltage and Current waveform obtained by fuzzy controller and synchronization block

Table 2 Comparison of results

Parameters	*Synchronous Switching*	*Asynchronous Switching*	*Closed Loop Control With two PI controllers*	*Closed Loop Control with two PI controllers and synchronization block*	*Closed loop Control using Hybrid Fuzzy logic PI controller*
Output Voltage	47.52V	48.8V	48.66V	47.55A	47.13A
Output Current	6.833A	6.112A	6.952A	5.923A	5.819A
Circulating Current	2.311A	2.565A	0.58A	0.48A	0.135A

References

1. Zachariah ME, Nisha GK, Babu J (2020) Circulating current of parallel connected converters. 2020 fourth international conference on inventive systems and control (ICISC)
2. Zachariah ME, Nisha GK, Babu J (2020) Synchronous and asynchronous switching of parallel connected converters. Int J Inform Technol Electr Eng
3. Shebani MM, Iqbal T, Quaicoe JE (2017) Synchronous switching for parallel-connected DC-DC boost converters. 2017 IEEE electrical power and energy conference (EPEC). pp 1–6, Saskatoon, Canada
4. Nagarajan R, Yuvaraj R, Hemalatha V, Logapriya S, Mekala A, Priyanga S (2017) Implementation of PV—based boost converter using PI controller with PSO algorithm. Int J Eng Comput Sci
5. Geetha MR, Suja Mani Malar R, Ahilan T (2016) Current sharing in parallel connected boost converters. J Eng 12:444–452
6. Augustine S, Mishra MK, Lakshminarasamma N (2015) An improved droop control algorithm for load sharing and circulating current control for parallel DC-DC converters in standalone DC Microgrid. IEEE Trans Sustain Energy 6(1):132–142
7. Augustine S, Mishra MK, Lakshminarasamma N (2013) Circulating current minimization and current sharing control of parallel boost converters based on Droop Index. 2013 9th IEEE international symposium on diagnostics for electric machines, power electronics and drives (SDEMPED), Valencia, pp 454–460
8. Anand S, Fernandes BG (July 2012) Modified droop controller for paralleling of dc-dc converters in standalone dc system. IET Power Electron 5(6):782–789
9. Anzurez-Marin J, Torres-Salomao LA, Lázaro-Castillo II (2011) Fuzzy logic control for a two tanks hydraulic system model. In: Proceedings 2011 IEEE electronics, robotics and automotive mechanics conference CERMA, Cuernavaca, Mexico, 201
10. Bull S (Aug 2001) Renewable energy today and tomorrow. Proc IEEE 89(8):1216–1226

Single Switch Non-isolated High-Gain Converter with Reduced Voltage Stress

Aakriti Pandey and Swapnajit Pattnaik

Abstract The necessity of high-gain DC–DC converter capable of operating at high voltage is gaining more and more importance for the utilization of renewable energy resources. It is a vital power conversion stage in renewable energy systems like grid-connected sources, solar PV cells, electric vehicles, etc. An efficient, low-cost, high-gain, high-efficient, compact and simple-control converter is a major need. This paper suggests an innovative non-isolated high-gain DC–DC topology with simple structure, limited component count, lower voltage stress and high efficiency. Here, the elevated gain ratio is attained without the inclusion of transformers or coupled inductors and thus free from higher switch voltage spikes during turn-off. The clamping circuit is also not required for a similar reason. There is only single switch that can easily be controlled and it is capable of providing higher gain without extreme value of duty ratio. The switch is subjected to lesser voltage stress with the help of output capacitor. A detailed analysis of the operation and simulation results is presented in the paper.

Keywords High-gain DC–DC converter · Non-isolated converter · Coupled inductor · Switch voltage stress

1 Introduction

In the present scenario, the employment of alternative energy resources is gaining substantial importance. Hence, designing better and efficient system of renewable energy utilization is gathering interests of scientists and scholars all over the globe. The use of solar photovoltaic (PV) cell, fuel cell, etc. yield an end voltage of 12–24 V. For utilization of these renewable resources, use of DC–DC high boost converter is indispensable [1–4]. It is used for providing the desired gain so that output from the resources can be fed to load by DC–DC converters or DC-AC converters.

A. Pandey (✉) · S. Pattnaik
National Institute of Technology, Raipur, India

A. R. Gupta et al. (eds.), *Power Electronics and High Voltage in Smart Grid*,
Lecture Notes in Electrical Engineering 817,
https://doi.org/10.1007/978-981-16-7393-1_27

There are several kinds and categories of topologies proposed in literature for the DC–DC high-gain ratio converters. For instance, it can be classified as voltage-fed or current-fed configurations. In the case of voltage-fed configuration, bulky electrolytic capacitors are required for balancing the high input ripple current of the energy source [5]. The other current-fed configuration having continuous current at the input side is used for high-voltage applications [6].

A broader and better classification of these converters is isolated type and non-isolated type of configuration [7]. In isolated type of configuration, the gain ratio is mainly provided by the turn-ratio present in the transformer. The usage of transformer increases the mass and dimension of the converter altogether, thus decreasing the power density [8, 9]. Moreover, the switches suffer surge in voltage spikes caused by the leakage inductance present in the isolation transformer. Thus, there is a requirement of clamping circuits for the control of switch voltage stress. Hence, there is an overall reduction in efficiency of the converter along with an increase in complexity and cost (Fig. 1).

In the non-isolated class of converters, the input and the load end are not isolated from each other [10–15]. The non-isolated kind of high boost converters are more compact, efficient, cost effective and simpler in operation. There are various techniques used in non-isolated type of converters presented in the literature [12–18]. The most commonly used technique is the use of coupled inductor [11]. In the DC–DC converters employing coupled-inductor, the leakage inductance present leads to high voltage spikes in the switch similar to isolated high-gain DC–DC converters. In this case also, there is a requirement of clamping circuit to suppress the voltage spikes. This increases the price and intricacy of the converter. The other techniques are use of switched inductor and switches capacitor. In switched inductor, the gain is increased

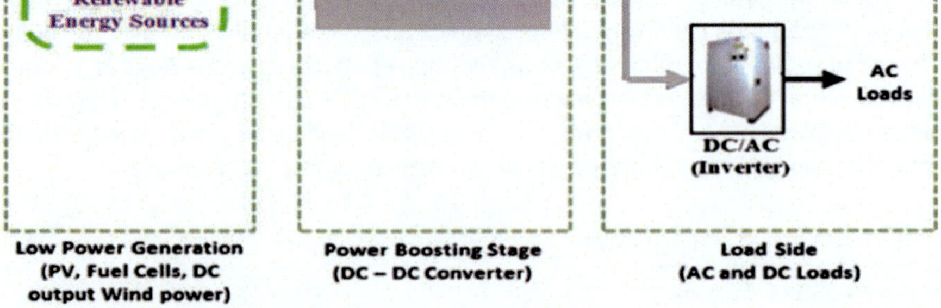

Fig. 1 Renewable energy system

by the energy stored during the charging time of the inductor [12, 13]. In switched capacitors, the problems arising due to the use of electromagnetic elements are eliminated. In such converters, the voltage regulation range is also reduced because the gain provided is generally the multiple of the capacitor voltage [14–19]. The drawback can be eliminated when the switched-capacitor-based converters are used in integration with other techniques. This causes the number of components to increase considerably. In [15, 21], voltage multiplier units (VMUs) are used with switched inductors or capacitors to provide the gain. In such types of converters, though the gain can be achieved by adding up more and more units of the voltage multiplier cells, the number of components also increases leading to increased losses and complexity of the circuit.

One other such technique reported is cascaded DC–DC converter [20]. In this, the voltage gain of each stage is multiplied together yielding a larger gain. This offers a simpler method but the aggregate component count increases. The higher number of the components can be reduced by "Quadratic Converters" [21] in which the common components can be shared by the different stages of the converter. The drawback of such topology is that the efficiency of all the converter stages is also multiplied and hence the overall efficiency is reduced. The converter in [22] shows converter derived from interleaved technique. In the proposed converter use of active clamping circuit with coupled inductor allows the recycle of the leakage energy and voltage stress regulation. The gain is achieved with the use of four switches and two pairs of coupled inductors. It is useful in high current applications but there is considerable increase in their number of switches. In [23], a three-state switching cell (3SSC) type converter is introduced. The gain is achieved with integration of 3SSC and voltage multipliers. The operation of such converters is more complex. Apart from all these above-mentioned techniques. The combination of two or more has been used in literature. A single switch DC–DC non-isolated converter is suggested in [24] where the coupled inductor is integrated with voltage-doubler cell and diode-capacitor technique. Here, the energy lost due to leakage inductance of the coupled inductor is restored and utilized. The combination of coupled inductor with intermediate capacitor is proposed in [27].

In the paper presented a higher gain with lesser voltage stress is achieved. Here, the gain is achieved without the utilization of transformer turn ratio or coupled inductors and thus is free from disadvantages of high voltage spikes and leakage energy losses. The gain to total component count ratio is higher to other converters in literature and the switch stress is also reduced. The detailed operation, simulation results and comparison with existing topologies are presented in this paper.

2 Proposed Circuit Topology and Operating Modes

2.1 Proposed Circuit Topology

In the proposed circuit topology, there are two inductors L_1 and L_2. The capacitors are four in number namely C_1, C_2, C_3 and C_4. There are five diodes D_1–D_4 and output diode D_0. There is only one single switch S.

For analytical simplification of the converter, a few presumptions are made as stated below (Fig. 2):

- All the components used are considered lossless in nature.
- The capacitor value is high enough, such that the voltage across can be considered constant.
- The inductor current is continuous and changes linearly.

The converter is operating at a frequency of 20 kHz.

2.2 Modes of Operation

There are two operating modes:

Mode 1. In this initial operating mode, switch S is conducting. The charging of inductors L_1 and L_2 is taking place through the diode D_1 and D_2. The diodes D_1 and D_2 are in conducting state. The capacitor C_1 voltage increases to input voltage

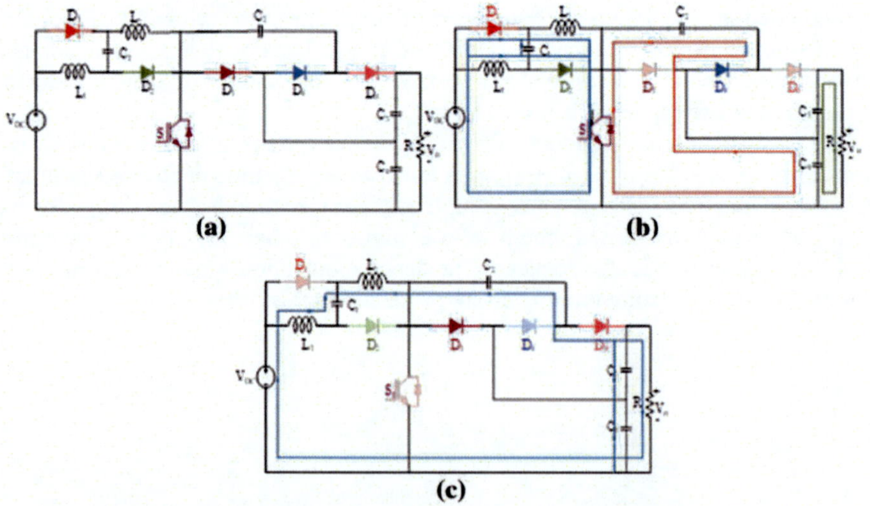

Fig. 2 **a** Proposed converter, **b** Mode 1 configuration, **c** Mode 2 configuration

V_1. The diode D_3 is in non-conducting state. The output capacitor C_4 is charging capacitor C_2 to voltage equal to half of the output voltage through diode D_4. The stress subjected to the switch is controlled by the capacitor C_4 and diode D_3 and the output capacitor D_0 is also in OFF state. The capacitors C_3 and C_4 at the output terminal are feeding the load.

$$V_{L_1} = V_{IN} \tag{1}$$

$$V_{L_2} = V_{IN} \tag{2}$$

$$V_{C_1} = V_{IN} \tag{3}$$

$$V_{C_2} = V_{C_4} = \frac{V_0}{2} \tag{4}$$

Mode 2. In the second mode of operation, the energy reserved in the inductor L_1, L_2, C_1 and C_2 is discharged to load. The inductors L_1 and L_2 change their polarity in discharge mode and reverse biases diodes D_1 and D_2. The output capacitor plays the key role in charging the capacitor C_2. It also performs the crucial function of limiting the switch stress to a maximum of half of the output voltage. The diode D_4 is OFF and the output is fed through D_0.

$$V_{IN} - V_{L_1} + V_{C_1} - V_{L_2} + V_{C_2} = V_0 \tag{5}$$

$$V_{L_1} = V_{L_2} \tag{6}$$

$$V_{C_2} = \frac{V_0}{2} \tag{7}$$

$$V_{L_1} = V_{IN} - \frac{V_0}{4} \tag{8}$$

Applying voltage-second balance equation for inductor voltage, the gain ratio for the given topology can be deduced as follows:

$$\text{Gain} = a = \frac{V_0}{V_{IN}} = \frac{4}{(1 - D)}, \tag{9}$$

where D denotes the duty ratio of the switch S.

Table 1 Value of parameters used in simulation

Parameters	Value (UNIT)
Rated power (P0)	120 W
Input voltage (V1)	10 V
Output voltage (V0)	120 V
Duty ratio	d1 = 65%
Switching frequency (FS)	20 kHz
Inductor	L1 = L2 = 360 μH
Capacitor	C1 = C2 = 100 μF; C3 = C4 = 470 μF

3 Simulation Result

For the verification of the working of the presented converter topology, it is simulated in MATLAB Simulink 2018. The frequency of operation is 20 kHz and the input.

Voltage fed is 10 V. The output of 120 V is derived as visible in the simulation results below (Fig.4).

The switch stress on the single switch S used is same as the desired value. The component description is listed in Table 1.

4 Comparison

A comparative study of the proposed circuit topology is done with various recent high-boost DC–DC configurations proposed in literature. The various parameters compared are as shown in Table 2. The aggregate component count and separate count of inductors, capacitors, switches and diodes are listed. The converter presented uses a total of 12 components. The gain equation is mentioned and the gain ratio is derived for all the converters for an optimum value of duty ratio of 0.65.

The suggested converter gives the maximum gain. For a better insight into the comparative analysis, the ratio factor of gain to total component count is compared and the presented topologies give the highest ratio as shown in Fig. 3. Apart from this, other vital comparison parameter is the ratio of gain to switch voltage stress (indicated as percent of output voltage). The given converter uses only single switch which is subjected to low stress of half of output voltage by capacitor voltage division at the load. The gain to switch stress ratio is also highest for this converter and hence, it can be useful in providing higher gain with lesser voltage stress subjected to the switch. Due to lessened stress, lower rating semiconductor device can be used and hence, the efficiency is enhanced.

Table 2 Comparative investigation of presented Converter with other converters

Parameters	[25]	[26]	[27]	[28]	Presented converter
Total no. of components	11	14	14	12	12
No. of Switches	3	2	1	1	1
No. of Inductors	4	4	4	3	2
No. of Capacitors	1	1	6	5	4
No. of Diodes	3	7	3	3	5
Gain	6.57	8.42	5.57	6.57	12
Gain Equation	$\frac{(1+2D)}{(1-D)}$	$\frac{(1+3D)}{(1-D)}$	$\frac{(3D)}{(1-D)}$	$\frac{(1+2D)}{(1-D)}$	$\frac{(4)}{(1-D)}$
Gain to component count ratio	0.55	0.602	0.397	0.55	1
Voltage Stress across switches (as percent of output voltage)	29.3%	50%	51.5%	43.5%	50%
Gain to switch voltage stress (as percent of output voltage) ratio	0.22	0.20	0.11	0.15	0.24

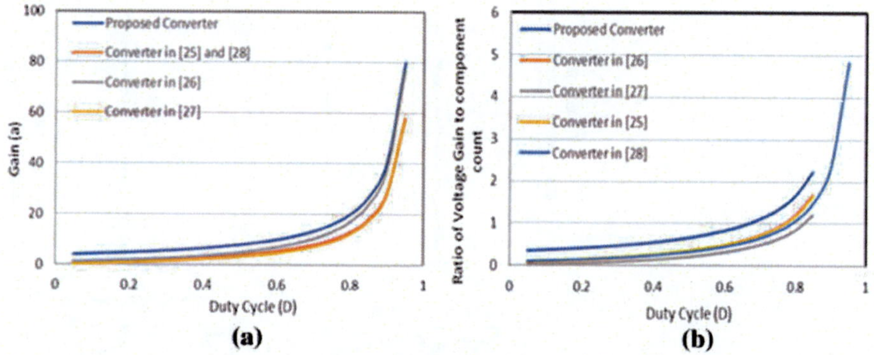

Fig. 3 **a** Variation of gain with component count of various converters, **b** Variation of ratio of gain to component count with duty cycle of various converters

5 Conclusion

In the present paper, an innovative non-isolated DC–DC high-boost topology is suggested. The prime benefit of the presented configuration is elevated gain ratio, single switch requirement, non-inverting output, low switch voltage stress and simple control system. The gain is acquired without employing any coupled inductor or transformers. Thus, there is no mutual inductance and hence, no severe voltage spike arising due to the same. Moreover, there is no requirement of clamping circuit.

An elaborated comparative investigation of the converter is done with different topologies in recent literature. The gain to the total component count ratio proves to be a key factor in proving the advantage and better functioning of the converter. The

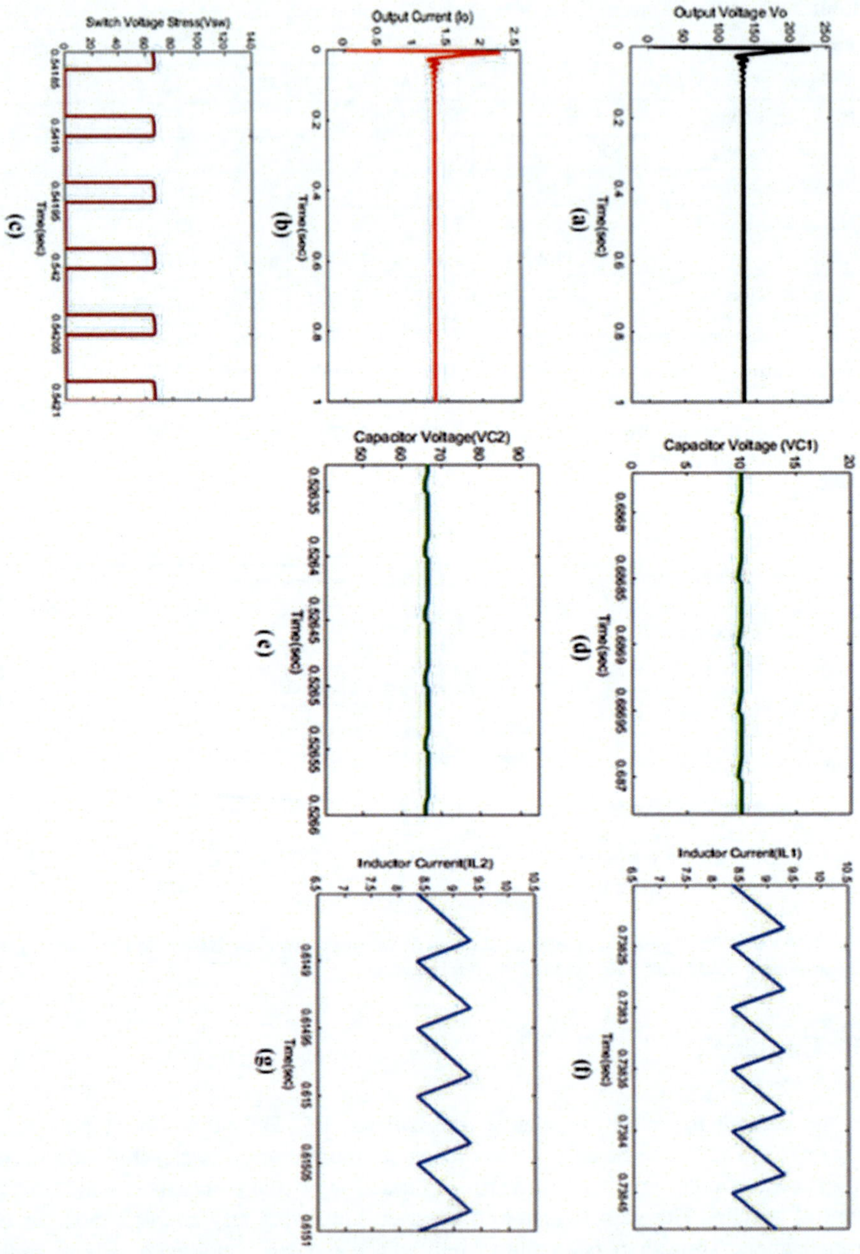

Fig. 4 MATLAB Simulation outcome **a** Output Voltage, **b** Output Current, **c** Switch Voltage Stress, **d** Voltage across intermediate capacitor $C_1(V_{C1})$, **e** Voltage across intermediate capacitor $C_2(V_{C2})$, **f** Inductor current I_{L1} and **g** Inductor Current I_{L2}

other factor, ratio of voltage gain to the percent voltage stress subjected to the switch is also highest for the presented converter. The lower switch stress allows the use of low rating semiconductor device and thus decreasing the conduction losses of the converter. Hence, the efficiency is improved and it can be inferred that this converter offers a better performance for high power, high-gain applications in renewable energy utilization.

References

1. Changchien S, Liang T, Chen J, Yang L (June 2010) Novel high step-up DC–DC converter for fuel cell energy conversion system. IEEE Trans Industr Electron 57(6):2007–2017
2. Das M, Agarwal V (2016) Design and analysis of a high efficiency DC–DC converter with soft switching capability for renewable energy applications requiring high voltage gain. IEEE Trans Ind Electr 63(5)
3. Pan CT, Lai C-M (2010) A high-efficiency step-up converter with low switch voltage stress for fuel-cell system applications. IEEE Trans Ind Electron 57(6):1998–2006
4. Agarwal V (2015) Novel high-performance stand-alone solar PV system with high-gain high-efficiency DC–DC converter power stages. IEEE Trans Ind Appl 51(6):4718–4728
5. Hsieh Y, Chen J, Liang T, Yang L (2012) Novel high step-up DC–DC converter with coupled-inductor and switched-capacitor techniques. IEEE Trans Industr Electron 59(2):998–1007. LNCS Homepage, http://www.springer.com/lncs. Accessed 21 Nov 2016
6. Leu C, Huang P, Li M (April 2011) A novel dual-inductor boost converter with ripple cancellation for high-voltage-gain applications. IEEE Trans Industr Electron 58(4):1268–1273. https://doi.org/10.1109/TIE.2010.2048835
7. Axelrod B, Berkovich Y, Ioinovici A (2008) "Switched capacitor/switched inductor structures for getting transformerless hybrid DC–DC PWM converters, IEEE Trans. Circuits Syst I Regul Pap 55(2):687–696
8. Forouzesh M, Baghramian A (2016) Galvanically isolated high gain Y-source DC–DC converters for dispersed power generation. IET Power Electron. 9(6):1192–1203
9. Padmanaban S, Bhaskar MS, Maroti PK, Blaabjerg F, Fedák V (Apr. 2018) An original transformer and switched-capacitor (T & SC)-based extension for DC–DC boost converter for high-voltage/low-current renewable energy applications: hardware implementation of a new T & SC boost converter. Energies 11(4):783
10. Wu B, Li S, Ma Smedley K (2016) A new single-switch isolated high-gain hybrid boosting converter. IEEE Trans Industr Electron 63(8:4978–4988. https://doi.org/10.1109/TIE.2016.2554542
11. Cheng T, Lu DD-C, Qin L (2018) Non-isolated single-inductor DC/DC converter with fully recongurable structure for renewable energy applications. IEEE Trans Circuits Syst II Exp Briefs 65(3): 351–355
12. Lakshmi M, Hemamalini S (Feb. 2018) Nonisolated high gain DC–DC converter for DC microgrids. IEEE Trans Industr Electron 65(2):1205–1212
13. Parastar A, Seok J (Feb. 2015) High-gain resonant switched-capacitor cell-based DC/DC converter for offshore wind energy systems. IEEE Trans Power Electron 30(2):644–656
14. Prudente M, Pfitscher LL, Emmendoerfer G, Romaneli EF, Gules R (March 2008) voltage multiplier cells applied to non-isolated DC–DC converters. IEEE Trans Power Electron 23(2):871–887
15. Yang L-S, Liang T-J, Chen J-F (2009) Transformerless DC–DC converters with high step-up voltage gain. IEEE Trans Ind Electron 56(8):3144–3152
16. Ismail EH, Al-Saffar MA, Sabzali AJ, Fardoun AA (2008) A family of single-switch PWM converters with high step-up conversion ratio. IEEE Trans Circuit Syst I Reg Papers 55(4): 1159–1171

17. Sadanala C, Pattnaik S, Singh VP , A novel switched capacitor based multilevel inverter with symmetrical and asymmetrical configurations. Electr Eng. https://doi.org/10.1007/s00202-020-01172-7
18. Park S, Choi S (May 2010) Soft-switched CCM boost converters with high voltage gain for high-power applications. IEEE Trans Power Electron 25(5):1211–1217
19. Wu H, Sun K, Chen L, Zhu L, Xing Y (May 2016) High step-up/step-down soft-switching bidirectional DC–DC converter with coupled-inductor and voltage matching control for energy storage systems. IEEE Trans Industr Electron 63(5):2892–2903
20. Baddipadiga BP, Ferdowsi M (Oct. 2017) A high-voltage-gain DC–DC converter based on modified dickson charge pump voltage multiplier. IEEE Trans Power Electron 32(10):7707–7715
21. Andrade AMSS, Martins MLDS (2017) Quadratic-boost with stacked zeta converter for high voltage gain applications. IEEE J Emerg Sel Topics Power Electron 5(4): 1787–1796. https://doi.org/10.1109/JESTPE.2017.2706220.
22. Muhammad M, Armstrong M, Elgendy MA (June 2016) A nonisolated interleaved boost converter for high-voltage gain applications. IEEE J Emerg Sel Topics Power Electron 4(2):352–362
23. Torrico-Bascope RP, Tofoli FL, Castelo Branco CG , Tavares Cruz CM (2018) Three-state switching cell (3SSC)-based non-isolated DC–DC boost-type converter with balanced output voltage and wide voltage conversion range. IET Power Electron 11(7):1217–1223. https://doi.org/10.1049/iet-pel.2017.0551
24. Hu X , Gong C (2014) A high voltage gain DC–DC converter integrating coupled-inductor and diode–capacitor techniques. IEEE Trans Power Electron 29(2): 789–800
25. Mashinchi Maheri H, Babaei E, Sabahi M , Hosseini SH (2017) High step-Up DC–DC converter with minimum output voltage ripple. IEEE Trans Industr Electron 64(5): 3568–3575. https://doi.org/10.1109/TIE.2017.2652395.
26. Babaei E, MashinchiMaheri H, Sabahi M, Hosseini SH (Dec. 2018) Extendable nonisolated high gain DC–DC converter based on active-passive inductor cells. IEEE Trans Industr Electron 65(12):9478–9487
27. Banaei MR, Sani SG (Dec. 2018) Analysis and implementation of a new SEPIC-based single switch buck-boost DC–DC converter with continuous input current. IEEE Trans Power Electron 33(12):10317–10325. https://doi.org/10.1109/TPEL.2018.2799876
28. Elsayad N, Moradisizkoohi H, Mohammed OA (May 2019) A Single-switch transformerless DC–DC converter with universal input voltage for fuel cell vehicles: analysis and design. IEEE Trans Veh Technol 68(5):4537–4549. https://doi.org/10.1109/TVT.2019.2905583

Torque Ripple Minimization of SRM Using Sliding-Mode Current Controller and Torque-Sharing Function

Stella Kurian and G. K. Nisha

Abstract The major drawback of high torque pulsations shadows the huge applications of switched reluctance motor (SRM). Quite a lot of control techniques are applied to decrease the amount of ripple in torque of SRM drives. Indirect control methods of torque control with torque-sharing functions (TSF) are used to reduce the amount of ripple content in SRM in this paper. Cosine and linear torque-sharing functions are derived for extracting torque reference signals. Sliding-mode current controller is implemented inner current control which generates gate pulses for the electronic converter. A comparative analysis of closed loop torque control using cosine and linear torque-sharing functions of the switched reluctance machine utilizing sliding-mode current controller is discussed in this paper.

Keywords Asymmetric bridge converter · Cosine torque-sharing function · Linear torque-sharing function · Sliding-mode current controller · Switched Reluctance motor

1 Introduction

The main two strategies in minimizing the amount of torque pulsations of SRM drives are either to improvise the magnetic structure of motor or to use electronic controlling methods. The pulsations in torque output can be minimized by varying the rotor and stator pole structures compromising certain motor output quantities. The electronic controlling methods are based on optimization of the controlling quantities like input voltage, triggering and commutating angles and output current.

S. Kurian (✉)
Department of EE, Government Engineering College, Palakkad, Kerala, India

G. K. Nisha
Department of EE, Mar Baselios College of Engineering and Technology, Trivandrum, Kerala, India

A. R. Gupta et al. (eds.), *Power Electronics and High Voltage in Smart Grid*, Lecture Notes in Electrical Engineering 817,
https://doi.org/10.1007/978-981-16-7393-1_28

The fundamental approaches of reduction of torque ripple in SRM are provided in Sect. 2. Section 3 enlightens the concepts of cosine and linear torque-sharing functions, conventional asymmetric bridge converter and sliding-mode current controller implemented in modulating total torque output in the motor [1]. Simulation strategy of torque control applying two torque-sharing functions and sliding-mode current controller of switched reluctance motor drive is described in Sect. 4. Simulation results of closed loop torque control of SRM with cosine and linear torque-sharing functions, implemented to generate the torque reference commands and sliding-mode current controller for inner current control in MATLAB SIMULINK are compared and presented in Sect. 5 [2].

2 Minimization of Torque Ripple

Torque production in switched reluctance machine happens the minimum reluctance path. The magnitude of torque produced by a phase varies proportionally with the rate of change of inductance and the square of the phase current. Polarity of torque can be varied depending upon slope of phase inductance as it is a function of square of magnitude of current and hence the direction of the current does not affect it [3]. Each stator phase winding is made by connecting the windings of the diametrically opposite stator poles serially or in parallel. When a phase winding is excited, the closest pair of rotor poles gets aligned towards excited stator poles in such a way that the minimum reluctance path is achieved. Phases are excited successively to develop constant torque is developed in both directions. The phase inductance profile is divided as region with constant inductance, region with increasing slope of inductance and region with decreasing slope of inductance. Figure 1 shows the inductance profile of SRM [4].

In the increasing region of inductance, rotor starts overlapping over the stator pole and torque generated is positive due to positive slope of inductance. Similarly, in decreasing region of inductance, rotor moves away from the stator pole and negative torque is generated due to negative slope. In region of constant inductance, the rotor poles and stator poles are either in unaligned position with minimum inductance or in completely aligned position with maximum inductance. In both positions, inductance will be constant and the slope will be zero. Since the torque is a function of slope of

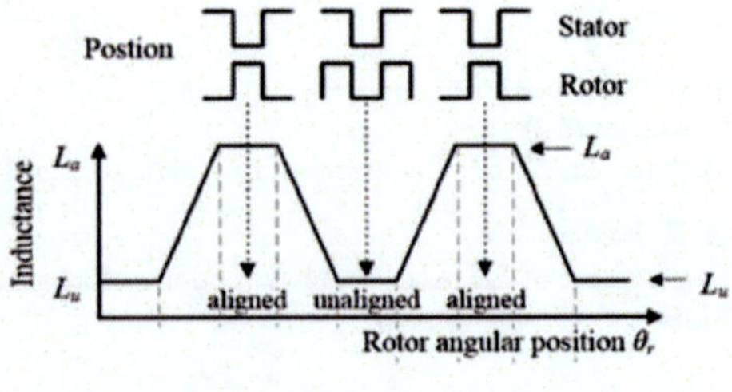

Fig. 1 SRM Inductance profile

inductance, both these regions do not contribute to the production of electromagnetic torque in SRM. Hence, the sequence in which phase windings are excited needs to be synchronized with the profile of inductance in SRM to realize effective output power.

The concentrated windings across stator, doubly salient stator structure, pulsating magnetic vector field acquired with exciting successive stator windings and different modes for magnetic saturation to increase torque to mass ratio contribute to the production of torque pulsations in SRM drives. The electronic controlling methods are based on optimization of the controlling quantities like input voltage, triggering and commutating angles and output current. The electronic control methods can be predominantly divided into direct and indirect torque control methods.

2.1 *Indirect Techniques for Controlling Torque*

The output torque of AC machines can be regulated by transforming the nominal torque values into equivalent phase current values and these techniques are referred to as indirect torque control methods. Similar approach can be implemented in switched reluctance motor, where phase current is modulated by indirectly regulating the output torque. The coupled non-linear function of current, rotor angular position and torque creates complicated mathematical relationship [5]. A hybrid torque ripple minimizing controller was proposed [6], where a torque-sharing function was integrated with a balanced commutator for a wide region. Variable range of speed was attained by changing the angle of commutation according to the speed. Ripple was minimized lesser than three percent for speed 300 rev/min and lesser than ten percent for speed 1,000 rev/min.

Two advanced TSF depending upon the overlap angle, turn-on angle and the expected torque for implementing torque ripple minimization was presented by Xue [7]. Genetic algorithm was used for the optimization of these torque-sharing functions. Exponential TSF yields more results when optimal speed is also evaluated. A novel family of torque-sharing functions was introduced by Vujicic [8]. Secondary objectives of reduction of variable losses and improved performance of motor were considered and optimal torque-sharing function was extracted evaluating the secondary objective. Of the two extracted functions, one was found to expand the possible speed range of ripple-less operation. The other TSF provided a better speed-torque curve. Direct torque control using sliding-mode controller was developed where torque references were directly realized using the sliding-mode controller to maintain the torque error within a limited range, a pulsating control voltage was combined with resultant voltage [9]. Torque ripple reduction up to a limited region was attained. The bandwidth of the controller varies according to the DC link voltage and frequency of sampling which decides the upper boundary.

3 Implementation of Torque Ripple Minimization

Indirect techniques by controlling torque with torque-sharing functions are implemented here in reducing the amount of torque pulsations in SRM. Torque-sharing functions produce the reference torque value for each phase, T_k^* from an input torque command, T^*. Reference torque of individual phases is then transformed as corresponding phase current reference signals, i_k^* in block 'Torque-to-Current' in accordance with the rotor angular position as shown in torque control block diagram in Fig. 2. The current reference command signals thus produced are compared to the actual output current values and are modulated by a current controller. The basic current regulators are pulse width modulated current controllers or hysteresis current controller. A sliding-mode controller is implemented here for current control [10].

Current controller outputs switching pulses for switches of the power electronic converter that excites the corresponding phases of the motor. The conventional asymmetric bridge converter with low current ripples is implemented to drive the motor [11].

3.1 *Torque-Sharing Functions*

Torque distribution or sharing function is a simple and powerful tool used in indirect torque control methods. Torque ripple is developed when multiple phases are in conduction during the commutation period. A suitable technique to generate constant and consistent torque is to correlate the torque produced by each phase such that the cumulative torque traces the command quantity. Total torque generated purely depends upon the square of the stator current and slope of inductance. Input reference torque, T^*, gets distributed in 'TSF block' into individual phase torque reference values, $T_k{}^*$, depending upon the rotor position which is developed by appropriately shifting pre-defined torque distribution functions. Phase current of two phases collectively delivers output torque in the overlapping region of inductances. The torque-sharing functions are separated as single conduction and commutation mode wherein a single phase contributes to torque production in single conduction mode and two consecutive phases collectively generate positive torque according to the discrete

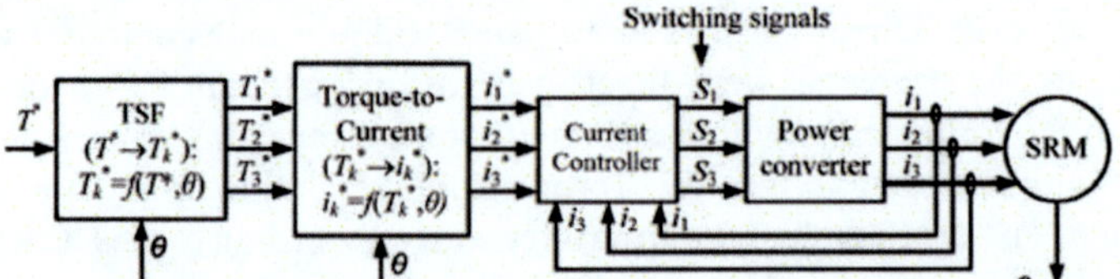

Fig. 2 Schematic diagram for torque control

functions in commutation mode. In overlap region, only two phases are energized simultaneously, such that negative torque is not produced [12].

The conventional torque distribution functions are differentiated based on torque-sharing curvature of the overlap region into cosine, linear, exponential and cubic torque-sharing functions. A phase is excited at the turn-on angle denoted by θ_{on} and is commutated at turn-off angle denoted by θ_{off}. According to the potential of torque generation, the angles need to satisfy the constraint $\theta \leq \theta_{on} \leq \theta_{off} \leq \theta_{al}$. Overlap angle, θ_{ov} specifies the time duration a phase collectively produces torque with the outgoing phase or incoming phase (overlapping period). The classic pattern of the cosine and linear torque distribution function of any phase k, considering $(k-1)$ as the outgoing phase and $(k+1)$ as the incoming phase is depicted in Fig. 3 [13]. The reference torque for any phase k is interpreted as,

$$T_k^*(\theta) = \begin{cases} 0, 0 \leq \theta \leq \theta_{on} \\ T^* \cdot f_{rise}(\theta), \theta_{on} \leq \theta \leq \theta_{on} + \theta_{ov} \\ T^*, \theta_{on} + \theta_{ov} \leq \theta \leq \theta_{off} - \theta_{ov} \\ T^* \cdot f_{fall}(\theta), \theta_{off} - \theta_{ov} \leq \theta \leq \theta_{off} \\ 0, \theta_{off} \leq \theta \leq \theta_{al} \end{cases} \quad (1)$$

During commutation intervals, f_{rise} must increase from zero to one and the function f_{fall} must fall from one to zero. For the linear torque-sharing function, the functions f_{rise} (θ) and $f_{fall}(\theta)$ are described as,

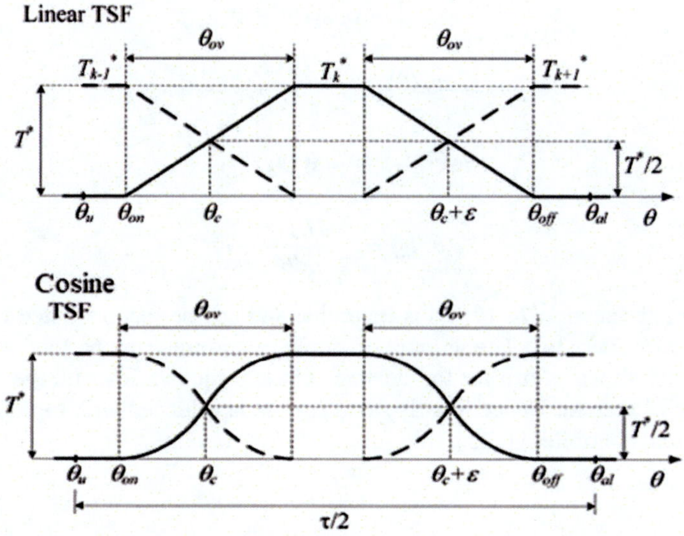

Fig. 3 Linear and cosine torque distribution functions

$$f_{rise}(\theta) = \frac{(\theta - \theta_{on})}{\theta_{ov}} \tag{2}$$

$$f_{fall}(\theta) = 1 - f_{rise}\left(\theta + \theta_{ov} - \theta_{off} + \theta_{on}\right) \tag{3}$$

Similarly, for cosine TSF,

$$f_{rise}(\theta) = \frac{1}{2}\left(1 - \cos\left(\frac{\pi}{\theta_{ov}}(\theta - \theta_{on})\right)\right) \tag{4}$$

$$f_{fall}(\theta) = \frac{1}{2}\left(1 + \cos\left(\frac{\pi}{\theta_{ov}}(\theta - \theta_{off})\right)\right) \tag{5}$$

3.2 *Current References Generation*

Artificial neural networks or static current, rotor position and torque (*i*-θ-*T*) characteristics are customarily utilized to generate current reference signals from torque commands. Reference torque values are translated into reference currents at a lower level of torque with the equation

$$\mathrm{i}_k^*(\theta) = \begin{cases} \sqrt{\frac{\mathrm{T}_k^*(\theta)}{\mathrm{a}_k(\theta)}}, \ \mathrm{a}_k(\theta) > 0 \\ 0, \ \mathrm{a}_k(\theta) \le 0 \end{cases} \tag{6}$$

$$a_k(\theta) = \frac{1}{2}\frac{dL_k(\theta)}{d\theta} \tag{7}$$

$$\varphi_k(\theta, i_k) = L_k(\theta, i_k) \cdot i_k \tag{8}$$

$$T_k(\theta, i_k) = \frac{1}{2}\frac{dL_k(\theta)}{d\theta} \cdot i_k^2 \tag{9}$$

The torque function T_k (θ, i_k) is invertible and can be reconstructed to current function i_k (θ, *Tk*) [14]. The amount of ripple in current must be low in order to produce smooth waveform for torque since phase torque varies with the square of magnitude of current. Hence less ripple in current is achieved with high frequency of the current controller [1, 15].

3.3 Sliding-Mode Current Controller

A second-order sliding-mode current controller (SMCC) operates with discontinuous control of second-order derivative of the control variable. The command U will be a continuous signal. The chattering problem can be eliminated in the second-order system while conserving the properties of convergence in finite time and robustness of single-order system. The sliding plane can be tracked more evenly and can be realized in finite time using second-order system [16]. Back-emf is derived from the electrical model of phase inductance, L, which depends on the rotor angular position and phase current. Electrical equation of individual phase is described by Eq. (10), where E the back-emf and L_{inc} is the incremental inductance of each phase.

$$U = Ri + L_{inc}\frac{di}{dt} + E \tag{10}$$

Hence, the first-order derivative of the constraint function $s = (i^* \ - i)$ is given by,

$$\frac{ds}{dt} = \frac{di^*}{dt} - \frac{U}{L_{inc}} + \frac{E + Ri}{L_{inc}} \tag{11}$$

Here 'i^*' denotes the reference value of stator current of each phase and 'i' denotes the actual current measured from the output of the motor. The command variable U is the average phase voltage and the constraint function 's' has a relative degree of 1 which is the required criteria for applying second-order sliding-mode control based on super-twisting algorithm (STA). It is required to consider certain assumptions to make sure that the sliding surface is accessible in the design of second-order sliding-mode controller. The sliding variable 's' has to be bounded. The average control voltage U is constituted of a continuous proportional term and a continuous integral term, which is independent of first time derivative of 's'[17]. The discontinuity happens to appear only in first-order derivative of the control input, U with integral term, U_i, and proportional term, U_p

$$U = U_p + U_i \tag{12}$$

$$\frac{dU_i}{dt} = \beta sign(s) \tag{13}$$

$$U_p = \alpha\sqrt{s}sign(s) \tag{14}$$

where $\beta < 0$, $\alpha > 0$ and sign(s) = 1 for s > 0 and −1 for s < 0.

The advantage of STA is that it integrates sign of error rather than integrating the error [18]. The proportional input term, U_p changes quickly than error, actuating a better responsive regulator since U_p is dependent on square root of error and value

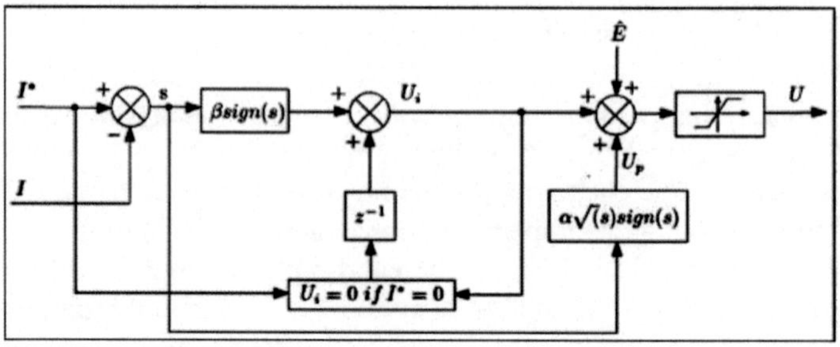

Fig. 4 Discrete SMCC with STA

of error is less near the reference. The convergence can be obtained on the sliding surface, S (s = $s^{\cdot}$ = 0) within finite time. A super-twisting sliding-mode current controller is depicted in Fig. 4. This scheme consists of an initialization block of the integrator with the current reference position returning to zero. The super-twisting algorithm promotes the motion control design and reduces the chattering problem at the outputs.

3.4 *Asymmetric Converter*

Asymmetrical bridge converters are the most conventional configuration of converters implemented with switched reluctance motors. Each stator phase winding has two semiconductor IGBT switches and two flywheel diodes connected across individual windings A, B and C as depicted in Fig. 5. The three phases of operation of asymmetric converter include magnetization mode, demagnetization and freewheeling mode [19]. In the course of or magnetization or energization mode, one IGBT switch from each phase is switched on while the other remains off. Corresponding stator current takes the path through excited IGBT switch and forward biased fly-wheeling diode. During the demagnetization mode or period of commutation, both switches remain in the off position. The magnetic energy stored in the phase inductances of machine is fed back to the supply through the fly backing

Fig. 5 Asymmetrical converter configuration

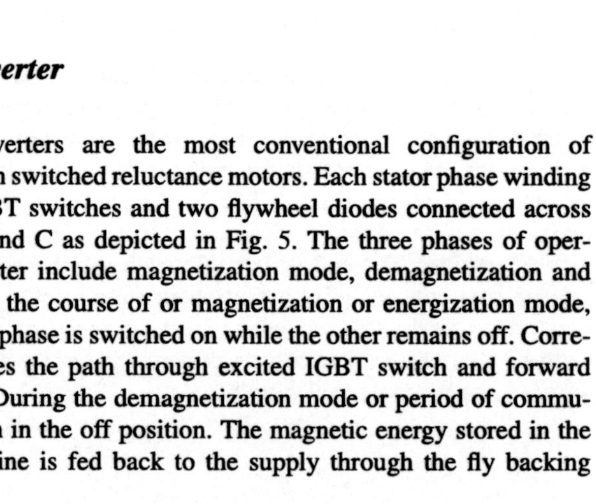

diodes. Phase current decays rapidly since the voltage appears reversed through the stator phase winding.

During the magnetization or energization phase, since both IGBT switches of a corresponding winding are switched on, phase current increases quickly through the phase winding. Voltage across phase winding will not be available in the fly backing mode as only one diode and an IGBT switch are excited and the magnitude of current is slowly decreasing. In this mode of operation, no energy transfer occurs through the supply [20].

4 Torque Control of Switched Reluctance Drive

A power electronic converter is necessary to run switched reluctance motor since the machine is electronically commutated. The classical asymmetric bridge converter navigates the machine in the desired direction depending upon the sequence in which phases are excited. The switching pulses for the exciting IGBT switches are produced by the sliding-mode current controller. The input DC supply is used to energize the stator phases. Simulation of closed loop torque control of a 3ϕ, four pole 60 kW switched reluctance drive using torque-sharing functions and sliding-mode current controller is implemented in MATLAB [21]. The mechanical load torque (Nm) applied to SRM is negative in case of generating mode and positive for motoring operation. The quantities available at the output of switched reluctance motor drive are voltage across stator windings (V), electromagnetic torque (Nm), stator phase current (A), speed of rotor (rad/s), rotor angular position (rad) and flux linkage (V-s). Discrete simulation is performed with constant 40° turn-on and 75° turn-off angles.

Energization of windings depends upon rotor position, teta in rad. The position of the rotor is sensed from the teta output available from the machine. The torque reference signal is distributed among the individual phases according to the cosine and linear torque distribution functions from the torque reference command input. Angular functions are generated from the torque reference commands for individual phases by calculating rate of change of phase inductance with rotor. Current command signals are derived from individual phase torque command signals and the angular functions. The sliding-mode controller differentiates the actual stator current from motor output and the current reference to generate required switching pulses for asymmetric bridge converter that rotates the motor.

5 Simulation Results and Discussions

The basic asymmetric bridge converter constitutes two diodes and two IGBT switches for individual phase winding. Each stator phase winding is energized in accordance to the sequence of switching signals produced from the sliding-mode controller. It is

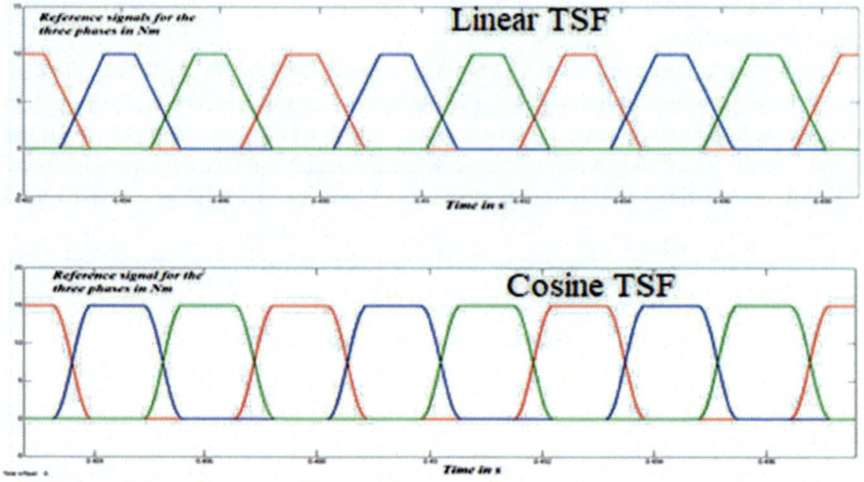

Fig. 6 Simulated linear and cosine TSF

ensured that each phase is energized periodically for every 90° rotation of the motor since four rotor poles are present making a total rotation of 360°. Same switching signals are applied to both the switches of an individual phase. Input voltage appears across the stator phase winding when the switching signals are high. A 240-V supply voltage is applied to energize the stator windings. The torque-sharing functions are defined depending upon the rotor angular position. It shows a linear profile for linear torque-sharing function and a sinusoidal variation for cosine torque-sharing function as depicted in Fig. 6, respectively, for all the three phases in the commutation region. Each phase has a phase shift of 30° considering the phase difference between three-phase windings. The cosine torque-sharing function seems to have better results than the linear function.

The period of commutation corresponds to decreasing inductance and increasing inductance interval of the respective outgoing and incoming phases. The sliding-mode current controller differentiates the stator current signal from reference signal to modulate the error as in the block diagram of the controller. A second-order controller applying super-twisting algorithm is implemented along with linear and cosine torque-sharing functions. Figure 7 presents gate signals from the sliding-mode current controller while simulating with the cosine torque-sharing functions. The integral and proportional gain parameters are tuned to obtain single pulse mode of the converter. Positive DC voltage is supplied across the phase winding when the rotor position is between the turn-on and turn-off angles.

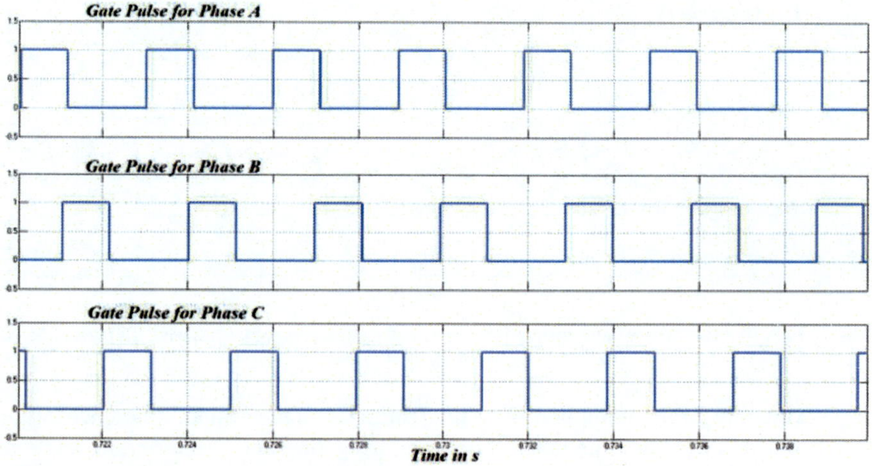

Fig. 7 Gate pulses generated from the SMCC

5.1 *Torque Control Using Sliding-Mode Current Controller with Linear Torque-Sharing Function*

In the closed loop operation of SRM drive with torque-sharing functions and sliding-mode controller, the torque reference signals are generated depending on the cosine and linear torque distribution functions. The motor performance is analyzed with the torque output, flux linkage and stator current of the machine as shown in Fig. 8. The torque pulsations and the peak current is reduced by implementing the torque-sharing functions with the sliding-mode controller. The reduction in peak current will improve the torque per ampere ratio of the motor drive. The decay time of the current reduces considerably thereby reducing the ripple content thereby making the phase current smoother. The amount of torque ripple is nearly 20 Nm which was 35 Nm in the current controlled drive with sliding-mode current controller alone.

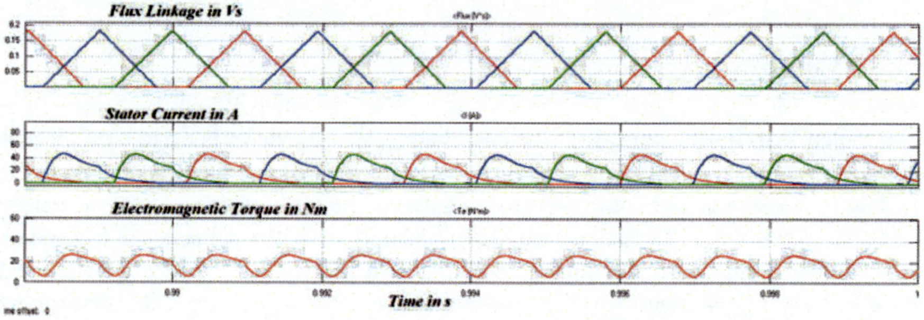

Fig. 8 Flux linkage, current and torque waveforms of torque control with linear TSF and SMCC

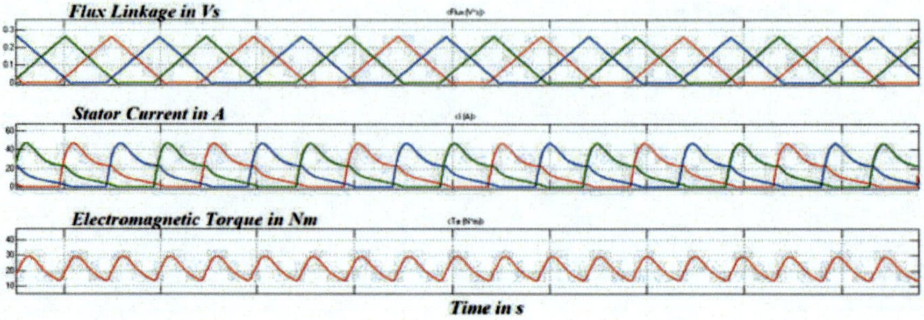

Fig. 9 Flux linkage, current and torque waveforms of torque control with cosine TSF and SMCC

Approximately 20 Nm is applied as the total torque reference value. Reference torque will be available in the time interval between triggering and commutating angle of the machine.

5.2 *Torque Control Using Cosine Torque-Sharing Function and Sliding-Mode Current Controller*

Figure 9 depicts the motor performance parameters with the SMCC and cosine TSF. Decay time is reduced in this system similar to the SMC-controlled system utilizing linear functions. The current controlled system with both linear TSF and cosine TSF achieves nearly same speed of rotation.

The ripple content in torque when the asymmetrical converter-driven SRM is simulated with sliding-mode current controller alone was 35 Nm. The torque ripple was reduced to 20 Nm when linear torque-sharing function was introduced with sliding-mode controller to distribute the total torque reference of 20 Nm into individual phases. The torque ripple is further reduced to 15 Nm with sliding-mode controller and the cosine functions.

6 Summary

The closed loop torque control of the switched reluctance drive is done with torque-sharing functions and sliding-mode current controllers. The control operation using the sliding-mode current controller is simulated using both cosine and linear torque distribution functions. The sliding-mode-controlled drive can achieve greater speed. Super-twisting algorithm is applied to implement the sliding-mode controller and the gains are finely tuned to obtain single pulse mode operation. The amount of ripple in torque seems to be lesser with the cosine torque distribution function and the linear torque distribution function. The torque ripple is reduced to 50% with the sliding-mode controller than the hysteresis current controller with the torque-sharing functions [1].

References

1. Kurian S, Nisha GK (2015) Torque ripple minimization of SRM using torque-sharing function and hysteresis current controller. International conference on control communication & computing, pp 149–154
2. Ahn J-W (2011) Torque control, InTech publisher, Chapter 8
3. Husain I, Ehsani M (1996) Torque ripple minimization in SRM drives by PWM current control. IEEE Trans Power Electron 11(1):83–88
4. Husain I (2002) Minimization of torque ripple in SRM drives. IEEE Trans Industr Electron 49(1):28–39
5. Electric and hybrid vehicles Design fundamentals (2003) I Husain. CRC Press, Boca Raton
6. Russa K, Husain I, Elbuluk M (1997) Torque ripple minimization in switched reluctance machines over a wide speed range. In: Proceedings IEEE IAS conference, pp 668–675
7. Xue XD, Cheng KWE, Ho SL (2009) Optimization and evaluation of torque sharing functions for torque ripple minimization in switched reluctance motor drives. IEEE Trans Power Electron 24(9):2076–2090
8. Vujicic VP (2012) Minimization of torque ripple and copper losses in switched reluctance drive. IEEE Trans Power Electron 27(1):388–399
9. Yahoo SK, Panda SK, Xue JX (2005) Direct torque control for SRM drive using sliding mode control. IEEE conference on power electronics, pp 1129–1134
10. Cunningham JD (2011) Switched reluctance motor drive circuit evaluation criteria for vehicle efficiency and responsiveness. MSE thesis, University of Texas, Austin
11. Suryadevara R, Fernandes BG (2013) Control techniques for torque ripple minimization in switched reluctance motor: an overview. IEEE conference on industrial and information systems (ICIIS)
12. Kurian S, Nisha GK (2014) State of the art of switched reluctance motor for torque ripple minimization. J Industr Electron Electr Eng 2(12)
13. Xue XD, Cheng KWE, Ho SL (2006) A control scheme of TRM for SRM drives based on flux linkage controller and torque-sharing function. International conf. on power electronics systems and applications, pp 79–84
14. Kurian S, Nisha GK (2015) Closed loop operation of switched reluctance motor with sliding-mode current controller. Int J Appl Eng Res 10(23):43407–43411
15. Rain X, Hilairet M, Talj R (2010) Second order sliding-mode current controller for the switched reluctance machine. IEEE Industr Electron Soc, 3301–3306
16. Inanc N, Obulur V (2003) Torque ripple minimization of a switched reluctance motor by using continuous sliding-mode control technique. J Electr Power Syst Res 66:241–251
17. Singh A (2010) Study of soft switching converter for switched reluctance drive. ME Thesis, Thapar University, Patiala, India
18. Ahn JW, Liang J, Lee DH (2010) Classification and analysis of switched reluctance converters. J Electr Eng Technol 5(4):571–579
19. Krishnan R (2001) Switched reluctance motor drives, Boca Raton. CRC Press, FL
20. Kurian S, Nisha GK (2014) Open loop control of switched reluctance motor using theta position sensing. Int J Innov Sci Eng Technol 1(10):64–68
21. Kurian S, Nisha GK (2015) Closed loop operation of switched reluctance motor with hysteresis controller. Int J Appl Eng Res 10(17):37645–37650

Blockchain Technology in Smart Grids and Microgrids: A Critical Review of Challenges and Opportunities

Avinash Pareek, Prashant Singh, and J. S. Lather

Abstract Our Electricity or energy market was designed a long time ago. The regulation that applies to it today that disciples the innovation were put in place about a 100 years ago has been lobbied to keep in place. Since then, there have been massive inefficiencies in the entire ecosystem and infrastructure, and the control and distribution of energy today are quite antiquated. Much regulation prevents new companies from implementing a new system, so it becomes necessary to work with the incumbents to integrate this system. Therefore, this paper aims to frame today's situation of the energy markets' state. Also, to find the opportunity for a new energy paradigm. We point out how the current public blockchain infrastructure is generally incompatible with today's energy infrastructure. Furthermore, it also enlightens the new infrastructures that will enable the transition and examples of applications in the energy infrastructure.

Keywords Blockchain · Demand side management · Microgrid · Renewable energy resources · Smart grids

1 Introduction

Smart energy networks are one of the main hopes for a sustainable future. After the complete depletion of fossil fuels, renewable sources will be the only source of energy left. Here we will talk about smart energy networks (SEN) and how they operate at low voltage levels. It works at real-time control with utmost management. The town's talk is how consumers can produce energy and become active prosumers. The earlier

A. Pareek
Electrical and Computer Engineering, Binghamton University State University of NY, Binghamton, NY, USA

P. Singh (✉) · J. S. Lather
Electrical Engineering Department, NIT Kurukshetra, Kurukshetra, India

J. S. Lather
e-mail: jslather@nitkkr.ac.in

A. R. Gupta et al. (eds.), *Power Electronics and High Voltage in Smart Grid*, Lecture Notes in Electrical Engineering 817,
https://doi.org/10.1007/978-981-16-7393-1_30

approach of using centralized methods is now not useful in dealing with smart grid management issues. Due to this, the decentralized methods and structures come into play. The infrastructure for SEN needs to implement the internet of things (IoT) and smart metering devices. The idea is to incorporate the above technologies into blockchain and develop a SEN that uses and transfers electric energy that generates minimal losses [1]. If the storage device is overcharged, it might become faulty and, this can result in grid failure and power outage. The renewable energy generation and the supply curve are not balanced during unexpected peak hours. The energy demand is not enough to meet the total generated energy. The Distribution System Operators (DSOs) are not capable enough to change the energy production that creates fault in grid operations. Demand-side management (DSM) is the answer to matching energy demand with the generation by making Distributed Energy Prosumers (DEPs) handle their energy demand during peak load hours [2, 3]. According to the demand response of the DSO, a regulation created to adjust the charging period and transfer energy during restricted hours [4, 5]. The DEP makes offers concerning vitality requests and, according to it, it increases or decreases their interest. After acknowledging the request, the DSO checks the compatibility between the incoming request and age at the network level is met.

A record is constructed from a set out of squares, chained to each other by connecting multiple hash pointers [8], as seen in Fig. 1. These have many legal exchanges of computerized resources. This connector run down is another form of information structure. Following these progressions will show up in the past enlisted squares with some irregularities. These irregularities will come up due to a change in that particular square's hash pointer. In this distributed system, these squares and exchanges are recreated. Registering more exchange will transfer it to its companion hubs that further will approve and multiply. In case of any logical irregularities, the exchange will not be transferred. A hub has an option not to advance exchange to maintain a strategic distance from circles in the systems. Shrewd contract execution can be done due to the new advances in the blockchain. The bits of code-brilliant contracts are used to execute diverse trade routines. These routines are analyzed and established by companion hubs from the system.

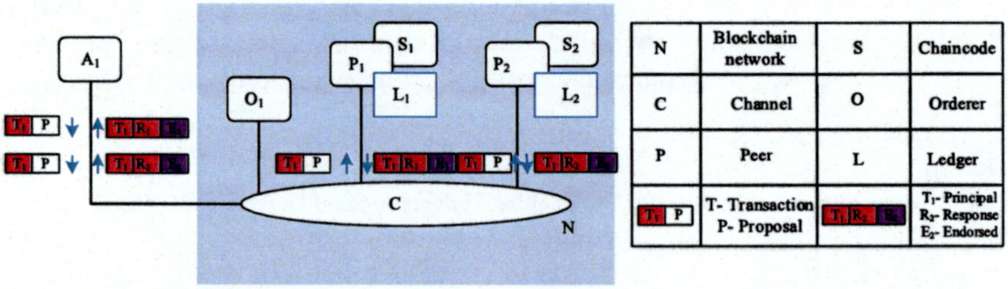

Fig. 1 Ledger distribution for Peer-to-peer network (P2P) [6, 7]

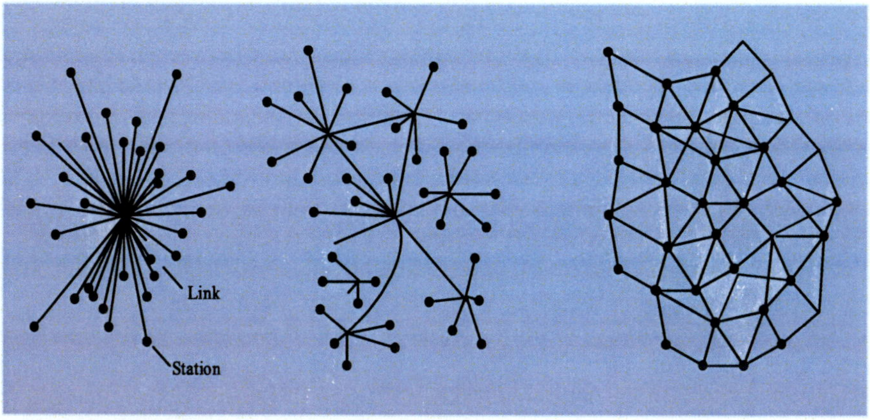

Fig. 2 Centralized, decentralized, and distributed systems

1.1 Background

Blockchain bandwidth demand is way higher than the real-time advanced metering systems, almost tenfold. Hence, it is essential to have a higher bandwidth channel to implement blockchain transactions [9]. Three parameters are critical for residential, industrial, and commercial customers. These technical issues are the following, high frequency of P2P trading decentralizing application for P2P power trading on blockchain, and developing and motivating the application for attracting more participants [10]. To optimize financial and energy flow in the transition toward active distributed networks, a new model for decentralized transactive energy management [11] is seen in Fig. 2.

This paper's main objective is to present a brief review of the existing algorithms based on blockchain. Blockchain technology enables the distribution of energy in a hybrid network. Some algorithms can track renewable generations' electricity consumption with peer-to-peer exchange methods.

2 Blockchain and Renewable Energy

In the network, the data communication works on duplicating data from one data block to another. Important point to note while transferring digital currency from one person's wallet to another in crypto world is to transfer the funds ensuring there is no double-spending problem. The conventional solution is to exchange the currency via a central entity like a bank or a neutral governing body that acts as a trusting body between two parties. Sometimes these mediators are not preferred by either one side or both. Adding them also causes an increase in the transaction cost [12]. One disadvantage of centralized systems is that they are prone to malicious attacks and can have a single point of failure [13]. Blockchain was introduced to defeat such intermediaries' purpose and replace it by implementing a distributed network of digital users who can have an association to approve the transactions and establish the integrity of the ledger. When the central management is eliminated, the main issue that remains is how to find a method to keep track of records and make multiple copies of the ledger to ensure integrity. There are various consensus algorithms [14] to ensure validation and ledger consolidations. Based on participation, blockchain is categorized as public or private [15]. To confirm the information on the network, specific policies are implemented that are also called protocols of agreement. Transaction velocity, safety, transparency, and scalability are the factors that achieve consensus.

The most widely used algorithm is proof-of-work (PoW) used by Bitcoin. Here the miners fight with other miners to solve a computational mathematical problem to add a new block to the current blockchain. To find this novel block, miners have the only option to try and solve the problem using trial and error. It is not possible to predict the blocks' values without solving them, hence the name—PoW. The addition of a novel block requires a large amount of computational and electric power. After this brute-force method, the Bitcoin network adds the new block if transactions are correct and unspent. According to Forbes 2020 report [16], the cost to mine one bitcoin is $8,206; so, the reward goes to the miner itself.

If a new block is constructed during mining, they accept it and work on a succeeding block. The network stores all resulting chains and, the members mining parallel blocks accept the longest and discard the shortest chain. Larger chains represent substantial computational power combined by a larger group of networks and hence seems most authentic. This feature helps the PoW suppress 51% of attacks, which means most nodes are honest. In the case of a 51% attack, the miners will have control over 50% of the network computational power and mining hash rate. The PoW requires much energy to run its operation and transactions and mostly comes from renewable resources (77%, according to a report by Coin Shares) [17]. Other than proof-of-work, proof-of-stake (PoS) is also an alternative algorithm to confirm transactions. These methods remove the need for the brute force of computational power and validate the block by the quantity of cryptocurrency that the miner has [18]. This method is a lot faster and energy-efficient. Instead of awarding them with crypto coins, the miners get a transaction fee. One of the leaders, Ethereum, is shifting its operation from PoW to PoS [19].

2.1 *Other Algorithms Based on Blockchain*

Other upcoming algorithms are Proof-of-Authority (PoA) and Proof-of-Participation (PoP). In PoA, a few authorized officials are selected to ensure nodes' validity. They can remove the flawed characters from the nodes. This makes the transactions faster; however, the disadvantage is what if the elected officials are corrupt? This PoP is used to work on guided decentralized rules and regulations, restricting them to add or delete new nodes individually. These algorithms are blooming in the energy markets like Aura, based on PoA implemented on an Energy Web Blockchain [20].

2.2 *Active Sites Working on This Platform*

Due to the blockchain's decentralization aspect, it can completely change the energy sector by supporting decentralization. Solar panels installed on building rooftops produce electricity that creates stress on electricity grids with surplus power and were initially designed as centralized power structures. Blockchain offers a way to transfer energy using peer-to-peer trading that incentives local consumers. When the consumers and prosumers pay to each other, all other spending procedures come into question. Some of the working projects are in Germany's Conjoule, New York's Brooklyn Microgrid [21], and Australia's Power ledger [22]. However, these operations are still in the native stage and still have a long way to go. Carbon credits or renewable energy certificates and Solar Coin are reward-based programs on electricity tracking tracked using blockchain. Small corporations or individuals with a shortage of financial resources but who have the zeal to jump into the renewable energy market can start a collective investment that ensures fair and transparent revenue sharing. New tools have to be put into place to successfully implement a decentralized system, as old electricity systems will become redundant in the future.

3 Blockchain Applications in Microgrids

Microgrid's decentralized business and control processes are compatible with the Blockchain's decentralized model. Following are the recently adopted concepts of microgrids and blockchain projects; hence they are being in the testing stage. Figure 3 shows the structure to trade energy using a distributed system.

- **PWR Company**

In microgrids, PWR Company [23] targets P2P trading of renewable energy. They have removed the idea of instant selling of energy, but rather, they encourage and install deep cycle batteries for energy storage so that the stress on the grid can be reduced. Based on Ethereum, it will convert its platform to PWR Token, which will be its kind of energy-based cryptocurrency. One MWh will be awarded one PWR Token, which can be traded into other crypto markets.

- **Power Ledger**

It is a blockchain-based market trading and clearing mechanism [24]. It brings the option to trade surplus energy at either the microgrid or to the distribution grid. The cost might vary at both ends. The energy traded by the distribution system operators at the distribution grid will attract income.

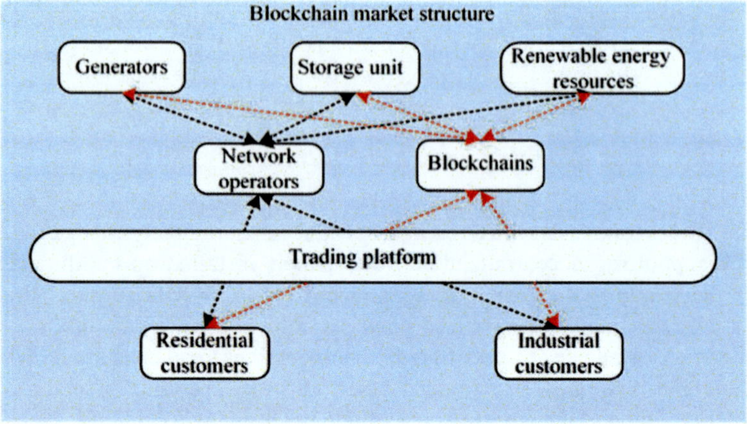

Fig. 3 Structure to TRADE ENERGY using a distributed system

- **Key2Energy**

One of the common approaches by other platforms is to trade the surplus power; however, Key2Energy [25] channels the Solar energy to the residents living in a multi-apartment system at a lower cost. In this way, the first agent works to sell the surplus power at a local market for the best possible cost. The second agent utilizes the energy for shared electric equipment in a multi-apartment setup such as water pumps, elevators, garden lights and other lighting. The blockchain model is responsible for all the related transactions and smart contracts.

- **LO3 Energy**

Ethereum and Smart contract-based transactive Grid platform is developed by the Transactive Grid and Brooklyn Microgrid LO3 Energy [26]. This platform focuses on various business models for the distribution grid and transactive energy space. Enabling P2P transactions and control of distributed energy resources for grid balancing also takes care of emergency management and demand response. TAG-e is also known as transactive Grid Elements, which measures energy production, consumption, and information sharing with an electric meter and computer. They are aiming to build a blockchain-based microgrid intelligence system.

- **Share and Charge** [27]

Instead of gas stations or fuel pumps, it aims at a network of electric vehicle charging stations. Traditional gas stations permit one-way transactions, selling fuel to the customer, whereas this model permits them to charge and discharge their batteries at the station. Once the consumer is registered to the station, the share and charge module ensures no unregistered units can be connected. It has a wallet-based system for transactions, and the layer of the transaction is Ethereum based.

4 Centralized Versus Decentralized Energy Infrastructure

Our current energy infrastructure was designed a long time ago. Much regulation that applies to it today that disciples the innovation was put in place about a 100 years ago and has been lobbied to keep in place since then. There are massive inefficiencies in the entire ecosystem and infrastructure. Today's control and distribution of energy are relatively antiquated. Many regulations prevent companies from implementing a new system, so it is super important to work with the incumbents to integrate this system. The current infrastructure is antiquated that favors a centralized, highly capitalized market system. In this, we have monolithic generators on a network that transmit power directly to a consumer. Total Energy units generated for the consumer, only ten of those get to the endpoint. This low percentage is due to inefficient power generation. Localizing the power generation with an infrastructure that enables the value which is to be transferred in that new ecosystem can increase the delivery flow.

Renewable technology is mostly a decentralized distributed power, small in size that currently does not have links to a network. A prosumer with a solar panel set up on the house can have a grid-tie that benefits the party from a feed-in tariff. However, a host of other tokenize derivatives is also available that can unlock a massive amount of economic abundance from already distributed systems with blockchain after implementation.

To do this, organize every device that can consume or produce energy and connect them as a node, and that node is an agent that is aware of all other nodes in its locality. They can react dynamically according to the current market conditions. Conditions like, how much energy is being sold for what time of day it is? Whether there is an abundance of solar power that needs to be consumed not to flood the market or flood the power lines. All these bits of hardware infrastructure get financed in the utility operator's interest. The challenge is finding a sustainable way of getting a return for investors on significant capital projects. Renewable energy is generally a distributed system. It brings about an unlock value, creating more incentive for people to either crowdsource or pull together resources or small groups of community investors into a big project. For example, biogas or wind project in Africa are being built by small community investors [28].

A platform is needed where people can come together to finance these distributed systems where renewable energy credit is a reward system. For example, the energy from a small solar installation on the house should be traded frictionlessly on the network in small or large pieces removing the need for intermediaries. In this model, smaller producers and participants in the network benefit by getting value from what they already have. The following Fig. 4 shows a simulation of a local energy grid in which each section of the network dynamically reacts to the other. It depicts a pricing model and how other houses in the network will react in the case of the turned-off power plant. In that case, the houses will dynamically adjust their consumption and production loads. In the event of an actual outage in a real power plant, the houses will still consume power locally and produce it and interact with each other. Once this is implemented, it is a considerable resiliency improvement for energy grids worldwide.

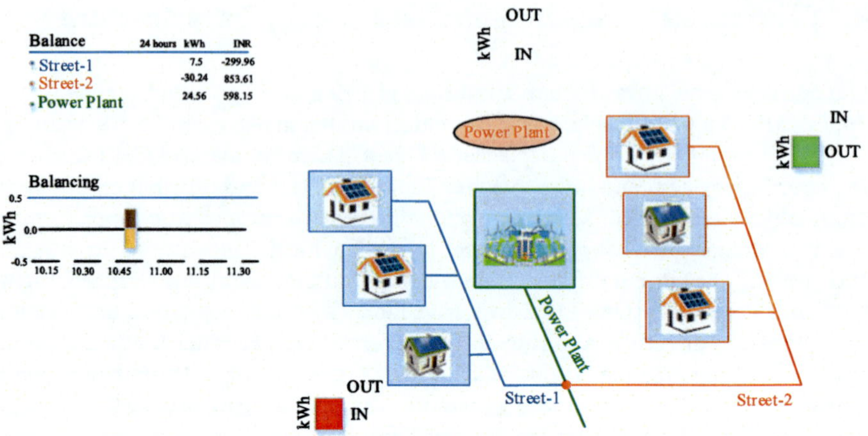

Fig. 4 Web of energy in a smart city using IoT and Blockchain

The current infrastructure does not work very well with this model, so using this with a new platform called the energy web blockchain will help. Another project known as the aurora network is an open-source hardware project where, through these mechanisms, people can deploy innovative hardware, which earlier would not be possible. A good example is biomass gasification. For the above household model, for PV energy, the supply–demand curve is unbalanced because of the massive abundance of solar energy during the day that is not sufficiently deployed or stored in storage batteries to be stored offline. Further the assessment of blockchain technology compared to conventional technology is given in Table 1.

Table 1 Evaluation between the conventional technology and blockchain technology

Event	Conventional technology	Blockchain technology
Failure in a single location	Yes	No
Energy plan uncertainty	Yes	No
Payment system	Centralized system	Peer-to-peer
Payment approval	Through central authority up to 60 days	It should be an agreement among all links Near real-time
Demand response arrangement	Through central authority	Independent signaling by link cooperation and smart agreements
Energy contract approval	Through central authority	Over contract amidst all links

5 Critical Review and Observations

- Power consumption demand is in the morning or the evening as most of the population works during the day. The dull period in between has an unbalanced demand–supply curve. Big power stations keep running as they cannot shut down as the cost of turning them again becomes very expensive. However, this system implementation can take energy on demand from core parts and synchronize to other equipment used at that particular time. For example, in a time of high demand, and equipment that costs more to run can have a variable pricing model than the usual, and that energy can be used and acquired to level across the network.
- These systems can be implemented over microgrid and brilliant grid as this vision is of a more distributed power network. For tokenizing energy on a network, high transaction throughput is needed. The rate and volume can be scaled as per the area's infrastructure. Scaling the global infrastructure can allow the transfer of value and flow of markets globally to transfer energy in abundance in one area where there may not be good infrastructure. Walnut shells are abundant in California that are supplied by walnut manufacturing companies [29]. However, due to a lack of infrastructure, those walnut shells are dumped or burned, and no value is derived [30].
- The smart way is to create a mechanism where the walnuts shells are put into a gasifier on-site and connect to the grid-tied back to the network. This energy can be sold directly to a consumer using tokenized methods. Also, linking the energy with smart meters, which clamps onto the incoming feed of the house and does a Fourier transform on the energy frequencies. The meter tells, about a month of learning, whether which equipment like a heater uses 50% of the power or a fan is using 5%. This model analyses this data, incentivizes good actors in the system, and overcharges the bad actors. If the person uses energy below a threshold, they will get paid for it, which incentivizes good behavior. This advises a better practice on the network where the participants contribute overall to lower demand in a fluid and organic way.
- The current power infrastructure that we are using right now is very static. During summer in California, an abundance of solar power is not utilized due to the insufficient market resources [31]. The utility companies ask the solar producers to turn off their panels so that the electrons' surge does not flood the network. This is a vast waste and ineffective utilization by the solar industry. Enabling all the network resources to be a lot more flexible will enable much higher utilization of all the electricity generated from renewables.

6 Conclusion

An extensive literature review on blockchain technology applications in smart grids and microgrids has been presented in this paper. Micro and smart grids have a decentralized framework that governs or administrates request reaction programs and to have a feedback mechanism to control or support the network. Integration of the lattice elements to blockchain blueprint with the associated brilliant contracts to authenticate the automatic meaning of expected vitality adaptability levels, the approval of demand response understandings, and balance between vitality creation and vitality request. Blockchain technology is being tried out by many companies like Grid Singularity, LO3Energy, Electron, active grid, and many more. This vision is for the more IoT-connected cities with feedback mechanisms that feed into a smart grid that understands precisely what to switch on and off. This Utopian vision of an efficient marketplace is where we unlock value in systems that are not currently used and use the resources elsewhere in the world to make an actual impact. Take an abundant station and produce revenue and use that revenue to develop other infrastructure elsewhere in the world. New development open-source hardware currently does not exist because it is hard to monetize and hard to justify costs. The ecosystem within energy applications seeks to use blockchain infrastructure, and some magnitude is expanding super rapidly.

References

1. Huang X, Zhang Y, Li D, Han L (2019) An optimal scheduling algorithm for hybrid EV charging scenario using consortium blockchains. Futur Gener Comput Syst 91:555–562
2. Su Z, Wang Y, Xu Q, Fei M, Tian Y, Zhang N (2019) A Secure charging scheme for electric vehicles with smart communities in energy Blockchain. IEEE Int Things J 6:4601–4613
3. Wang Y, Su Z, Zhang N (2019) BSIS: Blockchain-based secure incentive scheme for energy delivery in vehicular energy network. IEEE Trans Ind Inform 15:3620–3631
4. Liu C, Chai KK, Zhang X, Lau ET, Chen Y (2018) Adaptive Blockchain-based electric vehicle participation scheme in smart grid platform. IEEE Access 6:25657–25665
5. Yang X, Wang G, He H, Lu J, Zhang Y (2020) Automated demand response framework in elns: decentralized scheduling and smart contract. IEEE Trans Syst Man Cybern Syst 50:58–72
6. Kang J, Yu R, Huang X, Maharjan S, Zhang Y, Hossain E (2017) Enabling localized peer-to-peer electricity trading among plug-in hybrid electric vehicles using consortium Blockchains. IEEE Trans Ind Inform 13:3154–3164
7. Pop C, Cioara T, Antal M, Anghel I, Salomie I, Bertoncini M (2018) Blockchain based decentralized management of demand response programs in smart energy grids. Sensors (Switzerland) 18
8. Noor S, Yang W, Guo M, van Dam KH, Wang X (2018) Energy demand Side Management within micro-grid networks enhanced by blockchain. Appl Energy 228:1385–1398
9. Meeuw A, Schopfer S, Wortmann F (2019) Experimental bandwidth benchmarking for P2P markets in blockchain managed microgrids. Energy Procedia 159:370–375
10. Wang X, Yang W, Noor S, Chen C, Guo M, van Dam KH (2019) Blockchain-based smart contract for energy demand management. Energy Procedia 158:2719–2724
11. Li Z, Bahramirad S, Paaso A, Yan M, Shahidehpour M (2019) Blockchain for decentralized transactive energy management system in networked microgrids. Electr J 32:58–72
12. Grewal-Carr VMS (2016) Blockchain opportunity contents. Deloitte 27
13. Brandon D, Naucler C, Evans M, Bernard SA, Manning M (2016) Industrial Blockchain platforms : an exercise in use case development in the energy industry. Int J Acad Bus World 2420:0–25
14. Baliga A (2017) Understanding Blockchain consensus models. Whitepaper, 1–14

15. Zheng Z, Xie S, Dai H, Chen X, Wang H (2017) An overview of Blockchain technology: architecture, consensus, and future trends. In: 2017 IEEE international congress on big data (BigData Congress), pp 557–564
16. Young J (2020) Why the actual cost of mining bitcoin can leave it vulnerable to a deep correction
17. Bendiksen C, Beware of Lazy Research, https://coinshares.com/insights/beware-of-lazy-research-bitcoin-mining-update. Accessed 20 Feb 2021
18. Blinder M, Making Cryptocurrency more environmentally sustainable [online], https://hbr.org/2018/11/making-cryptocurrency-moreenvironmentallysustainable. Accessed 12 Jan 2021
19. Kim C, Take Two: Ethereumis getting ready for the Constantinople hard fork redo [online], https://www.coindesk.com/take-two-ethereum-isgetting-readyfor-the-constantinople hard-fork-redo. Accessed 12 Jan 2021
20. J. Bentke, Proof of Authority [online], https://energyweb.atlassian.net/wiki/spaces/EWF/pages/11993089/Proof+of+Authority,last accessed 2021/01/12.
21. Brooklyn Microgrid, https://www.brooklyn.energy/. Accessed 20 Feb 2021
22. Power Ledger, https://www.powerledger.io/. Accessed 20 Feb 2021
23. Price water house Coopers: Blockchain – an opportunity for energy producers and consumers?, https://www.pwc.com/gx/en/industries/assets/pwc-blockchain-opportunityfor-energy producers-and-consumers. Accessed 12 Jan 2021
24. Power Ledger - Where Power meets Blockchain, Power Ledger – A New Decentralised Energy Marketplace [Online], https://powerledger.io/. Accessed 12 Jan 2021
25. Komendantova N, Schwarz MM, Amann W (2018) Economic and regulatory feasibility of solar PV in the Austrian multi-apartment housing sector. AIMS Energy 6:810–831
26. Transactive Grid, LO3 Energy [online], http://lo3energy.com/transactivegrid/. Accessed 12 Jan 2021
27. Share & Charge, FAQs – easily charge your electric car with reliable charging station providers [Online], https://shareandcharge.com/en/faqs-2/. Accessed 12 Jan 2021
28. Klagge B, Nweke-Eze C (2020) Financing large-scale renewable-energy projects in Kenya: investor types, international connections, and financialization. Geogr Ann Ser B Hum Geogr 102:61–83
29. Hirsch J, Totally Nuts: This Farm Runs on Walnut Shells. https://modernfarmer.com/2013/08/the-farm-powered-by-walnut-shells/. Accessed 20 Feb 2021
30. Xie J, Should the U.S. Be Harvesting More Energy From Nuts?, https://www.bloomberg.com/news/articles/2014-05-12/should-the-u-s-be-harvesting-more-energy-from-nuts. Accessed 20 Feb 2021
31. Reaching boiling point: When solar panels overheat, https://www.opusenergy.com/blog/when-solar-panels-overheat/. Accessed 20 Feb 2021

A Hybrid Islanding Detection Technique for Synchronous Generator Based Microgrids

Santosh Kumar Singh, Mayank Rawal, Mahiraj Singh Rawat, and Tripurari Nath Gupta

Abstract Recently, microgrid has been emerged with an optimum solution for control and operation of distributed generations (DGs) in large power system. The islanding detection is one of the critical issues in grid connected microgrid. The hybrid islanding detection technique (IDT) uses the advantages of both passive and active techniques. Among passive techniques, the voltage unbalance (VU) IDT is the most effective technique and can be implemented comfortably for systems with load fluctuations. In this paper, a novel hybrid IDT has been developed using VU and rate of change of frequency (ROCOF) for synchronous generators based microgrid system. The proposed hybrid technique is compared with existing hybrid technique based on VU and frequency set point (FSP). In contrast to the current hybrid technique (0.21 s), the proposed technique detects the islanding in 22 ms, according to simulation results. Moreover, the proposed hybrid IDT can be applied for Inverter based DGs such as Solar Photovoltaic, wind energy, etc. in microgrid. The simulation results are obtained using MATLAB/Simulink 2019b.

Keywords Voltage unbalance · ROCOF · Islanding detection technique · Frequency set point · Hybrid technique

1 Introduction

As per IEEE 1547 standard [1], "An island is defined as a condition in which a portion of an area electric power system (EPS) is energized solely by one or more local EPSs through the associated PCCs while that portion of the area EPS is electrically separated from the rest of the area EPS". The islanding phenomenon can be categorized into unintentional and intentional islanding. Unintentional islanding can be hazardous to construction personnel operating on the main grid and can also result in utility equipment destruction. In literature, various IDTs have been proposed and implemented to detect unintentional islanding. The IDTs are generally classified into

S. K. Singh · M. Rawal · M. S. Rawat · T. N. Gupta (✉)
National Institute of Technology, Uttarakhand, India
e-mail: mt19eee009@nituk.ac.in

A. R. Gupta et al. (eds.), *Power Electronics and High Voltage in Smart Grid*,
Lecture Notes in Electrical Engineering 817,
https://doi.org/10.1007/978-981-16-7393-1_31

remote and local techniques. The implementation of local techniques is performed on distributed generation (DG) side. Further, the local IDTs are categorized into active, passive, and hybrid. The hybrid IDTs merge the advantages of active and passive IDTs or two passive/active IDTs and hence more efficient. A detail review of IDTs has been presented in [2–4]. In [5–7], the phase locked loop (PLL) is used to calculate ROCOF at point of common coupling (PCC). The ROCOF based IDT is sensitive to load fluctuations and hence the setting of threshold becomes difficult. For islanding detection, the VU based passive IDT utilizes the measurement of negative (NSV) and positive (PSV) sequence of voltage. This passive technique suffers from the disadvantages such as difficult to set threshold, unresponsive to network disturbances, and undesired harmonics in negative sequence component [8]. Recently, the hybridization of two IDTs has been utilized to improve islanding detection time, NDZ, etc. [9]. The hybrid technique using ROCOF over reactive power and d-axis current injection has been utilized to improve islanding detection time. This hybrid technique is best suited for mixed DG environment [10]. The hybrid techniques i.e., Q-f droop and reactive power variations (RPV) [11]; power control loop and signal processing [12]; VU and voltage phase angle (VPA) [13]; wavelet transform and artificial neural network (ANN) [14]; voltage and real power shift [15]; Sandia frequency shift and ROCOF [16]; "wavelet packet transform and probabilistic neural network" [17]; "adaptive neuro-fuzzy inference system" [18]; Voltage unbalance factor [20]; "rate of change of reactive power and Rate of change of active power" [21]; Pattern-recognition method [22]; VU/THD and bilateral reactive power variation [23]; Discrete fractional Fourier transform [24] have been proposed and implemented by various researchers.

The passive technique has several disadvantages such as large NDZ, failed to detect balance islanding and difficult to implement in multiple DG systems. Moreover, the active techniques have also several disadvantages such as degradation in power quality, long detection time, and synchronization issues. In this Paper, an improved hybrid IDT is proposed using VU and ROCOF. The two passive methods are combined together due to easy implementation and have negligible influence on power quality and having fast detection time. The following is how the remainder of the paper is organized: The second section delves further into the passive and active strategies employed. The proposed hybrid technique using VU and ROCOF is explained in Sect. 3. Section 4 is on focused simulation results and analysis. At last, the paper is concluded in Sect. 5.

2 Passive Methods

In this section, the passive methods i.e., VU and ROCOF for islanding detection have been reviewed.

2.1 Voltage Unbalance

The DGs mounted in the microgrid can match the load demands if the power grid is isolated. The voltage unbalance can be observed at terminals of DGs. This passive technique identifies the islanding dependent on the VU of three-phase voltages at PCC. In this method, the ratio of NSV to PSV is utilized to identify islanding event. The voltage unbalance at any time t can be determined from the expression given below:

$$VU_t = \frac{V_{NS_t}}{V_{PS_t}}, \tag{1}$$

where V_{NS} and V_{PS} represent the NSV and PSV component of voltages at the DG's output terminal. The detection time is 53 ms. The disadvantage of this technique is that it is influenced by undesirable harmonics when removing the negative sequence portion of voltage, difficult to set the threshold value.

2.2 Rate of Change of Frequency (ROCOF)

In this passive ID, ROCOF i.e., *df/dt* over a limited cycle at PCC is observed. If the value of df/dt is larger than threshold value, the inverter should cutoff the power output. The islanding detection time is around 24 ms. From the literature, the optimum threshold value for ROCOF technique is found to be 0.3 Hz/sec with 0.7 s islanding detection time. This technique may lead to erroneous results due to load fluctuations and switching. The *df/dt* can be determined by Eq. (2).

$$\frac{df}{dt} = \frac{f(t_k) - f(t_k - \Delta t)}{\Delta t}, \tag{2}$$

where $f(t_k)$ represents the frequency at time of the kth sample, $f(t_k - \Delta t)$ represents the value of determined frequency before the time of the kth segment, i.e., $t_k - \Delta t$. This method fails to recognize balanced islands. The setting of threshold values is very critical in this technique to distinguish islanding and switching events.

3 Proposed Hybrid IDT

The proposed hybrid IDT consists of the advantages of VU and ROCOF. Figure 1 displays the flow chart for the proposed algorithm. The voltage and frequency at PCC are monitored using this technique. ROCOF (*df/dt*) and voltage unbalance is determined for each DG. For any disturbance such as switching of loads or

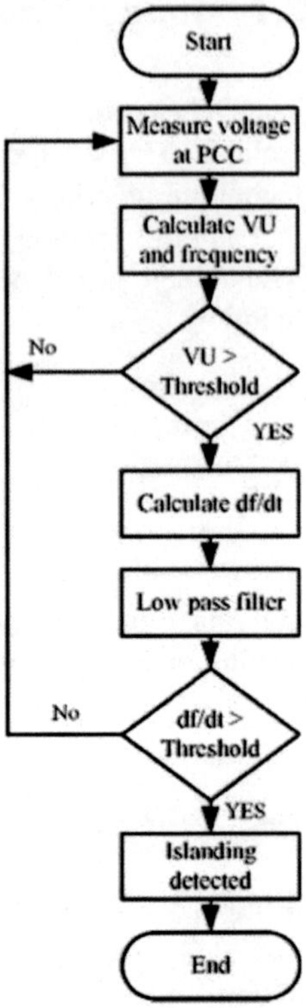

Fig. 1 Flow chart of proposed hybrid islanding method

islanding, a spike can be observed in VU. Whenever VU spike crosses the threshold value, the *df/dt* value at PCC is observed for 2 to 50 cycles and sent to the low pass filter (LPF). The aim to use LPF is to remove high frequency transients generated by components of the power system. The islanding case is observed if the ROCOF value still exceeds the threshold value. The maximum permissible VU spike value can be determine from [19].

4 Results and Analysis

The proposed hybrid method has been verified on test system as shown in Fig. 2. A diesel generator of 3 MVA is connected through the main grid via 25 kV/2.4 kV, Y-Δ transformer. Moreover, a 5 MW load is connected at main grid. The short circuit capacity (SCC) of main grid is 1000 MVA. The 1 MW load is connected near diesel generator. The Simulink model of test system is shown in Fig. 3.

The voltage and current profile at PCC before islanding and load changing is shown in Fig. 4. The following disturbances were simulated to study the islanding event. First, at 0.25 s, 0.8 MW load is inserted at PCC (near DG) and 0.9 MW load

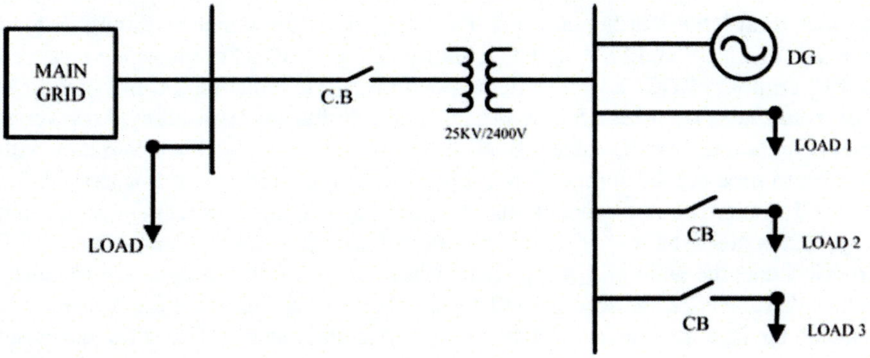

Fig. 2 Single line diagram of test system

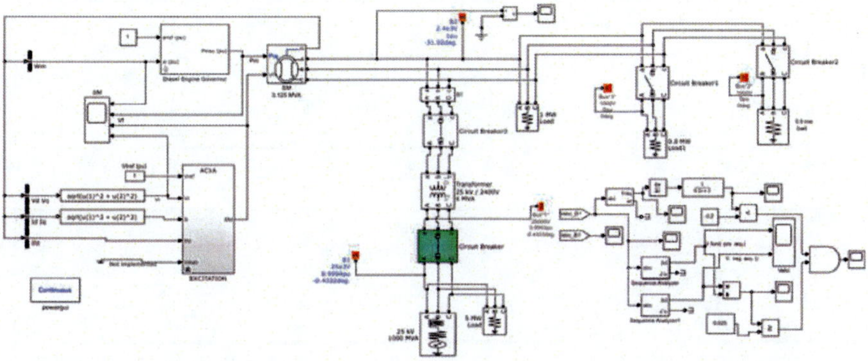

Fig. 3 Simulink model of test system

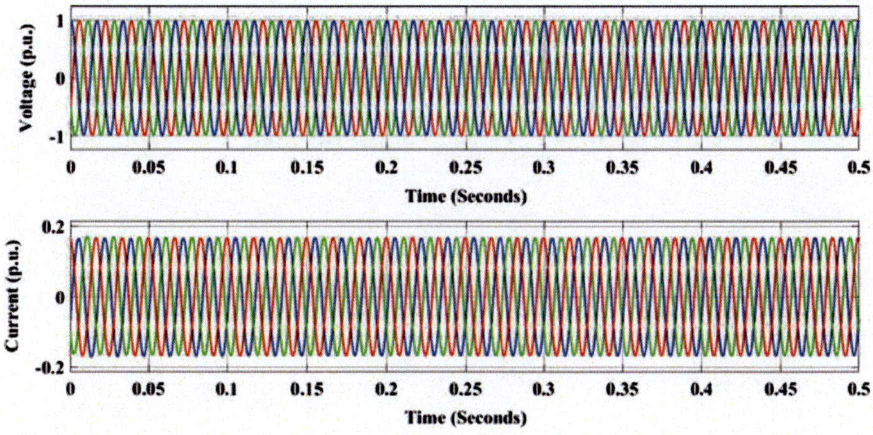

Fig. 4 Voltage and current profile at PCC before islanding

is inserted at 0.5 s. Further, at t = 1.0 s, main grid is disconnected through circuit breaker. The NSV and PSV at PCC is shown in Fig. 5. The VU spikes are observed at the terminal of DG at load switching and main grid isolation. From Fig. 6, it is observed that VU spikes are low in magnitude during load switching; however at main grid failure, the VU value is 0.04 pu. The frequency and ROCOF variation with respect to time can be observed from Figs. 7 and 8 respectively. It is observed that ROCOF is higher at the event of islanding compared to load switching event. From Fig. 9, it is observed that when VU spikes and *df/dt* both crosses the set threshold, to disconnect the main grid, a trip signal is sent to the circuit breaker. The islanding detection time is 22 ms, according to the simulation data. However, for the same test system, the islanding detection time for hybrid method using VU and frequency set point was found to be 0.21 s.

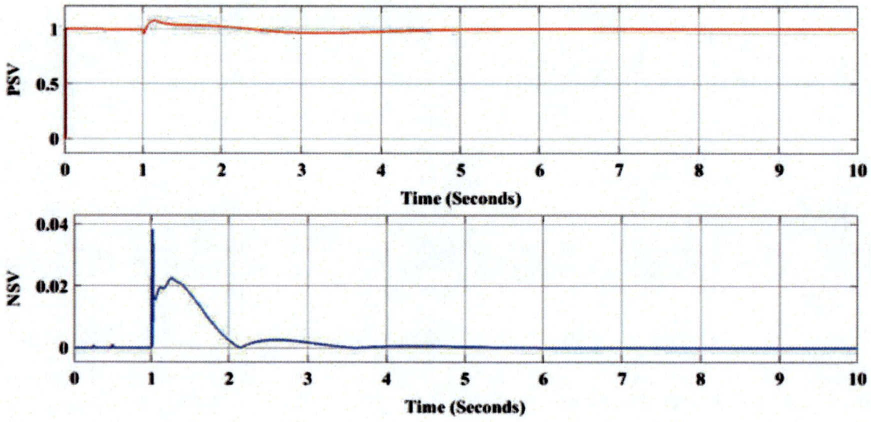

Fig. 5 Positive and Negative sequence voltages

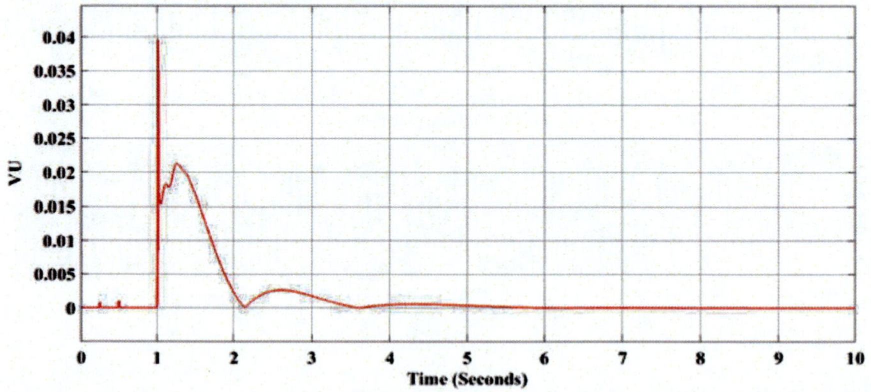

Fig. 6 Voltage unbalance at DG terminal

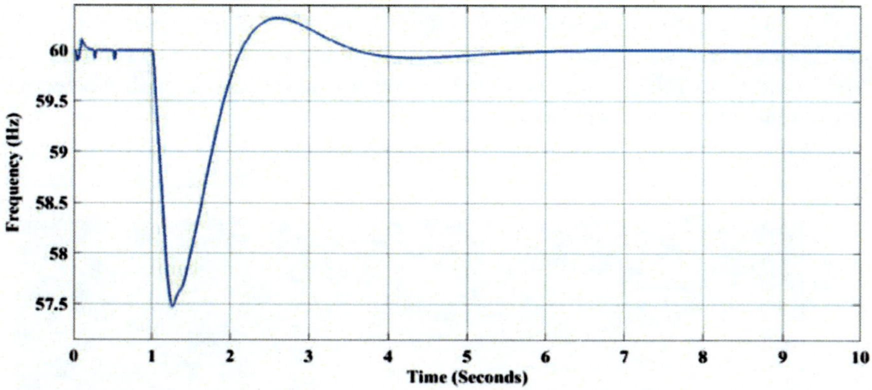

Fig. 7 Frequency variation at PCC

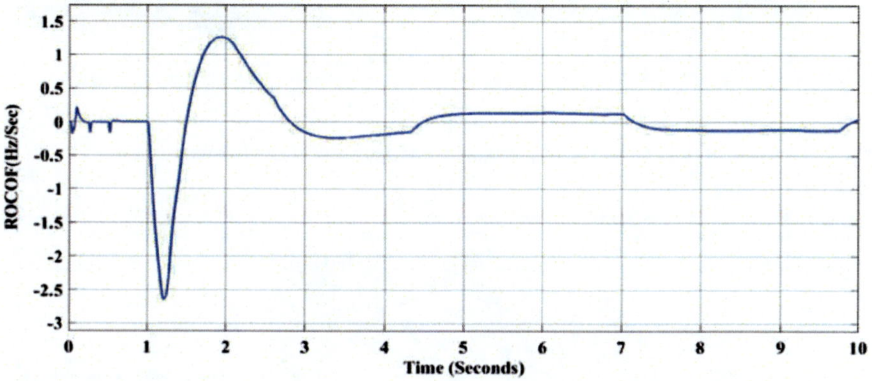

Fig. 8 ROCOF at PCC

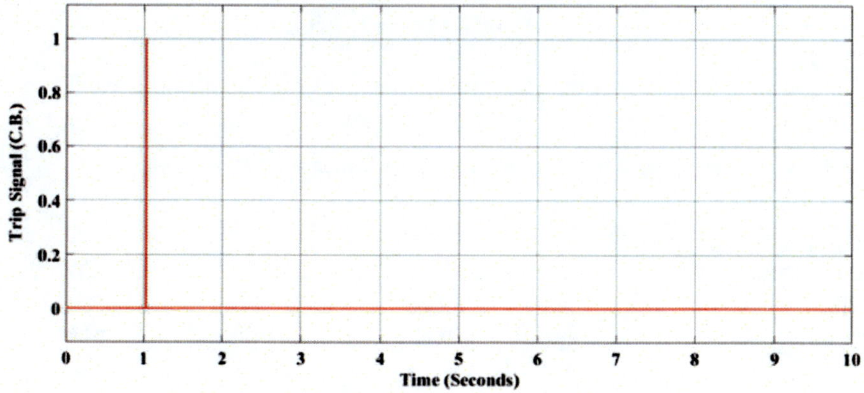

Fig. 9 Trip Signal input to circuit breaker at PCC

The voltage variation at DG terminal and current at PCC are shown in Figs. 10 and 11 respectively. A comparison between discussed IDTs is shown in Table 1. It is observed from Table 1 that the proposed method has lowest islanding detection time compared to other techniques.

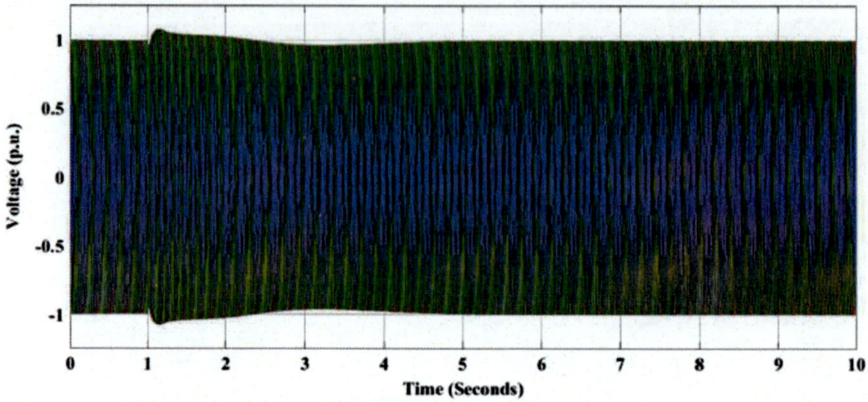

Fig. 10 Voltage during load switching and main grid isolation at DG terminal

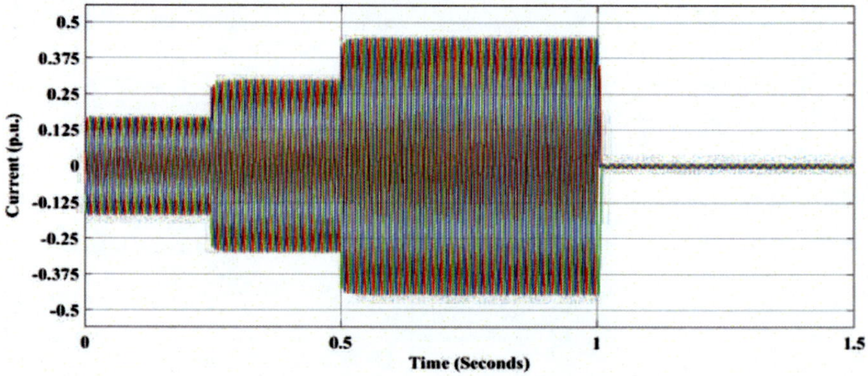

Fig. 11 Current during load switching and main grid isolation at PCC

Table 1 Evaluation of various IDTs

Methods	Detection time	NDZ	Power quality
VU	53 ms	Large	No impact
ROCOF	24 ms	Small	No impact
VU and FSP	0.21 s	Small	Degrade
VU and ROCOF	22 ms	No information	No impact

5 Conclusion

In this paper, a novel hybrid IDT using VU and ROCOF is presented for diesel generators based microgrid system. The proposed hybrid technique is compared with hybrid IDT based on VU and frequency set point (FSP) given in literature. The suggested hybrid system detects islanding in 22 ms, according to simulation results. Moreover, the islanding detection time for hybrid method using VU and frequency set point was found 0.21 s. This proposed method can also be applied on the inverter based DG systems. The simulation results are obtained using MATLAB/Simulink. In future, the proposed method can be investigated for hybrid microgrid system.

References

1. IEEE application guide for IEEE Std 1547(TM) (2009) IEEE standard for interconnecting distributed resources with electric power systems. IEEE Std 1547.2–2008, 15, 1–217
2. Mahat P, Chen Z, Bak-Jensen B (2008) Review of islanding detection methods for distributed generation. In: 3rd international conference on electric utility deregulation and restructuring and power technologies (DRPT). IEEE, pp 2743–2748, Nanjing, China
3. Khamis A, Shareef H, Bizkevelci E, Khatib T (2013) A review of islanding detection techniques for renewable distributed generation systems. Renew Sust Energ Rev 28:483–493
4. Kunte RS, Gao W (2008) Comparison and review of islanding detection techniques for distributed energy resources. In: 40th North American power symposium. IEEE, pp 1–8, Calgary, Canada
5. Guha B, Haddad RJ, Kalaani Y (2015) A passive islanding detection approach for inverter-based distributed generation using rate of change of frequency analysis. In: 1st international conference on power electronics, intelligent control and energy systems (ICPEICES), pp. 1–6, Delhi
6. Marchesan G, Rossato M, Cardoso G, Mariotto L, De Morais AP (2016) Passive method for distributed generation islanding detection based on oscillation frequency. IEEE Trans Power Del 31(01):138–146
7. Tzelepis D, Dysko ABooth C (2016) Performance of loss-of-mains detection in multi-generator power islands. In: 13th IET international conference on developments in power system protection (DPSP). IET, pp 1–6, Edinburgh, United Kingdom
8. Jang SI, Kim KH (2004) An islanding detection method for distributed generations using voltage unbalance and total harmonic distortion of current. IEEE Trans Power Del 19(2):745–752
9. Bower W, Ropp M (2002) Evaluation of islanding detection methods for photovoltaic utility-interactive power systems. International Energy Agency (IEA), Albuquerque, NM, USA
10. Nougain V, Prakash S, Mishra S (2018) Hybrid islanding detection method based on ROCOF over reactive power and d-axis current injection. In: 8th India international conference on power electronics (IICPE). IEEE, pp 1–4, Jaipur, India
11. Raipala O, Mäkinen A, Repo S, Järventausta P (2014) A novel anti-islanding protection method based on the combination of a Q-f droop and RPV. In: IEEE PES innovative smart grid technologies (IEEE), pp 1–6, Europe
12. Zamani R, Golshan MEH, Alhelou HH, Hatziargyriou N (2019) A novel hybrid islanding detection method using dynamic characteristics of synchronous generator and signal processing technique. Electr Power Syst Res 175:105911
13. Seyedi M, Taher SA, Ganji B, Guerrero JM (2009) A hybrid islanding detection technique for inverter-based distributed generator units. Int Trans Electr Energy Syst 29(11):1–21

14. Kumar SA, Subathra MSP, Kumar NM, Malvoni M, Sairamya NJ, George ST, Suviseshamuthu ES, Chopra SS (2020) A novel islanding detection technique for a resilient photovoltaic-based distributed power generation system using a tunable-q wavelet transform and an artificial neural network. Energies 13(16):1–22
15. Mahat P, Chen Z, Bak-Jensen B (2009) A hybrid islanding detection technique using average rate of voltage change and real power shift. IEEE Trans Power Del 24(2):764–771
16. Khodaparastan M, Vahedi H, Khazaeli F, Oraee H (2015) A novel hybrid islanding detection method for inverter-based DGs using SFS and ROCOF. IEEE Trans Power Del 32(5):2162–2170
17. Ahmadipour M, Hizam H, Othman ML, Radzi AMM (2018) An anti-islanding protection technique using a wavelet packet transform and a probabilistic neural network. Energ 11(10):1–31
18. Hashemi F, Ghadimi N, Sobhani B (2013) Islanding detection for inverter-based DG coupled with using an adaptive neuro-fuzzy inference system. Int J Electr Power Energy Syst 45(1):443–455
19. Menon V, Nehrir MH (2007) A hybrid islanding detection technique using voltage unbalance and frequency set point. IEEE Trans Power Syst 22(1):442–444
20. Nayak AM, Mishra M, Pati BB (2020) A Hybrid Islanding Detection Method Considering Voltage Unbalance Factor. international symposium on sustainable energy, signal processing and cyber security (ISSSC). IEEE. Gunupur Odisha, India, pp 1–5
21. Jhuma UK, Mekhilef S, Mubin M, Ahmad S, Rawa M, Alturki Y (2020) Hybrid islanding detection technique for malaysian power distribution system. In: 5th international conference on computing communication and automation (ICCCA). IEEE, pp 785–790, Greater Noida, India
22. Chaitanya BK, Yadav A, Pazoki M (2020) Reliable islanding detection scheme for distributed generation based on pattern-recognition. IEEE Trans Ind Informat, pp 1–9
23. Wang G, Gao F, Liu J, Li Q, Zhao Y (2020) Design consideration and performance analysis of a hybrid islanding detection method combining voltage unbalance/total harmonic distortion and bilateral reactive power variation. CPSS Trans Power Electron Appl 5(1):86–100
24. Dutta S, Olla S, Sadhu PK (2021) A secured reliable and accurate unplanned island detection method in a renewable energy based microgrid. Int J Eng Sci Technol, 1–14